Protein-Protein Interactions

Krishna Mohan Poluri • Khushboo Gulati •
Sharanya Sarkar

Protein-Protein Interactions

Principles and Techniques: Volume I

Springer

Krishna Mohan Poluri
Department of Biotechnology
Indian Institute of Technology Roorkee
Roorkee, Uttarakhand, India

Khushboo Gulati
Department of Biotechnology
Indian Institute of Technology Roorkee
Roorkee, Uttarakhand, India

Sharanya Sarkar
Department of Biotechnology
Indian Institute of Technology Roorkee
Roorkee, Uttarakhand, India

ISBN 978-981-16-1593-1 ISBN 978-981-16-1594-8 (eBook)
https://doi.org/10.1007/978-981-16-1594-8

This Springer imprint is published by the registered company Springer Nature Singapore Pte Ltd.
The registered company address is: 152 Beach Road, #21-01/04 Gateway East, Singapore 189721, Singapore

Dedicated to all our teachers, mentors, and parents for their invaluable guidance and unconditional support throughout our research careers

Preface

Mother Nature has bestowed a variety of functions to the cell by sculpting distinct variety strings of proteins in which the amino acids were systematically positioned as beads of a chain. Proteins perform their functions either alone or in conjunction with several biomolecules. Most of the cellular processes are conducted via complex interactions among proteins. These interactions are well known as protein–protein interactions (PPIs). To obtain deep insights into such functional processes of cell and modulate them in a variety of pathophysiological conditions, it is essential to understand the basic concepts underlying PPIs, which opened an emerging era of research that focuses on PPIs.

PPIs are highly prominent in terms of their magnitude and diversity. It has been established that proteins interact with multiple protein partners and these differentially interacting proteins together form an interaction network. This is an outcome of evolvement of high throughput experimental techniques that are able to decode the criteria followed by the proteins during the course of their interactions with other protein partners. This phenomenon has also stimulated computational biologists to formulate novel computational approaches for the prediction of PPIs, tools to predict missing connections in PPI network, to identify distinctive roles of each PPI network component, and databases to store and organize the ample amount of PPI data generated. Hence, studying PPIs is a multidisciplinary endeavor. To gain better understanding of PPIs and for the development of drugs targeting specific PPIs, it is essential to analyze basic physicochemical properties of PPIs, their evolutionary profiles, thermodynamic and kinetic parameters associated with them, and additionally the experimental and computational means to unravel them.

Keeping in mind the contemporary requirements for studying PPIs, the current book is designed. This book comprises six chapters, each of them discusses a unique perspective about this briskly evolving field of protein–protein interactions (PPI). Chapter 1 presents an overview of the fundamental concepts pertaining the protein's structure–function–folding paradigm and also provides an introduction to the evolution of proteins and glimpse of the PPI field. Chapter 2 presents a comprehensive knowledge about the classification of PPIs, delineates their functional significance for the sustenance of life, and discusses all the parameters employed for analyzing a specific PPI. Chapter 3 provides an outline about the basic principles of energetics and majorly aims at describing the thermodynamic and kinetic parameters associated with PPIs. Several mutagenesis related case studies have also been discussed to provide glimpse of high affinity protein–protein interactions. Chapter 4 sheds light on the protein complexes that are formed as a resultant of interactions among the proteins. This chapter describes the mechanism of formation of protein complexes, their structures, and applications. Evolutionary concepts involved in the formation of these protein complexes in the living cell are also outlined. Chapter 5 focuses on the wide spectrum of experimental techniques employed to decipher interactions among the proteins. Such techniques not only aid in discovering novel protein–protein interactions but also assist in validating the hypothesized interactions. This chapter also offers one to understand the basic principle of each technique along with its strengths and limitations. Chapter 6 is devoted to discuss computational approaches currently being employed for the prediction of PPIs and their analysis. It also provides a comprehensive view of databases designed to embrace differential aspects related to a variety of PPIs.

This book is intended to provide thorough understanding of the egressing field of protein–protein interactions. The book is designed to be self-contained as it offers an outline about the fundamental concepts related to PPIs, an in-depth knowledge of various experimental and computational approaches for the prediction and development of PPI libraries. Hence, this book is of great interest for variety of researches, students, and academicians who are seeking to gain comprehensive understanding and an up-to-date knowledge about PPIs.

Roorkee, India

Krishna Mohan Poluri
Khushboo Gulati
Sharanya Sarkar

Acknowledgements

KMP greatly acknowledges the support of Grants CRG/2018/001329 from Science and Engineering Research Board (SERB), DBT-IYBA fellowship-BT/07/IYBA/2013-1 from Department of Biotechnology, and GKC-01/2016-17/212/NMCG-Research from NMCG-MoWR, Government of India (GoI). The authors are also immensely thankful to Ms. Nupur Nagar and Ms. Shweta Tripathi of Department of Biotechnology, IIT Roorkee, for their sincere assistance in preparation of figures and/or proof reading the chapters contained in this book.

Contents

About the Authors

Krishna Mohan Poluri is currently working as an Associate Professor at the Biotechnology Department, IIT-Roorkee. He completed his PhD at Tata Institute of Fundamental Research, TIFR-Mumbai and Postdoc at Rutgers University and the University of Texas Medical Branch (UTMB-Texas). His areas of expertise are protein biochemistry and biophysics, bio-molecular interactions, structural biology, glycoimmunology, bionanotechnology, and algal biotechnology. He has published more than 110 articles and editorials in various respected international journals, books, and book chapters. He is also a guest editor/editorial board member for several international research journals. He won numerous awards and fellowships for his research work including; young scientist awards from the Indian Science Congress Association (ISCA) in 2009, National Academy of Sciences India (NASI) in 2014, and the DBT Innovative Young Biotechnologist Award (IYBA) in 2013.

Khushboo Gulati holds a PhD in Protein Biophysics from the Department of Biotechnology, Indian Institute of Technology (IIT), Roorkee. She completed her bachelor's in Bioinformatics and master's in Biotechnology at Punjab Panjab University, India. Her research interests are in the area of protein–protein interactions, protein engineering, molecular biophysics, and structural biology. She has authored 18 peer-reviewed research articles in SCI indexed international journals, 6 book chapters, and a book on protein engineering techniques. She is a recipient of the ICMR young scientist award.

Sharanya Sarkar is currently working as a Research Associate at the Mechanistic Biological Chemistry Laboratory, Department of Biotechnology, Indian Institute of Technology (IIT), Roorkee. She completed her master's degree in Biotechnology from IIT Roorkee in 2019 and her bachelor's in Microbiology from St. Xavier's College, Calcutta, in 2017.

Structural and Functional Properties of Proteins

1

Abstract

Proteins are one of the most fundamental biomolecules present in living organisms. They are indispensable to the sustenance of these organisms owing to the multitude of functions that they perform. They serve as structural constituents of cells and are involved in cell signaling, transport of biological components, cell adhesion, eliciting immune responses, catalyzing chemical reactions, etc. In order to get better insights into these functional aspects, it is quintessential to understand not only the physicochemical features of proteins, but also their structural organization (starting from primary sequences and ending in quaternary structures), and the various interactions that stabilize the higher-order structures. Many biological functions in the cell are mediated when a protein interacts with one or two other proteins in a phenomenon known as protein–protein interaction (PPI). The chapter concludes by providing an overview of PPIs including their functional importance.

Keywords

Transcription · Translation · Ramachandran map · Structural organization · Energy landscape · Biomolecular engineering · Protein folding · Non-covalent interactions

1.1 Introduction

Proteins are the ultimate players in the biochemical factory of cell. They are the work horses of cell which actively participates in almost all the cellular events. Proteins are highly interesting molecules as they are the most generous macromolecules in cell.

K. M. Poluri et al., *Protein-Protein Interactions*,
https://doi.org/10.1007/978-981-16-1594-8_1

Moreover, they are also highly diverse in contrast to any other macromolecules. Distinct variety of proteins can range from small or large, fibrillar or globular, monomers or oligomers, hydrophilic or hydrophobic, fixed conformation or alternative conformations, existing alone or associated with other macromolecules including carbohydrates and/or ligands, etc. Such diversity in proteins is the direct outcome of differential arrangements of amino acids. Since amino acids are the building blocks of proteins, they act as the beads of the protein string. Further, cellular events such as post-translational modification and gene splicing also add complexity to the diversity of these entities. Such a variation results in different structures of proteins which ultimately make them amicable to perform differential functions (Goodsell 1991).

1.2 Functions of Proteins

Proteins play numerous biological roles in cell including maintaining cell shape, structural organization, act as catalysts, molecular sensors, help in cell interactions, cell movement, provide defense mechanism to cell, etc. (Fig. 1.1). Different proteins exhibiting variety of functional roles are discussed in the following paragraphs.

Enzymes are the proteins that exclusively catalyze the biochemical processes in a given living organisms. Reaction is catalyzed when there is a binding of substrate in the protein's active site. In case of large substrates, there is more than one binding site in the enzyme. Each binding site binds to the different part of the substrate. However, catalysis occurs in one of those sites. For many reactions, such as for transferring a chemical group to the substrate, transferring electron, or substrate polarization, etc., enzymes require small molecules or metallic elements. When these molecules are permanently attached to the enzymes, these groups are known as prosthetic groups. In contrast, when attached temporarily, they are known as co-enzymes. Enzymes are broadly classified into six different categories, namely: oxidoreductases, hydrolases, transferases, lyases (syntheses), isomerases, and ligases (synthetases) (Silverman 2000). Each of these enzymes carries out specific chemical reaction.

Oxidoreductases aid in catalyzing redox reactions that mediate transfer of electrons from one molecule to the other. Such reactions are mostly involved in extracting out energy from the food stuff or in utilization of energy in the synthesis of molecular complexes. *Transferases* are involved in transferring a chemical moiety from one molecule to the other molecule. For instance, the transfer involves variety of chemical groups such as carbonyl, acyl, amino, glycosyl, methyl, sulfate, phosphate, or selenium containing groups. *Hydrolases* use water as an attacking group to cleave covalent bond in the substrate. Several types of bonds are attacked such as ester, amide, glycoside, and C-C bond. These enzymes exhibit low specificity toward the substrate. For instance, phosphatases remove the phosphate group from the substrate. *Lyases* catalyze the formation or cleavage of covalent bonds in the substrate without utilizing water. When the reaction involves the formation of bond the enzyme is known as synthase. *Isomerases* catalyze the movement of the

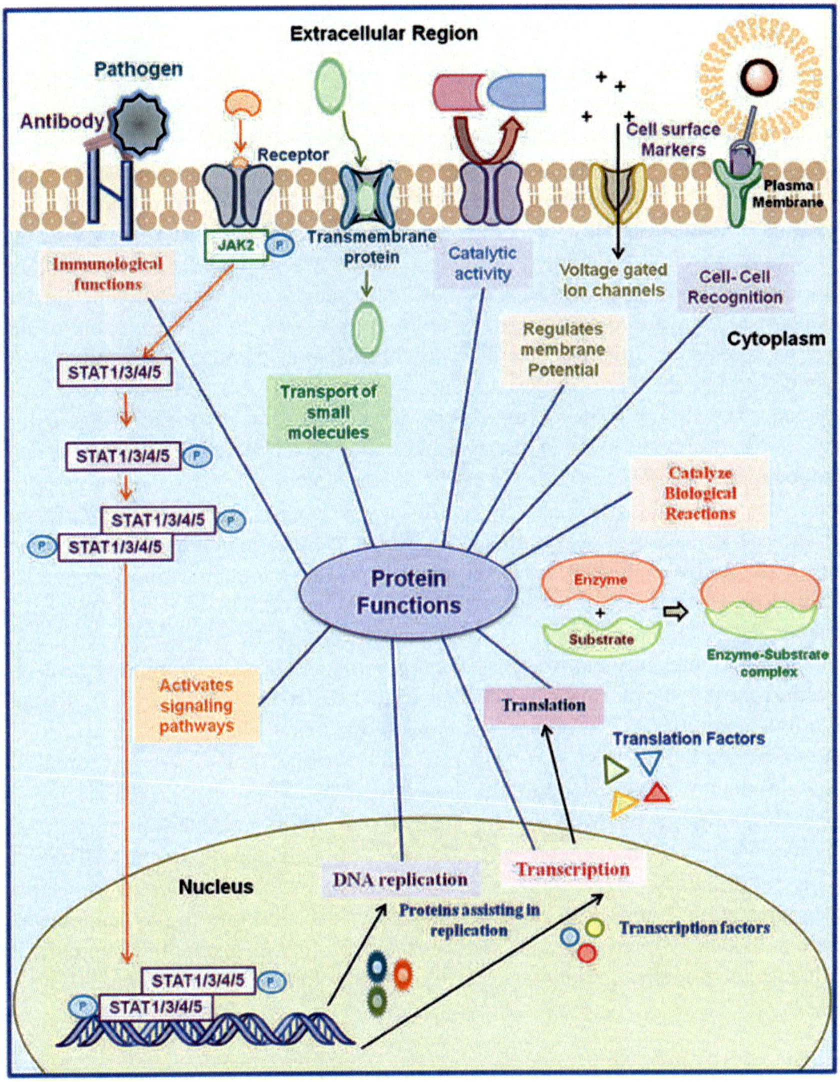

Fig. 1.1 Schematic showing the different functional roles of proteins

chemical moieties from one position to the other in the substrate. Isomerases are termed in accordance with the specific reaction they carry out. For example, epimerase, phosphomutase, and racemase belong to isomerase group, however they carry out specific reactions. *Ligases* catalyze the covalent binding of two substrate molecules for the formation of a single product. Such enzymes have traditional naming, for example, synthase, carboxylase, synthetase. DNA ligase is involved in ligating the two DNA strands via catalyzing the formation of phosphodiester bond (Moss 2020).

Membrane proteins are broadly divided into two categories, namely: integral (intrinsic) or peripheral (extrinsic) membrane proteins. Most of the bio-membranes encompass both types of membrane proteins. Intrinsic proteins, also termed as integral membrane proteins, anchor via two or more fragments across the phospholipid bilayer of the membranes. These proteins comprise hydrophobic amino acids, whose side chains are involved in interaction with the fatty acyl groups in the membrane. Most of the proteins completely span across the phospholipid bilayer and also exhibit domains flanking on both extracellular and cytosolic sides of the membrane. They are known as *transmembrane proteins*. In contrast, some of the integral membrane proteins are attached through one membrane via covalently bound fatty acids. Peripheral membrane proteins are also termed as extrinsic proteins; they do not span the membrane and are also not involved in interaction with the hydrophobic core of the phospholipid bilayer. They are present on the surface of the membrane. They are either directly interacting with the lipid's polar head groups or indirectly interacting with integral membrane proteins. Membrane proteins carry out various diversified roles. Some of them maintain cell shape, while others are involved in cell adhesion and cell-to-cell communication. Membrane proteins also assist in triggering the signaling pathways once they receive the signals (Tan et al. 2008).

Antibodies are also known as immunoglobulins. They are Y-shaped proteins secreted mostly by plasma cells that neutralize the pathogens. They are divided into five major classes on the basis of types of the heavy chains in their structures. They are namely IgG, IgM, IgA, IgD, IgE. All of them exhibit differential immunological functions (Schroeder and Cavacini 2010).

Variety of other types of proteins such as translation machinery proteins, transcription factors, polymerases, helicases, etc. are also present in the cell which participates in differential cellular processes such as vesicular trafficking, transcription, replication, translation, etc. Such flexibility and diversity in the functions of proteins is due to their structural features that in turn depend on their amino acid sequence, oligomerization, dynamic features, and folding (Voet and Voet 2011).

1.3 A Journey from Genes to Proteins

DNA (de-oxyribonucleic acid) is a long linear polymer of nucleotides that carries information of the cell and can be transferred from one generation to the other. Nucleotides are made of three components, namely: deoxy-ribose sugar, nitrogenous bases, and phosphate group. They have been classified into four different types on the basis of their constituting bases. These nitrogenous bases are of four different types, namely: adenine (A), cytosine (C), guanine (G), and thymine (T). Nucleic acid chain is formed by the sequential arrangement of these nitrogenous bases which encode the genetic information. This genetic information is coded to RNA and that process of formation of RNA is known as *transcription*. RNA (ribose nucleic acid)

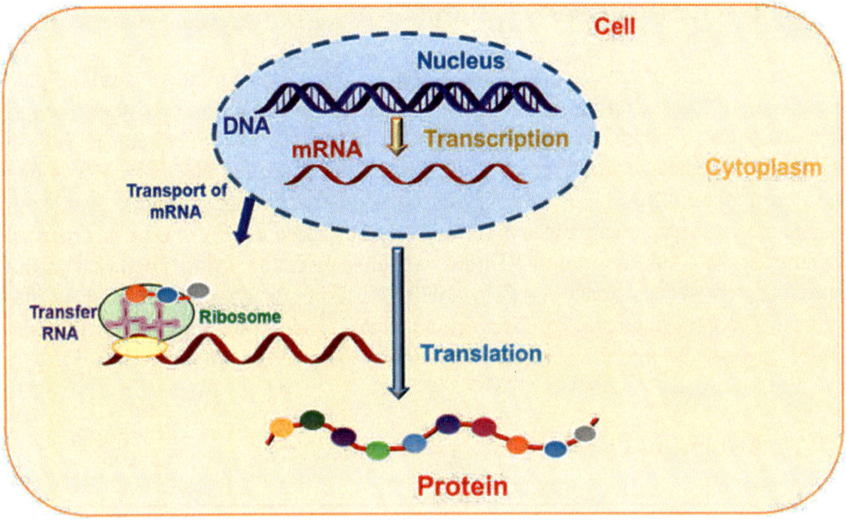

Fig. 1.2 Illustration depicting the flow of genetic information form DNA to RNA to protein termed as "central dogma" of molecular biology

uses codons to carry this information and ultimately decodes this message in the process of *translation*, whereby the RNA codes for distinct amino acids that results in the formation of the polypeptide chain of protein. The whole process of transfer of information from DNA to RNA and RNA to proteins is termed as "central dogma" (Fig. 1.2). There are 64 possible codons which code for 20 distinct amino acids. This difference in number of codons and amino acids results in degeneracy in the genetic code. This degeneracy implies that more than one codon is encoding for an amino acid. Among 64 possible codons, three codons including UAA, UAG, UGA do not code for any of the amino acid and are known as stop codons. These codons are added at the end to terminate the growth of polypeptide chain during the translation process.

Translation involves three distinct forms of RNA, namely: messenger RNA (mRNA), ribosomal RNA (rRNA), and transfer RNA (tRNA). mRNA contains the information that directs the amino acid sequence of a polypeptide chain. The association of amino acid into a polypeptide chain is performed in rRNA. tRNA aids in transferring amino acids from cytosol to the developing chain of polypeptide in accordance with mRNA's triplet code. The process of adding amino acids continues till the stop codon has appeared in mRNA. Finally, the ribosome releases the complete polypeptide chain for further processing (Crick 1970).

1.4 Amino Acids: Beads in the String of Proteins

Proteins are made of combination of one or more polypeptide chains. A polypeptide chain is a polymer of amino acid residues which are linked with each other through the formation of peptide bond which is also known as CO-NH linkage. Such a linkage is made as a result of removal of water elements from α-amino moiety of one of the amino acids and α-carboxyl group of the other amino acid residue. There are around 20 naturally occurring amino acids that are frequently present in proteins. Asparagine was the first amino acid that was discovered in 1806. The last was the threonine which was discovered in 1938. Different polypeptides differ in the sequence and composition of their amino acids. These polypeptides or proteins can be hydrolyzed into their constituting amino acids (Nelson and Cox 2013).

1.4.1 Structural Characteristics of Amino Acids

All the 20 amino acids are also termed as α-amino acids since each of them exhibits a carboxyl moiety and an amino moiety that are bonded to the single α-carbon atom. Amino acids vary because of their R group or side chain groups (Fig. 1.3). R groups differ in their structure, charge, size, and also affect the amino acid's solubility in water. In addition to 20 amino acids, some other amino acids that are either modified after the formation of protein have also been observed or they are present in the organism but are not a part of proteins. All the standard 20 amino acids are assigned with a 3-letter abbreviation and a single code letter symbols (Table 1.1 and Fig. 1.3).

All the amino acids except glycine have four distinct groups, namely: carboxyl group, amino group, hydrogen atom, and R group. Glycine in place of R group contains hydrogen atom. Hence, the α-carbon in rest of 19 amino acid residues is a chiral center which makes them optically active. The four different groups can be organized in two different manners that are non-superimposable mirror images of each other. These two forms are class of stereoisomers (D and L forms of amino acids) known as enantiomers. All the amino acid residues that are observed in proteins are in L form. D form amino acids are present in minute quantities mostly in small peptides, such as those found in cell wall of bacteria and some of the antibiotics that are based on peptides.

1.4.2 Classification of Amino Acids

Amino acids are divided into three different groups on the basis of the nature of their R groups. They include: (a) non-polar, (b) uncharged polar, and (c) charged polar amino acids.

1.4.2.1 Non-Polar Amino Acids

Nine amino acid residues encompass non-polar R groups. Six amino acid residues, namely: alanine, glycine, methionine, isoleucine, leucine, proline, valine, exhibit

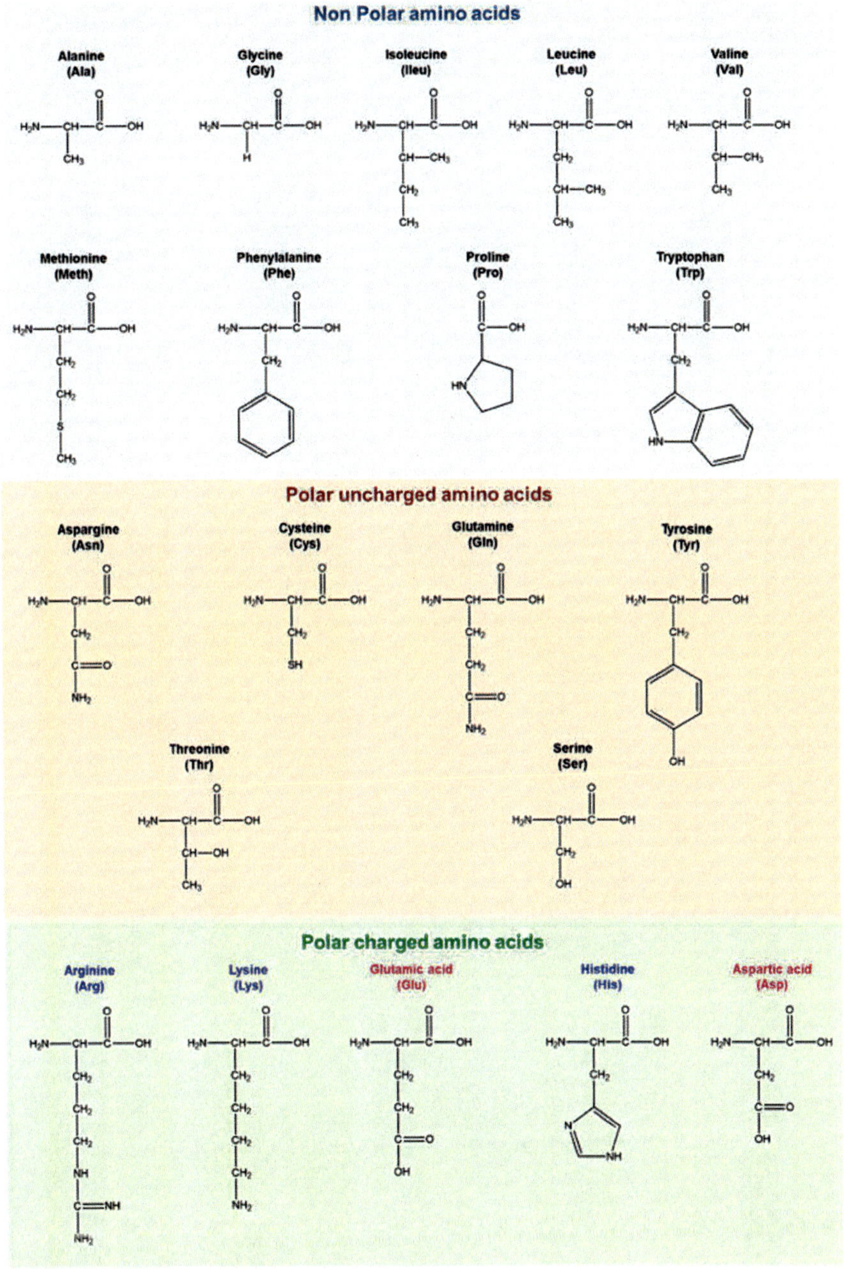

Fig. 1.3 Schematic depicting the structures of 20 standard amino acids, and their three-letter codes are mentioned in parenthesis

Table 1.1 Demonstrating the distinctive physiochemical characteristics of 20 standard amino acids which are categorized on the basis of their charge. Single letter codes of each amino acid are also shown in parenthesis

Nature	Amino acid	Codons	Molecular weight	Hydropathy	pKa	P(α)	P(β)
Non-polar	Alanine (A)	GCA, GCG, GCU, GCG	71.09	1.8	–	1.45	0.97
	Phenylalanine (F)	UUC, UUU	147.18	2.8	–	1.12	1.28
	Glycine (G)	GGA, GGC, GGG, GGU	57.05	−0.4	–	0.53	0.81
	Isoleucine (I)	AUA, AUC, AUU	113.16	4.5	–	1.00	1.60
	Leucine (L)	UUA,UUG,CUA, CUC, CUG,CUU	113.16	3.8	–	1.34	1.22
	Methionine (M)	AUG	131.19	1.9	–	1.2	1.67
	Proline (P)	CCA, CCC, CCG, CCU	97.12	−1.6	–	0.59	0.62
	Valine (V)	GUA,GUC, GUG, GUU	99.14	4.2	–	1.14	1.65
	Tryptophan (W)	UGG	186.21	−0.9	–	1.14	2.29
Polar (charged)	Aspartic acid (D)	GAC, GAU	115.09	−3.5	3.9	0.98	0.80
	Glutamic acid (E)	GAA,gag	129.12	−3.5	4.07	1.53	0.26
	Histidine (H)	CAC, CAU	128.17	−3.2	6.04	1.24	0.71
	Lysine (K)	AAA, AAG	128.17	−3.9	10.79	1.07	0.74
	Arginine (R)	Aga,AGG,CGA,CGC,CGG,CGU	156.19	−4.5	12.48	0.79	0.9
Polar (uncharged)	Cysteine (C)	UGC,UGU	103.15	2.5	8.35	0.77	1.30
	Asparagine (N)	AAC, AAU	114.11	−3.5	–	0.73	0.65
	Glutamine (Q)	CAA,cag	128.14	−3.5	–	1.17	1.23
	Serine (S)	AGC, AGU, UCA,UCC,UCG,UCU	87.08	−0.8	16	0.79	0.72
	Threonine (T)	Aca,ACC,ACG,ACU	101.11	−0.7	16	0.82	1.2
	Tyrosine (Y)	UAC, UAU	163.18	−1.3	–	0.61	1.29

aliphatic non-polar side chains. The side chains of alanine, isoleucine, leucine, and valine are involved in stabilizing the proteins via hydrophobic interactions. Glycine with simple hydrogen as R group does not really make any contribution to the hydrophobic shell of the protein. Methionine contains thio-ether group in its side chain. Proline contains pyrrolidine ring which contributes to the reduction in the structural flexibility of the protein. Phenylalanine, tryptophan, tyrosine are aromatic amino acids with relatively non-polar side chains. Tryptophan and tyrosine are relatively more polar than phenylalanine, owing to the tryptophan's nitrogen-containing indole ring and hydroxyl group of tyrosine. Tryptophan, tyrosine, and to some extent phenylalanine absorb UV light, this feature is utilized by many researchers to characterize the protein's structure-stability-functional aspects.

1.4.2.2 Uncharged Polar Amino Acids

The amino acids, namely: asparagine, glutamine, threonine, serine, and cysteine, fall under this category. These amino acids show more water solubility than the non-polar amino acids. This is due to the presence of groups that can involve in hydrogen bonding with water. Hydroxyl groups of threonine and serine, amide groups of glutamine and asparagine, and sulfhydryl group of cysteine contribute to the polarity of these amino acids. Cysteine gets readily oxidized to form cystine by forming disulfide bonds. Cysteine is highly hydrophobic and is mainly involved in joining different parts of the proteins or for connecting two different polypeptide chains by a covalent disulfide linkage.

1.4.2.3 Charged Polar Amino Acids

Charged amino acids are highly hydrophilic in nature. Out of 20, there are five amino acids that are charged. Out of five, there are three residues with positive charge, namely: histidine, lysine, and arginine. Histidine with its imidazole moiety, lysine with its aliphatic side chain containing amino group at ε position, and arginine with guanidino group are highly positively charged at pH 7.0. Remaining two amino acids, namely: glutamic acid and aspartic acid with side chain carboxyl groups, are the negatively charged amino acids that remain negative at pH 7.0.

1.4.2.4 Uncommon Amino Acids

Beside 20 amino acids, many other unusual or uncommon amino acids have also been observed. Some of these amino acid residues are the modifications of the common amino acids. For instance, 5-hydroxylysine and 4-hydroxyproline are derived from lysine and proline, respectively. 4-hydroxyproline is present in cell walls of protein. However, both the 4-hydroxyproline and 5-hydroxylysine are present in collagen, which is an abundant fibrous protein found in connective tissues. Carboxyglutamate is present in prothrombin and also in other calcium-binding proteins. Desmosine is observed in elastin. Selenocysteine is introduced in the protein during the synthesis of proteins rather than by the post-translational modifications. In this amino acid sulfur in cysteine is replaced by selenium 6-N-methyllysine which is present in contractile muscle protein known as myosin. Many other uncommon amino acids with variety of functions are present in cell but they are not present as the part of proteins. For example, citrulline and ornithine are important intermediates during the arginine biosynthesis and urea cycle.

1.4.3 Acidic and Basic Properties of Amino Acids

Amino acids dissolved in water exit as dipolar ions or zwitter ions in the solution. This implies that they can act as both acid and base (proton donor as well as proton acceptor). Such dual nature entities are amphoteric in nature and are known as ampholytes. pk1 denotes the α-carboxylic group values which lie in the range of 2.2 indicating that it stays in the form of carboxylate ion when the pH is above 3.5. In contrast, pk2 refers α-amino group values which lie in the range of 9.4 implying that it remains as ammonium ions when the pH is below 8.0. pkr denotes the values for different side chains of amino acids. pKa values of all the 20 amino acids are summarized in Table 1.1.

1.4.4 Hydrophobicity Profiles of Amino Acids

Hydrophobicity is a measure of affinity of interaction between water/polar solvents and side chain of amino acids. Different nature of side chains of proteins attributes their differential interaction with polar solvents. Hydropathy is the measure of exact relative values of hydrophobicity or hydrophilicity of amino acids. All the non-polar amino acid residues with lower affinities for water are hydrophobic in nature and exhibit higher positive values in the hydrophobicity scale as compared to other polar charged/uncharged amino acids. Hydropathy scores of all the amino acid residues are briefly described in Table 1.1 (Kyte and Doolittle 1982).

1.5 Structural Hierarchy of Proteins

Proteins are the large complex macromolecular entities. They are made of different types of polypeptides with distinct amino acid sequences which are folded in a fashion to give biologically and functionally active proteins (Edison 2001). Hence, the structures of these complex entities have been defined at four different levels, namely: primary, secondary, tertiary, and quaternary structures (Poluri and Gulati 2017f). Primary structures define the amino acid sequence. Secondary structure denotes the arrangement of amino acids in a repetitive structural fashion. Tertiary structure implies the folding of polypeptide chains in three dimensions. Quaternary structures imply the arrangement of polypeptide chains in space.

1.5.1 Primary Structure of Proteins

Primary structure of protein is defined as the linear organization of amino acid residues that are linked to each other through the formation of covalent peptide bonds (Fig. 1.4). These peptide bonds are rigid and planar owing to the resonance or partial sharing of pair of electrons between amide nitrogen and carbonyl oxygen. These peptide bonds are devoid of free rotation due to the partial double bond character. Peptide bonds mostly exist in "trans" conformation. Polypeptide backbone is allowed to rotate about two bonds, namely: Cα-N and Cα-C. The angles about these bonds Cα-N and Cα-C are known as Φ (phi) and Ψ (psi) torsion angles. These angles dictate the conformational state of a polypeptide chain. When the conformation of polypeptide chain is fully extended, the values for torsion angles are 180°. G. N. Ramachandran pioneered a two-dimensional graph termed as Ramachandran plot which defines the energetically favorable or sterically allowed values of Ψ and Φ that a polypeptide can adapt (Fig. 1.5). 75% of the region in graph denotes the inaccessible conformations of the polypeptide. Ψ and Φ values for all the secondary structures in proteins lie in the conceded regions of the Ramachandran plot. Some residues that are involved in protein folding compensate for steric clashes, hence their value lie in the forbidden regions of Ramachandran plot. Ψ and Φ values for all the secondary structure are represented in Table 1.2.

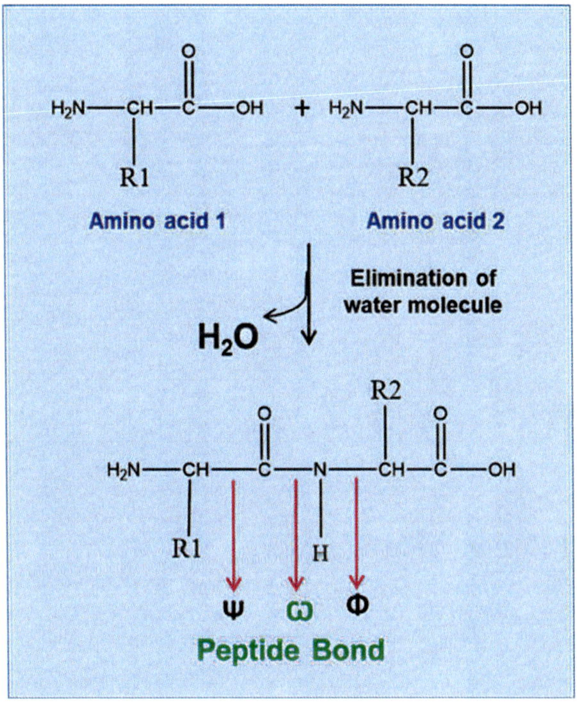

Fig. 1.4 Schematic showing the covalent peptide bond formation between the two amino acid residues; the backbone torsion angles (Ψ and Φ) are also shown

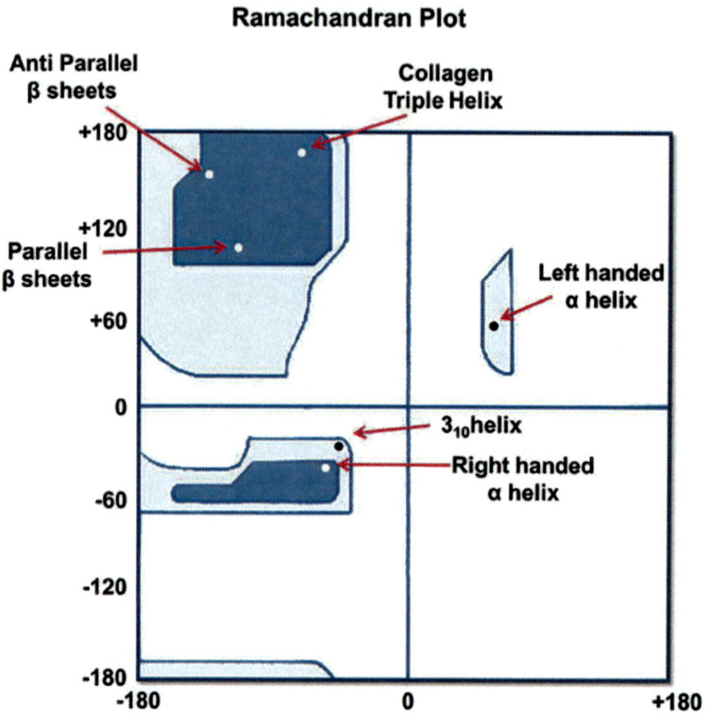

Fig. 1.5 Ramachandran plot showing sterically allowed values of Ψ and Φ. Conformational regions for differential secondary structures are also marked

Table 1.2 List of Ψ and Φ values for different secondary structure elements

Secondary structure	Φ (in degrees)	Ψ(degrees)
2.27 –ribbon	−78	59
Antiparallel β-pleated sheet	−139	135
Left-handed α helix	57	47
Left-handed polyglycine	−79	150
Polyproline type I helix	−75	160
Polyproline type II helix	−75	150
Parallel β-pleated sheet	−119	113
Right-handed α helix	−57	−47
Right-handed 310 helix	−49	−26
Right-handed π helix	−57	−70
Type I β-bend (residue2)	−60	−30
Type I β-bend (residue3)	−90	0
Type II β-bend (residue2)	−60	120
Type II β-bend (residue3	90	0

1.5.2 Secondary Structure of Proteins

Secondary structure represents the next degree of organization of structure in proteins. It basically refers to the local conformation attained by various parts of protein. It is a regular folding pattern followed by the polypeptide backbone. Existence of these structures was first predicted by Pauling and Corey in 1951. Different types of secondary structures were observed among the proteins; however, few of them including α-helices and β-sheets are highly stable and commonly present in proteins. Different secondary structures are described in the following paragraphs.

α-helices—polypeptides acquire helical structure when they twist by the equal quantities around each of its Cα atom. *α-h*elices are made as a resultant of formation of hydrogen bonds between amide groups and carbonyl groups found in close vicinity in the sequence. Hence, the polypeptides with helical structures are fully involved in hydrogen bonding except the amide group of the initial (first) residue and the carbonyl group of the amino acid present at the end of the polypeptide chain. This type of exhaustive hydrogen bonding cylinders the protein structures in which the middle axis is represented by the fully hydrogen bonded backbone and all the R groups or side chains are protruding outwards from the polypeptide for interacting with differential regions of other polypeptide chains or to some other different protein (Fig. 1.6a).

In the right-handed α-helices, the repeating unit is the single turn of helix with pitch (which is defined as the distance that the helix rises along its axis per turn) of 5.4 Å, torsion angles $\Psi = -47°$ and $\Phi = -57°$ and around 3.4 amino acid residues per helical turn (n). Hydrogen bonds are formed between amide (N-H) group of nth amino acid residue and carboxyl (C=O) group of n-fourth amino acid residue. The common notation for helices is n_m, where n displays the number of residues per

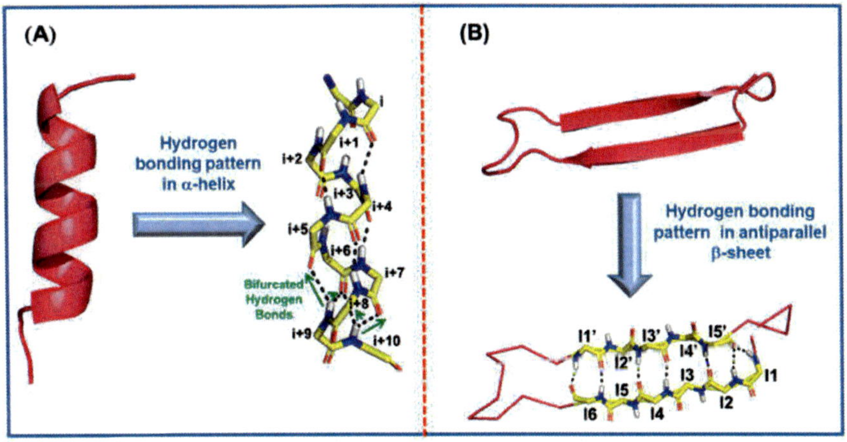

Fig. 1.6 Hydrogen bonding patterns followed by amino acids in (**a**) α-helix and (**b**) antiparallel β-strands

helical turn and m denotes the number of atoms that are indulged in hydrogen bonding, respectively. Differential forms of helices are observed in proteins such as α-helix, 4.4_{16} or π-helix, 2.2_7 ribbon, and 3_{10}-helix. These helices have been categorized on the basis of differential torsion angles, pitch, and number of residues per helical turn (n) (Table 1.2). α-helices are one of the most commonly observed helical structures in globular and fibrous proteins in contrast to other helices such as 4.4_{16} and 3_{10}-helix, which are found only in small segments of proteins or in turns. In general, ¼ of the amino acids acquire helical structure in all the polypeptides, the exact fraction varies from protein to protein. The preference for formation of helices in the protein structures is majorly due to their stability and the involvement of hydrogen bonding as a major stabilizing force. L-amino acids, that occur naturally, can form both the right-handed and the left-handed α-helices. However, left-handed α-helices are never observed in proteins. Formation of α-helices and their stability are highly dependent on sequence and identity of amino acids in a polypeptide. For instance, the presence of a block of acidic amino acids like Glu in a segment of protein will not be able to form α-helix at pH -7.0, as the negative charge of carboxyl groups of Glu residues will make them repel from each other. Similarly, presence of block of Lys/Arg residues will inhibit the formation of α-helix due to repulsion of positively charged NH groups of Lys/Arg. Proline because of its ring structure introduces a kink in the α-helix. Hence, Pro is rarely observed in helices. Gly is observed infrequently in α-helices owing to its conformational flexibility.

β-*sheets*—these are the second type of repetitive structures that are acquired by the proteins. β-sheet structure was confirmed using X-ray analysis. It is a more extended conformation of polypeptide chains with a zigzag structure. This structure resembles pleats when polypeptide chains are arranged side by side. β-sheets are made due to formation of hydrogen bonds between the neighboring segments of polypeptide chains. Segments forming β-sheets can be the segments that belong to the same polypeptide chain that can be present nearby or far from each other. The other possibility is that they can be the segments of different polypeptide chain. The amino acid side chains point outwards from the zigzag structure in the opposite directions (Fig. 1.6b).

β-sheets are of two types, namely: parallel or antiparallel depending on the orientation of adjacent polypeptide chains. Parallel β-sheets are formed when both the adjacent polypeptides have same amino to carboxyl orientations. Antiparallel β-sheets are formed as a result when adjacent polypeptides run in opposite direction. Structures for both parallel and antiparallel are quite similar; however, their repeat distances and hydrogen bonding patterns are different. Repeat distance for parallel and antiparallel β-sheets is 6.5 Å and 7.0 Å, respectively. C=O and NH groups from the two polypeptides that are facing each other form hydrogen bonds (Fig. 1.6b). Other NH and C=O groups that are at the ends can be involved in hydrogen bonding with water molecules that are present on the surface or these groups can also be involved in interaction with side chains in α-helices to pack against them. In case of β-barrel formation, NH and C=O groups from the adjacent polypeptides curve around one another. β-sheets that are antiparallel are more common among the proteins in comparison with parallel β-sheets, as antiparallel β-sheets are more stable than parallel β-sheets. This is due to the presence of distorted patterns of hydrogen bonding in parallel β-sheets. β-sheet strands are either connected by β-turns or hairpin turns or by complex helical turns. Knotted topologies have never been observed in β-sheets. β-sheets exhibit a structure that is fully extended and have phi and psi values ($\Phi = -119°$ and $\Psi = 113°$ for parallel β-sheets and $\Phi = -139°$ and $\Psi = 135°$ for antiparallel β-sheets) that belong to sterically permitted regions of Ramachandran plot. β-sheets are present on the surface of the protein owing to the presence of adjacent hydrophobic and hydrophilic amino acids in the adjoining β-strands of the β-sheets.

In addition to these regular secondary structures, globular proteins also exhibit some irregular and non-repetitive structures. These structures are known as loops/turns/coil conformations. Loops and turns allow the polypeptide chain to make a turn in reverse direction and also aid in connecting other regular secondary structures. β-turns are most commonly found in linking the segments of antiparallel β-sheets. β-turn involves four residues in which carbonyl group of the first amino acid is involved in forming hydrogen bond with the amino group of the fourth amino acid. The primary two residues are not involved in formation of any inter-residue hydrogen bonding. Proline and glycine are most commonly found in β-turns because of the small size of glycine and cyclic structure of proline. Turns with glycine and proline are tight turns. β-turns are usually observed on the surface of the protein in which the peptide groups of central amino acid residues can involve in hydrogen bonding with water. Three-residue turn known as γ-turn in which hydrogen bond forms between the first and third amino acid residue is less common among the proteins.

Amino acids do have the preferences for the secondary structures to adapt. Some of them, for instance, proline and glycine prefer to be present in turns than in regular secondary structures including β-sheets or α-helices. This is due to the several constraints in secondary structures. In 1974, Chou and Fasman performed an analysis on several protein crystal structures and determined the preferential scores of all the 20 standard amino acids for differential secondary structures including α-helices, β-sheets, and reverse turns (Fasman 1990). Based on which they categorized residues as strong helix breaker or former and strong β-sheet breaker or former. Table 1.1 enlists the conformational parameter scores for each of the standard amino acid.

1.5.3 Tertiary Structure of Proteins

Tertiary structure is defined as the spatial arrangement or three-dimensional arrangements of all the atoms in the protein with an aim to form a compactly folded conformation. In secondary structures, the spatial arrangement of adjacent amino acids has been defined. In contrast, spatial arrangement of long range amino acids will be defined in tertiary structure. Amino acids that are present far from one another in polypeptide sequence or the residues present in differential secondary structure may interact in a folded protein structure. Residues may interact via weaker interactions or by covalent bonds such as disulfide bonds. Proteins acquire natively folded structures rapidly; however, in some cases, proteins remain in their partially folded or natively unfolded state.

In globular proteins, the side chains of non-polar residues like Ile, Leu, Met, Phe, and Val occur in the protein's hydrophobic interior. These non-polar side chains can interact to each other and form a cluster in the water repelling interior of the protein and thereby resulting in a hydrophobic core in the interior of a protein. In contrast, the side chains of charged polar residues like Asp, Arg, Glu, His, and Lys reside on the surface of globular proteins from where these residues are accessible to the polar solvents or aqueous environment. In case of large polypeptides, the tertiary structural arrangement is in the form of domains. Different segments of polypeptide form different globular domains which can either directly interact with one another or via a connecting polypeptide. Such domains can have different functions and conformations which can alter by the interaction with each other by means of cooperativity and allostery. In most of the cases, the interactions between the domains are highly flexible, as the interface forms the binding site for both biomolecules and small molecules (Voet and Voet 2011).

1.5.4 Quaternary Structure of Proteins

Numerous proteins comprise multiple polypeptide chains/subunits and their association performs variety of functions. Multi-subunit proteins are involved in regulatory functions. Small molecule binding may affect the interaction between the subunits of the protein, thereby greatly affects their activity. In many cases, different subunits perform differential but related functions. Large proteins serve as the site of various multi-step complex reactions (Petsko and Ringe 2004).

Multi-subunit proteins are known as "multimers." They are made of either two subunits or several hundreds of subunits. If a protein is composed of only few subunits, then it is known as oligomer. Protein is said to be asymmetric, if it is composed of non-identical subunits. In multimers, mostly the subunits are identical or they constitute repeating groups of non-identical subunit arranged in symmetric fashion. Hemoglobin is first oligomeric protein whose 3D structure was defined by employing X-ray crystallography in 1959. It is composed of four polypeptide chains along with four heme prosthetic groups with iron atoms in ferrous state. These polypeptide chains are arranged symmetrically.

Oligomers can range from dimers, trimers, tetramers, etc. Identical subunits interact in symmetric manner due to the presence of complementary binding surfaces. In contrast, non-identical subunits interacting asymmetrically will have different binding surfaces. For instance, a subunit with X binding surface will interact with the subunit having X' binding surface. The asymmetric subunit is also referred to as the protomer. Subunits that have two complementing interacting surfaces result in formation of more complex from of protein. For example, monomers interact to form dimers, dimers can interact to form tetramer (Fig. 1.7a–c). Further, interaction of subunits can also lead to hexamer and octamer (Fig. 1.7d, e) and other higher order structures like decamers and dodecamers. Some of the

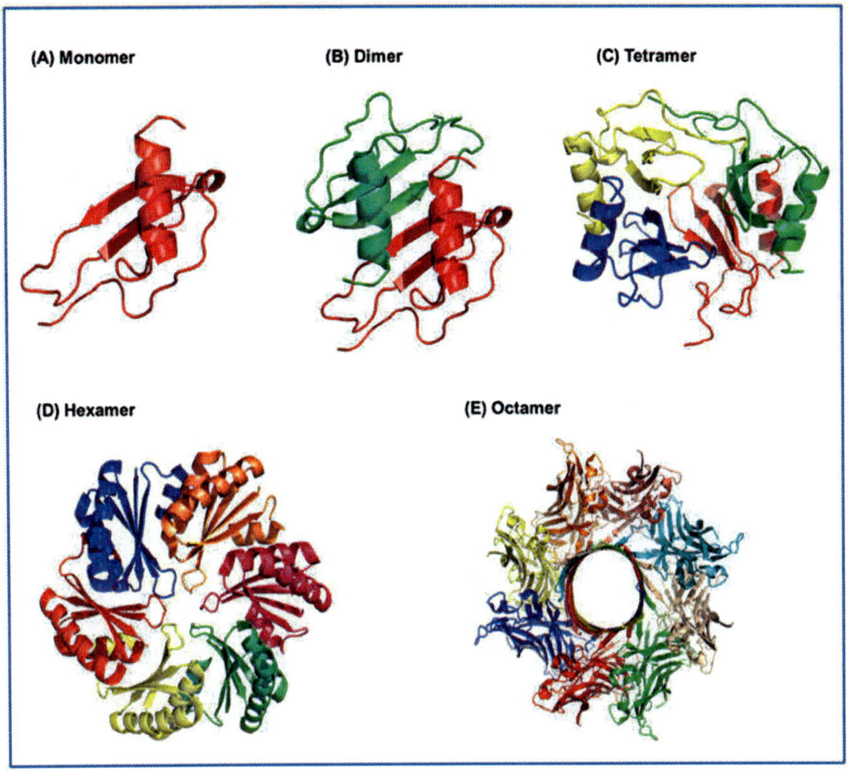

Fig. 1.7 Molecular representation of the differential oligomeric structures: (**a**) monomer, (**b**) dimer, and (**c**) tetramer structures of human CXCL7 (PDB id: 1NAP), (**d**) Aeromonas hydrophila's microcomparment protein forming hexamer (PDB id: 4QIV), (**e**) Staphylococcus aureus's gamma hemolysin forming an octameric pore (PDB id: 3B07)

proteins are pseudo symmetric in nature. Proteosome, a giant and complex multi-subunit protein degrading machinery is pseudo-symmetric.

1.5.5 Motifs/Super Secondary Structures

Super secondary structures also termed as motifs are the intermediate structural entities between the levels of secondary and tertiary structural organization of proteins. They are formed as a consequence of amalgamation of several secondary structure elements and are found in globular proteins. Such secondary structures are compactly packed, and the hydrophobic side chains are clustered together in the interior of protein that leaves behind only nominal space in the interior. Additional interactions between polar side chains (hydrophilic interactions) and non-polar side

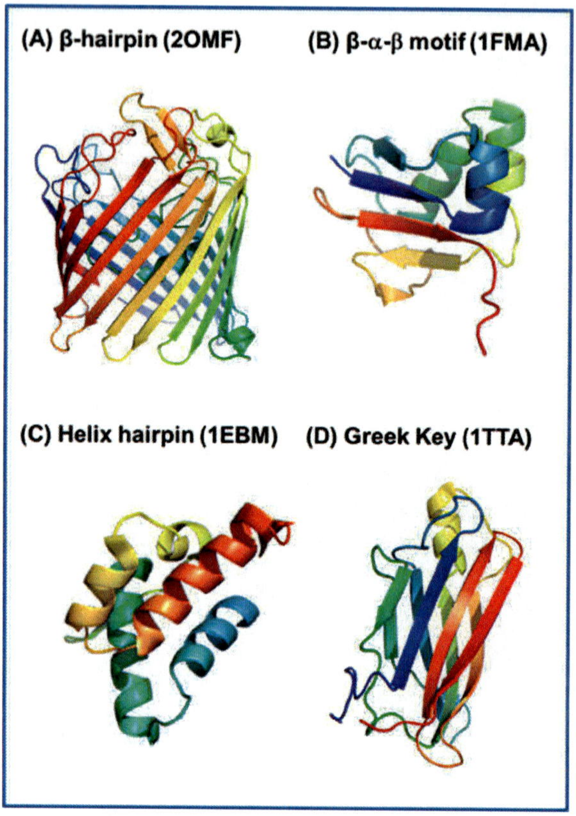

Fig. 1.8 Cartoon representation of commonly observed folds in differential proteins: (**a**) E. coli outer membrane protein (PDB id: 2OMF) with β-hairpin (**b**) Molybdopterin synthase (MOAE subunit) (PDB id: 1FMA) with β-α-β motif, (**c**) 8-oxoguaninglycosylase (1EBM) with helix-hairpin (chain A central domain (135–221) is displayed), (**d**) Pre-albumin (PDB id: 1TTA) with Greek key motif

chains (van der Waal interactions) also persist to further stabilize these secondary structures.

Many different motifs are present, for instance, β-hairpin, which is formed when two antiparallel β-strands are joined together by reverse turn (Fig. 1.8a). βαβ motif is formed when there is right-handed crossover connectivity between β-strands that contain α-helix (Fig. 1.8b). αα-hairpin motif is formed, when there is a connectivity between the two antiparallel α-helices via a loop (Fig. 1.8c). Greek key motif is formed, when antiparallel β-strands fold upon itself (Fig. 1.8d). These motifs together form different types of domains, which have been further categorized based on their secondary structure content. α-domains are those which predominantly comprise α-helices and β-domains are those that contain essentially the β-sheets. α/β-domain contains both the sheets and helices.

Globin is the common α-domain found in hemoglobin and myoglobin (Vojtechovsky et al. 1999). It involves the folding of 8 helices in the form of two layers (Fig. 1.9a). 4-helix bundle is also found in proteins such as cytochrome b_{562}, where the helices are arranged in up-down-up-down connectivity (Fig. 1.9b)

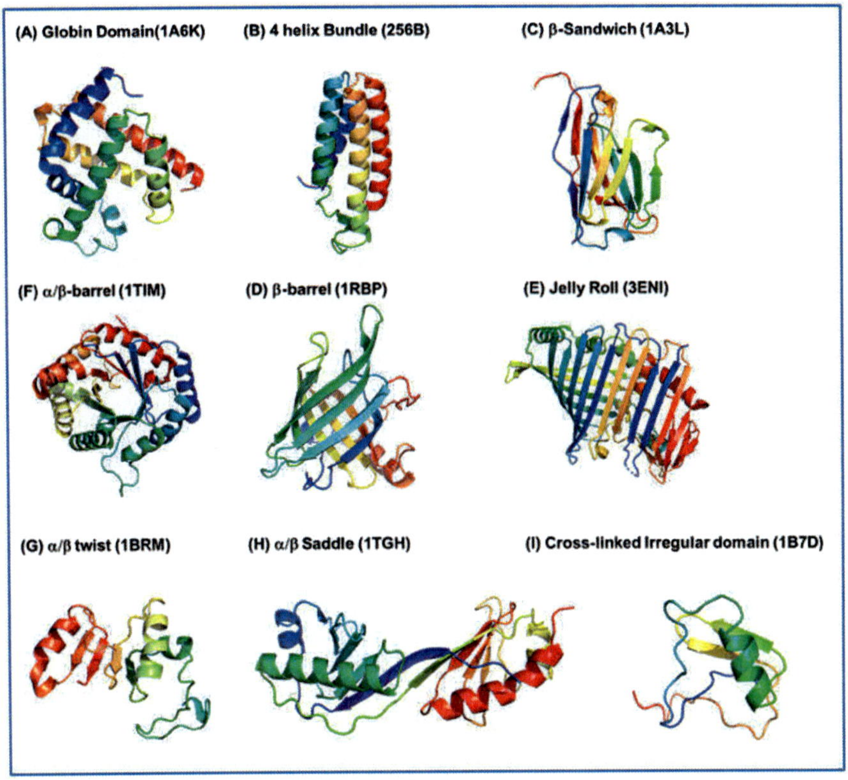

Fig. 1.9 Differential proteins containing distinct variety of domains (**a**) Myoglobin (PDB id: 1A6K) with single globin domain (**b**) E. coli Cytochrome b562 (PDB id: 256B) with four helix bundle (**c**) Immunoglobulin A protein (PDB id: 1A3L) with β-Sandwich (only light chain is shown), (**d**) Retinol binding protein (PDB id: 1RBP) with β-barrel, (**e**) Fenna-Matthews-Olson Protein from Chlorobaculum tepidum (PDB id: 3ENI) with Jelly roll, (**f**) Triosephosphate isomerase (PDB id: 1TIM) with Alpha/beta barrel, (**g**) Aspartate beta-semialdehyde dehydrogenase (PDB id: 1BRM) with alpha/beta twist, (**h**) TATA binding protein (PDB id: (1TGH) with alpha + beta saddle, (**i**) Scorpion toxin (PDB id: 1B7D) with cross-linked irregular domain

(Lederer et al. 1981). 4-helix bundle with dissimilar connectivity such as up-up-down-down connectivity has also been observed in different proteins.

Different β-domains have also been reported for the proteins (Fig. 1.9c–e). One is β-sandwich structure in which the two sheets are packed against each other, most usually at the right angles to one another. Sheets can be made of 4–10 β-strands in antiparallel trend. β-sandwich fold is found most commonly in immunoglobulins and is termed as immunoglobulin fold. This type of fold is made of two sheets packed against each other, with one sheet made up of four antiparallel β-strands and other sheet comprises three antiparallel β-strands (1A3L) (Heine et al. 1998). The second is β-barrel, which is made up of single β-sheet folded in a cylindrical form, and all the strands in the β-sheet are hydrogen bonded to each other. The first β-strand is hydrogen bonded to the last one. β-barrel structure is observed in human retinol binding protein (1RBP) with up and down connectivity. β-barrel having Greek key motif is found in pre-albumin (Cowan et al. 1990). Jelly roll also known as Swiss roll is a β-barrel variety that comprises eight antiparallel β-strands with up and down connectivity. Jelly roll structure was observed in bacterial chlorophyll A protein (Li et al. 1997).

α/β domains contain two most common motifs, one is α/β barrel which is a closed structure with β-α-β-α motif repeated four or more times (Fig. 1.9f–i). This structure was observed first in enzyme triosephosphate isomerase (TIM). It comprises eight stranded parallel β-barrel surrounded by a barrel of eight helices. This barrel is also termed as TIM barrel (Banner et al. 1976). The internal hydrophobic region that forms a core is formed by β-strands and enclosed by helices. Hydrophobic face of helices was packed against β-sheet and the polar side chains pointing toward the aqueous environment. TIM barrel structure is predominantly observed in enzymes. Around 10% of enzymes display TIM barrel structure.

Other is α/β twist comprising six antiparallel β-strands connected by five helices (Hadfield et al. 1999). Dinucleotide binding domain also known as Rossmann fold contains α/β twist structure which is common among nucleotide binding proteins. In α/β saddle, two β-sheets form an open sheet structure which gets twisted to form a saddle (Juo et al. 1996). Two β-sheets are arranged in two halves and the helices lie on those sheet halves. Cross-linked irregular domain is a distinctive small domain present in a few intra- or extra-cellular proteins. Covalent interactions between different parts of domains are a major stabilizing force for these domains. Small disulfide linked irregular domain is common among various toxin proteins such as scorpion toxin, cobra venom neurotoxin, and secreted protease inhibitor (Leon and Roth 2000).

1.6 Major Driving Forces for Protein Structural Stability

Protein structure undergoes stabilization via the additive effects of various interacting forces at different levels of structural organizations (Voet and Voet 2011). Such forces can be the non-covalent interactions playing in the protein or the interactions of protein with external environment. Additive effect of these forces leads to conformationally stable protein structures. Non-covalent interactions prevalent in proteins include: electrostatic or attractive/repulsive ionic interactions, hydrogen bonding, hydrophobic interactions which are described below in the following sections.

1.6.1 Ionic Interactions

Ionic interactions also known as electrostatic interactions are set into play via the interaction of two oppositely charged groups. The interactions form salt bridge or ionic pair. Amino acid residues such as arginine, histidine, and lysine with positive charge interact with amino acid residues (like aspartic acid and glutamic acid) containing negative charge present in their close proximity. Changes in solution's pH and concentration of salt can vary the charge on the residues' side chains and thereby result in disruption of these ionic bonds.

1.6.2 Van der Waals Interactions

Electrically neutral molecules form the permanent or induced dipoles which interact with each other to give rise to weak van der Waals forces. Amide and carbonyl groups in the backbone are the permanent dipoles in the protein which interact with one another or with other amide or carbonyl groups resulting in dipole–dipole interactions. Further, permanent dipoles possess a tendency to trigger the dipoles in the non-polar molecules and thereby results in weak dipole-induced dipole interactions. Induced dipoles can also in turn result in dipole induction in the non-polar molecules and then interact with those induced dipoles and result in the weaker London forces of interaction. These interactions are imperative during the interaction of protein with other complementary proteins or small molecules.

1.6.3 Hydrogen Bonding

Electrostatic attraction between the electron acceptor and donor groups results in the formation of hydrogen bond. Their association energies vary from −12 to −40 kJ/mol that essentially range between that of van der Waals interactions and covalent bonds. Hydrogen bonding is notably significant in holding the protein's structure. It is a major driving force of interaction during the formation of secondary structure such as β-sheets and α-helices. Bifurcated hydrogen bonds are present in both β-sheets and α-helices. Since the hydrogen bonding observed between the amide group of nth and carbonyl group of n-fourth residue is also additionally observed between the amide group of nth and carbonyl group of n-third residue. 3-centered hydrogen bonding pattern is also common among β-sheets. In β-sheets, bifurcation is observed among 40% of hydrogen bonds approximately.

1.6.4 Hydrophobic Interactions

Hydrophobic interactions are highly important in discerning the fold of proteins since these interactions are the major driving force during the folding of proteins. The side chains of hydrophobic residues are also involved in interaction with each other and cluster together to form the protein's hydrophobic core and thereby eventually results in the folding of protein. The relative hydrophobicity values of amino acids known as hydropathies have been estimated by Kyte and Doolittle which are mentioned in Table 1.1 (Kyte and Doolittle 1982). Higher positive values denote the higher hydrophobicity of an amino acid. Hydropathies are highly important in discerning the hydrophobic and hydrophilic parts of protein which ultimately give an outlook for the particular fold of the protein.

1.7 Protein Folding

Amino acids assemble to form proteins at a very high rate in living cells. For instance, a biologically active protein with 100 amino acids is formed in 5 s at 37 °C. The question arises that how a polypeptide arrives at its native conformation. If it is assumed that each amino acid takes up ten different conformations, then polypeptide will take up 10^{100} conformations. Now, if it is assumed that protein folds spontaneously in a random process, then it takes up around 10^{77} years for a protein to try all the possible conformations to finally reach its native state conformation. Hence, the protein folding cannot be a trial/error or random process. This problem was initially marked by Cyrus Levinthal in 1968 and is also known as Levinthal's paradox (Levinthal 1968).

The folding mechanism of long polypeptide chains is a complex process. Various models have been suggested to describe the rapid folding process in comparison with random search mechanism. They are: (1) sequential protein folding model or framework, (2) nucleation growth mechanism, (3) diffusion–collision mechanism, (4) hydrophobic collapse model, and (5) energy landscape or folding funnel model.

According to sequential protein folding model, some parts of the protein fold first. Some amino acids fold readily to form helices or sheets, followed by the long-range interactions, for instance, between two helices that will interact to form super secondary structures. The process will continue until the completely folded polypeptide is formed. In a hydrophobic collapse model, the hydrophobic interactions among the non-polar side chains result in hydrophobic collapse of the polypeptide in a compact structure. This state is termed as molten globule state of protein in which there is a high content of secondary structures comprising many amino acids side chains which do not possess particular conformation. Many proteins fold by the mechanism which has features of both the above-described models.

Among all the models, the energy landscape model is the latest and most prominent model (Fig. 1.10) (Bryngelson et al. 1995; Onuchic et al. 1997). In this model, the folding process is viewed as a thermodynamic process. Folding is regarded as free energy funnel in which the conformationally stable state exhibits lower entropy and lower free energy. In contrast, the unfolded state exhibits higher entropy and relatively higher free energy. The conformations at the top of the funnel have higher free energies than those at the bottom of the funnel. Small depressions that are observed along the walls of the funnel denote semi-stable intermediate states which can slow down the folding process briefly. At the end of the funnel, there exists a single fully folded native conformation of the protein or a small set of native conformations.

It has been observed that the thermodynamic stability does not have even distribution about the whole protein; however, there are several regions of higher stability which are separated by the segment of lower stability. The low stability

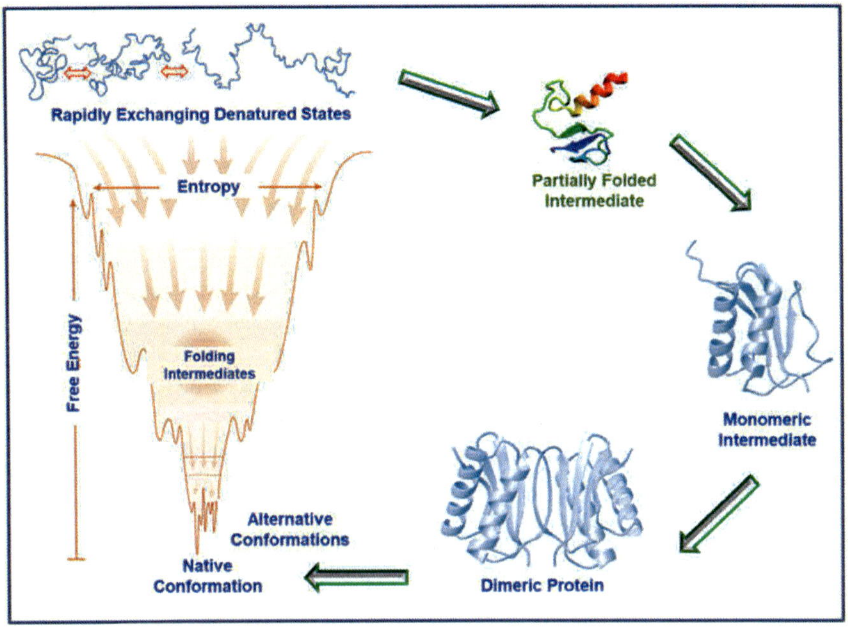

Fig. 1.10 Illustration of protein folding energy landscape: Funnel surface denotes a complete range of conformations of protein ranging from the multiple denatured conformations to the unique native conformation. The ordered state represents the protein structure which is natively folded and that exhibits precisely defined secondary and tertiary structure. Alternative conformations show the high energy native state conformations that comprise all the secondary and tertiary structure features of folded state (Krishna Mohan et al. 2008; Mohan et al. 2008). Molten globule states are intermediate entities with compact structures in the protein folding pathway mechanism that possesses significant content of secondary and nonspecific tertiary structures, with compelling structural flexibility (Sharma et al. 2016; Sharma et al. 2018). Random coils refer to the highly unstructured and denatured protein states (Chakraborty et al. 2012). Differential intermediate conformations along the folding trajectory of protein (dimer) are sculpted on the right side of the funnel (Mohan et al. 2010; Mohan and Hosur 2009; Mohan et al. 2009)

regions allow the protein to try one or more alternating conformational states. Such regions with differential stability are essential for the protein to function. Defects in the protein folding is associated with many genetic disorders, one is cystic fibrosis which occurs as a result of defect in cystic fibrosis trans-membrane conductance regulator (CFTR) protein which serves as a channel for chloride ions (Hull 2012). Most commonly, the mutation involving the deletion of Phe508 in CFTR causes an erroneous folding of the protein (Wang et al. 2000).

1.8 Evolution of Proteins

"How proteins have evolved?" The answer to this question is hidden in their amino acid sequence. Protein's amino acid sequence not only provides information related to its structure, function, or its cellular location, but also illuminates its evolutionary pathway. Thousands of protein sequences are stored in databases (Xu and Xu 2004). The sequence of newly discovered protein can be compared with the stored protein sequences and the relationship between them can be enlightened. Proteins with similar sequences, structure, and function have evolutionary relationship with each other; together these proteins constitute a protein family. An individual protein is assigned to some family based on the degree of its sequence similarity. Proteins within the family are 25% or more identical in their sequences and also share some functional and structural elements. Some protein families are defined on the basis of identity of few amino acids that are crucial for their function or forming special substructures or domains. Hence, evolutionary relationship of proteins can also be deciphered on the basis of structural and functional characteristics (Gangele et al. 2020; Gangele et al. 2019; Gulati et al. 2017; Joshi et al. 2020). Function of a protein depends on the structural fold which finally depends on the amino acid sequence of a protein and also on their composition. Thus, protein sequence carries wealth of information about how proteins have evolved and how the life has started on this planet (Voet and Voet 2011).

During the 1960s, Emile Zuckerkandl and Linus Pauling traced molecular evolution field employing nucleotide and protein sequences. It was proposed that if there is a close relationship between the two organisms, then it is assumed that their nucleotide and protein sequences should also be closely related or similar (Morgan 1998; Zuckerkandl and Pauling 1965b). Divergence among the sequences greatly increases as the distance between the organisms increases at the evolutionary levels. Vast amount of genome sequencing data has been generated for numerous species that geared up the efforts in revealing the historical events involved in evolving the proteins (Batley and Edwards 2009).

Evolution of proteins has not followed a simple straight pathway. Several complexities are associated in a path for deciphering the details of evolutionary events that are hidden amino acid sequence of a protein (Graur and Li 2000; Gupta 1998). For instance, the amino acids that are essential for the functioning of the protein remain conserved over the time of evolution. In contrast, less essential amino acids undergo changes, such amino acids help in tracing the evolutionary events. Substitution of amino acids is also not a random process. In a particular protein, a specific amino acid which is important for the functioning of protein can tolerate only few or some specific amino acid substitutions (Ng and Henikoff 2001). Hence, different proteins with variable rates of amino acid substitutions evolve at different rates.

A gene or group of genes, when transferred in a rare manner from one organism to other, leads to lateral gene transfer and further adds another layer of complexity in tracing the evolutionary history. The genes that are transferred are quite similar in contrast to the other distantly related genes present in the organisms. Transfer of antibiotic resistance gene in the bacterial population is an example of lateral transfer of gene (Bello-Lopez et al. 2019). The proteins obtained from such transferred genes serve as great candidates to unravel the bacterial evolutionary events, since they share a little evolutionary history with their host organism.

Molecular evolutionary studies mainly focus on families of closely related proteins also known as multigene family proteins (Ohta 1991). Mostly, the families that are involved in essential function of cells are chosen for the evolutionary analysis since there will be lesser chances that such genes are the resultant of lateral gene transfer. For example, the proteins involved in translation are present in bacterial cells as well as in all the eukaryotes. EF-1α, an elongation factor which is present in all eukaryotes, a similar protein EF-TU is also found in bacterial cells, which performs same function as EF-TU (Merrick et al. 1993). Such proteins with similar sequence and function indicate that they share same ancestor, hence are known as homologs. Homologs have been further classified into paralogs and orthologs. Paralogs are related proteins which are found in same species. In contrast, orthologs are those homologs which are found in different species. Evolutionary analysis involves first searching for the homologous proteins, followed by their utilization to rejuvenate the evolutionary paths. Gulati et al. have unraveled the evolution structure and function relationship of growth-related oncogene (GRO) chemokines (Gulati et al. 2017).

Marvelous amount of genome sequencing data has been generated for a number of species and stored in large databases that can be accessed to trace evolutionary events followed by the protein. With an advent in computational power, numerous bioinformatics tools have been designed to concede the evolutionary concepts of proteins (Batley and Edwards 2009; Lee et al. 2012). Such tools include multiple sequence alignment, substitution rates, phylogenetic analysis, and coevolutionary analysis. Comparative analysis of protein sequences both at nucleotide and amino acid levels forms the basis of such tools, since the protein's structural and functional details are masked in their sequences. Hence it is quintessential to comprehensively investigate the protein sequences in order to draw major evolutionary conclusions (Koonin et al. 2002; Rokas et al. 2003; Zeldovich and Shakhnovich 2008; Zuckerkandl and Pauling 1965a).

Indeed, several models have been suggested to study the evolution of multigene protein families. They include: (1) divergent evolution, (2) concerted evolution, and (3) birth and death model (Bielawski and Yang 2003b; Eirin-Lopez et al. 2012). According to divergent evolution, the genes acquired after gene duplication event from their parental sequences acquire novel functions independently. In contrast, the concerted evolution proposes that genes obtained after duplication from their parental sequences evolve in a coincidental way. This implies that the mutation that occurs in a repeat will divulge into all the gene family members as a consequence of repeated occurrence of gene conversion or unequal crossover, thereby resulting in DNA sequence homogenization in all the members. Birth and death model proposes that new gene arose as a result of gene duplication event from the parent sequences. Those new genes can either persist for long time or get deleted or become deactivated due to deleterious/dysfunctional mutations. This model has been

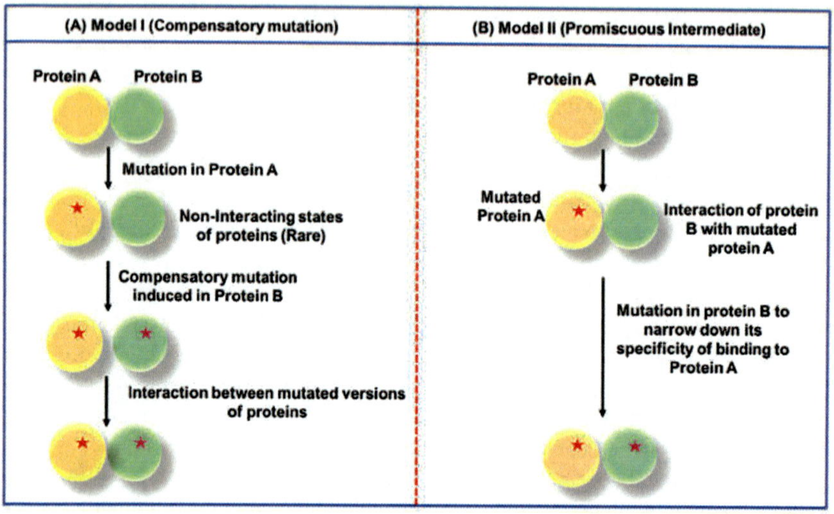

Fig. 1.11 Schematic depicting the differential models of evolution of PPIs specificity (**a**) Compensatory mutation, (**b**) Promiscuous intermediate state model of protein coevolution

followed by many genes, namely: major histocompatibility (MHC) genes, disease resistance genes, and immunoglobulins (Nei and Rooney 2005; Smith 1974).

Divergence in the genes occurs as a resultant of mutations in the coding sequences which exert great effects on their expression patterns, specificity, and interaction with their binding partners. Immunoglobulin heavy chain variable region (VH) genes underwent divergent evolution (Ota and Nei 1994). In contrast, concerted evolution pattern has also been observed for many genes, namely: histone genes (Matsuo and Yamazaki 1989) and ribosomal RNA genes (Rooney and Ward 2005).

Mutations play a significant role in evolution of proteins. After duplication of genes, newly formed genes undergo mutations. Such mutations can be either selected or lost. Some of them can be synonymous or non-synonymous. Synonymous mutations are those in which the changes occurring in the coding region of protein are not altering the protein sequence. In contrast, the non-synonymous mutations are those in which the nucleotide substitutions in the coding region of protein will affect the encoded amino acid sequence. Several studies have demonstrated that proteins are highly susceptible to non-synonymous substitutions in order to attain modified or novel functions (Bielawski and Yang 2003a).

Studies have proposed that interacting protein partners often coevolve. Coevolution implies that a mutation occurring in one of the proteins often drives a compensatory mutation in its interacting partner protein (Ovchinnikov et al. 2014; Skerker et al. 2008). This model involves the formation of a non-interacting or non-functional intermediate state. This model is known as compensatory mutation model (Fig 1.11a) (Aakre et al. 2015). However, the formation of such non-interacting state is highly unusual, if that interaction is highly important for the viability of cell. An alternative model has been proposed in which one of the interacting protein partners passes through a promiscuous intermediate state. This model is known as promiscuous intermediate state model (Fig 1.11b) (Aakre et al. 2015). This involves first the mutation in one of the proteins, say "A" will broaden its specificity, which induces the mutation in its interacting protein partner "B" to disrupt its interaction with ancestral form of protein A. Further, mutation occurs in A to narrow its specificity for the derived form of B and disrupt its interaction with ancestral form of A. In this model, the specificities of interacting protein partners have evolved without the formation of a non-interacting/non-functional intermediate (Kuriyan and Eisenberg 2007) (Fig 1.11b). Aakre et al. have demonstrated the plausibility of promiscuous model using bacterial toxin–antitoxin systems. They screened large libraries of interface variants and showed that the reprogramming of toxin antitoxin specificity passes through the formation of promiscuous variants. They proposed that the evolution of toxin–antitoxin systems and other paralogous protein families involve the formation of various promiscuous variants (Aakre et al. 2015).

1.9 Protein Engineering

Protein engineering is a fledgling field which has fascinated global scientific community for promising proteins with desired properties. Protein engineering allows the researchers to design proteins imbibed with features of their own choice. These features include specificity, activity, stability, solubility of proteins (Lutz and Iamurri 2018; Sinha and Shukla 2019). Engineered proteins have gained momentum due to the advancements in various diversified fields including recombinant DNA technology, screening strategies, and bio-computing techniques. Diverse varieties of proteins have been engineered in both industrial and health care sectors and hence made a remarkable progress in these eras. According to market and market film report, the protein engineering market is expected to rise by $ 4.77 billion by the year 2027, at a CAGR of 12.3%.

Protein engineering has provided us various valuable approaches for designing the novel proteins or improving the existing proteins by attenuating their undesired activities and inculcating desired characteristics in them (Chen and Zeng 2016; Faber and Whitehead 2019; Singh et al. 2018; Wilding et al. 2019). Protein engineering tool box has enabled proteins to work smartly to accomplish their actions. Protein engineering approaches have been broadly classified into two categories, namely: directed evolution/experimental-based approaches and computational/rational designing of proteins (Gulati and Poluri 2019; Gulati and Poluri 2016; Poluri and Gulati 2016; Qu et al. 2019). These approaches are being exploited either alone or in combination to each other to attain their dream proteins.

1.9.1 Directed Evolution (DE)

Directed evolution (DE) is an experimental method of tailoring proteins (Fig. 1.12). DE-based experimental approaches are exploiting the nature's idea of evolution to create or redesign proteins of one's wish (Poluri and Gulati 2017b). However, DE is quite different from natural evolution majorly in one aspect. Natural evolution of proteins took millions of years; in contrast, DE takes few days or months to evolve proteins. DE explores the sequence space that has not been dissected by nature. DE methods have flourished not only to return proteins with desired traits for numerous biological applications. However, they are also being utilized to engineer operons, viruses, novel pathways, and even other whole microbes. DE has also cropped up as

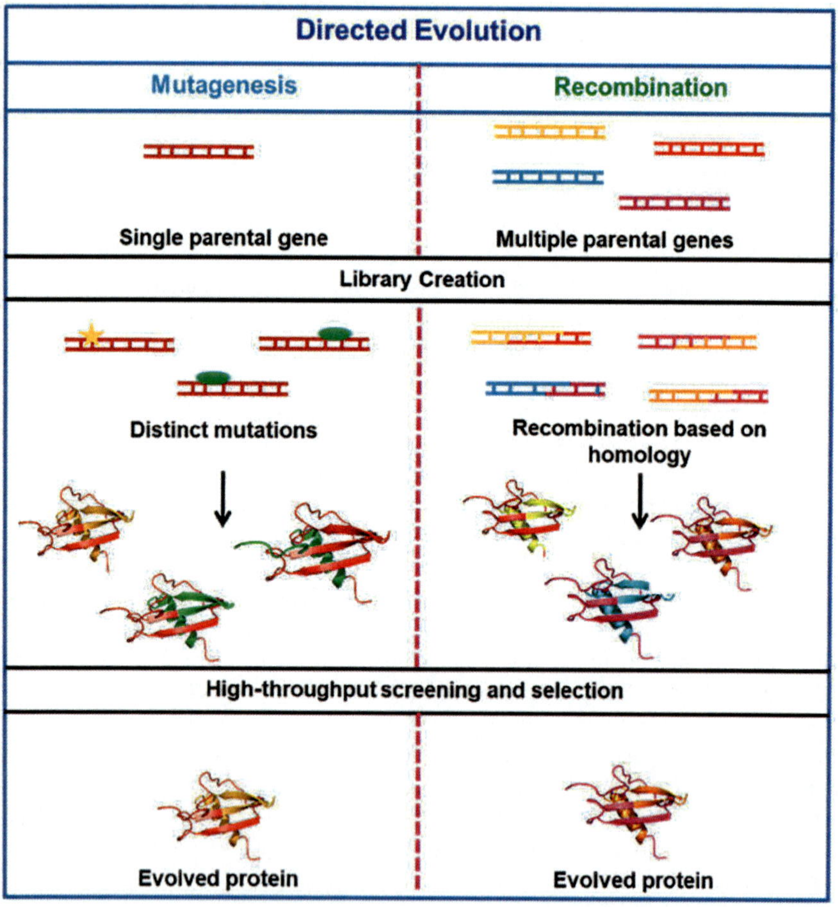

Fig. 1.12 Schematic representing an overview of directed evolution techniques

imperative tool in assessing and modulating the complex protein–protein interactions (Cobb et al. 2013a; Cobb et al. 2012; Cobb et al. 2013b).

DE strategies involve majorly two steps. The first step involves creation of diverse libraries of proteins. The second step involves high-throughput screening process, in which the protein with desired property is selected. These steps are iterated until the protein with interested trait is attained. DE relies majorly on two aspects. Firstly, the quality of protein libraries plays a major role. These libraries should be large containing numerous protein variants with refurbished properties that can be screened easily. Secondly, DE depends on the screening strategy followed (Arnold 1998; Bloom and Arnold 2009).

Methodologies to generate libraries for the DE of proteins have been segregated into two categories, namely: (1) *asexual methods* and (2) *sexual methods*. The choice of method (asexual/sexual) depends on several factors including how much information is available about the parent protein and the extent of novelty that has to be imbibed in the protein (Chen and Arnold 1993; Dalby 2011; Moore and Arnold 1996; Stemmer 1994b; You and Arnold 1996). Further, the availability and accessibility of various resources and techniques also play a significant role in selecting the appropriate strategy. Asexual methods involve single parental gene to generate diverse libraries. These methods have been segregated into two, namely: random mutagenesis and focused mutagenesis.

Random mutagenesis is a robust technique for generating libraries of mutant proteins. These techniques involve introduction of mutations like insertion, deletion, transitions, and transversions in the genes randomly. These methods include: (1) chemical mutagenesis (Deshler 1992; Labrou 2010), in which several chemicals such as mitomycin C, hydroxylamine, sodium bisulfite, etc. are used as chemical mutagens that can randomly mutate the DNA. (2) Error-prone PCR (Cadwell and Joyce 1992; Patrick et al. 2003), which generates libraries by means of erroneous activity of Taq DNA polymerase lacking 3′-5′ exonuclease activity. (3) Mutator strain (Greener et al. 1996) involves transforming the plasmid in the strain lacking DNA repair pathways. *Epicurian coli XL1*-red is the most commonly employed strain which introduces errors in plasmid during replication as it lacks DNA repair mechanism. Several other methods have also been developed in this category which include: rolling circle error-prone PCR (Fujii et al. 2014), dual approach to random chemical mutagenesis (DuARCheM) (Mohan and Banerjee 2008), mutagenesis by random insertion and deletion (Murakami et al. 2002), targeting glycosylases to embedded arrays for mutagenesis (TaGTEAM) (Finney-Manchester and Maheshri 2013), random mutagenesis methods that alter the target DNA length (Fujii et al. 2013; Kipnis et al. 2012), transposon-based random mutagenesis (Tee and Wong 2013).

Focused mutagenesis involves mutations focused at the hotspot residues of proteins. Hence, small libraries are generated in these methods that can be screened rapidly. Site saturation mutagenesis is the most widely used focused mutagenesis method which tries all the 20 amino acids against targeted site at one go. Site-directed mutagenesis (Alcalde et al. 2006; Matsumura and Rowe 2005) and overlap extension PCR (Gratz and Jose 2008) belong to site saturation mutagenesis method.

Several other focused mutagenesis methods such as single-primer reactions in parallel (SPRINP) (Edelheit et al. 2009), sequence saturation mutagenesis (SeSAM) (Wong et al. 2004), cassette mutagenesis (Reidhaar-Olson and Sauer 1988), Ω-PCR (Chen et al. 2013), OSCARR (Hidalgo et al. 2014), PFunkel (Firnberg and Ostermeier 2012), OmniChange (Dennig et al. 2011), trimer–dimer mutagenesis (Gaytan et al. 2009), synthetic saturation mutagenesis (Patwardhan et al. 2009), etc. have been developed which are described in detail elsewhere (Poluri and Gulati 2017d).

In contrast to asexual methods, *sexual methods* are based on recombination. These methods were first originated by Stemmer and his coworkers. They involve recombining homologous or non-homologous parental sequences in vitro (Stemmer 1994a; Stemmer 1994b). The outcome of these recombination events will acquire characteristic of all the recombining parental sequences. Such methods have been divided into two categories, namely: (1) homologous and (2) non-homologous recombination methods based on parental sequences used for recombination.

Homologous recombination methods involve recombination among those gene sequences which share high sequence homology. The resultant chimeric gene inherits features of all the recombining genes. These methods have been segregated into two, namely: in vitro and in vivo homologous recombination methods. DNA shuffling is the in vitro homologous recombination method developed by Stemmer and his coworkers (Stemmer 1994b). Further improvements in DNA shuffling method resulted in other methods, namely: family shuffling (Crameri et al. 1998) and DOGs (degenerate oligonucleotide gene shuffling (Gibbs et al. 2001). Other homologous recombination methods include truncated metagenomic gene-specific PCR (TMGS-PCR) (Wang et al. 2010), random priming in vitro recombination (RPR) (Shao et al. 1998), synthetic shuffling (Ness et al. 2002), random chimeragenesis on transient templates (RACHITT) (Coco 2003). In vivo homologous recombination methods include: phage-assisted continuous evolution (PACE) (Esvelt et al. 2011), mutagenic organized recombination process by homologous in vivo grouping (MORPHING) (Gonzalez-Perez et al. 2014), cloning performed in yeast (CLERY) (Abecassis et al. 2003).

Non-homologous recombination methods involve recombination of non-homologous parental sequences sharing little sequence identity. Kolkman and Stemmer developed exon shuffling method for the recombination of non-homologous sequences (Kolkman and Stemmer 2001). Other non-homologous recombination methods which include sequence homology independent protein recombination (SHIPREC) (Sieber et al. 2001), Degenerate Homoduplex Recombination (DHR) (Coco et al. 2002), incremental truncation for the creation of hybrid enzymes (ITCHY) (Ostermeier et al. 1999), Golden Gate Shuffling (GGS) Recombination (Engler et al. 2009), Y-Ligation Based Shuffling (YLBS) (Kitamura et al. 2002), Integron (Bikard et al. 2010) have also been developed that are described elsewhere (Poluri and Gulati 2017d).

1.9.2 Computational Designing

Directed evolution methods have made immense progress in designing proteins. However, the creation of diverse libraries and their screening to obtain desired protein is a highly daunting task. Hence, researchers have formulated a new, advanced, and brisk way to engineer proteins. Advancements in the era of computers as well as development of various databases and other bioinformatics tools have made it easy to engineer proteins meeting the structural and functional demands. An increased understanding of biophysics and structural biology has further motivated researchers to design novel proteins with high therapeutic potencies or enzymatic activities. The major goal is to redesign proteins with improved features such as high binding affinity, altered specificity for substrate, enhanced catalytic efficiency, etc. The other goal is to develop a completely novel protein with contemporary structure and function (Khan et al. 2016; Saven 2011; Tiwari et al. 2012). Computational designing of proteins tests our knowledge about the various aspects of existing proteins. The two approaches to computationally design proteins are (1) rational design, (2) de novo design (Fig. 1.12b and c) (Poluri and Gulati 2017c; Poluri and Gulati 2017e). Rational design mainly involves reconstituting the protein with desired traits. In contrast, de novo designing implies creating a protein from scratch. Hence, de novo design examines our knowledge about both structural and functional aspects of proteins (Lutz 2010). Both rational and de novo designing of proteins have been successfully employed to design numerous proteins with distinguished features.

Rational designing of proteins requires comprehensive knowledge about the proteins. As all the information regarding their evolution, structure, and function are hidden in their sequences, an exhaustive scan has been carried out to discern hotspot residues essential for the functionality of protein or that aid in adopting a specific fold. Numerous bioinformatics tools have emerged to attain all such information about the proteins. There are three major approaches, namely: MSA (multiple sequence alignment), coevolutionary analysis, and structure-based designing of proteins that are being exploited to rationally design a protein.

Multiple sequence alignment is a simplest way for redesigning a protein for which structural information is not available. Target protein sequence is aligned to its homologs using MSA tool which provides information about the residues essential for the functionality of protein (Fig. 1.13a). Those residues can be mutated to alter/develop the new functionality of proteins (Steipe et al. 1994; Sullivan et al. 2011; Watanabe et al. 2006).

Structure-based designing of proteins is based on the notion that proteins share structural similarity despite huge differences in their sequence. Hence, proteins can be redesigned using the structure of an existing protein as a template (Choi et al. 2009; Choi et al. 2015; Qi et al. 2012). There are numerous proteins whose structure is not known due to experimental discrepancies, structures of such proteins can be obtained using in silico methods such as protein threading (David et al. 2000), homology modeling (Xiang 2006), ab initio (Bonneau and Baker 2001), or fragment-based methods (Zhang 2007) (Fig. 1.13b).

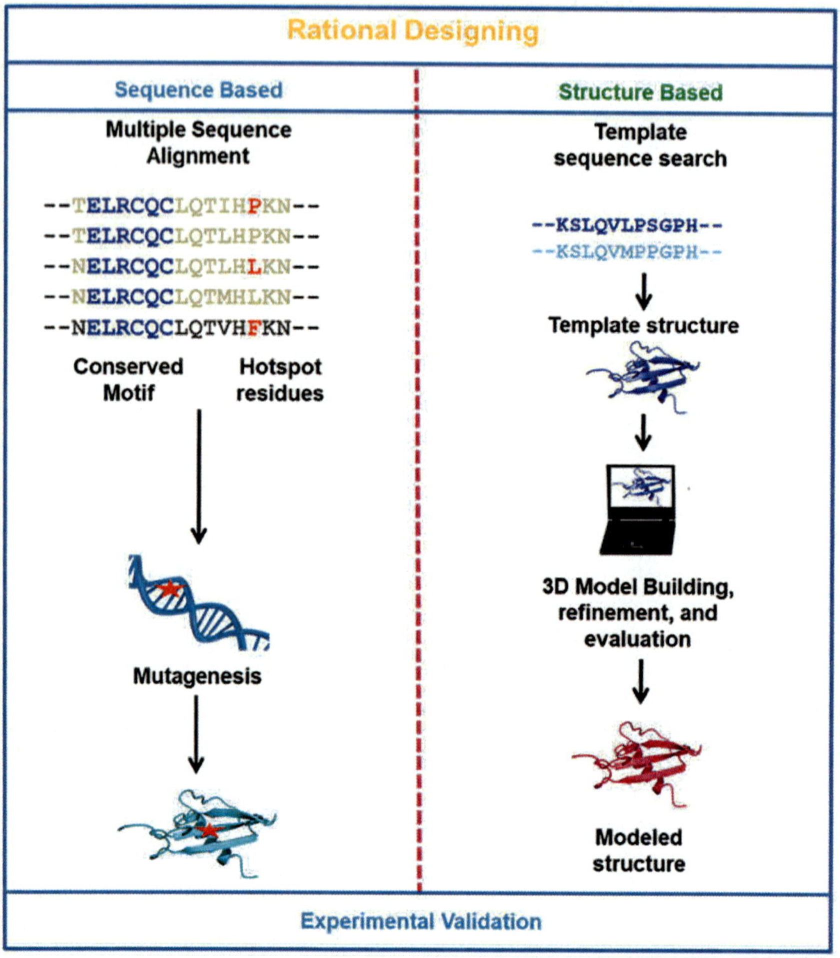

Fig. 1.13 Illustration of rational approaches for protein engineering (**a**) sequence-based and (**b**) structure-based methodologies

Coevolutionary analysis involves decoding the residues that are evolving simultaneously in the protein (Lovell and Robertson 2010). These residues are mostly far away in the sequence but come closer to interact with each other during the structural folding of a protein. These coevolving residues are targeted by several research groups to inseminate new features in the existing proteins (Xiang 2006) (Fig. 1.14).

De novo designing is a highly demanding approach for designing a protein to which no prior information about the desired protein is available. De novo designing is also referred to as inverse protein folding problem as a protein sequence is fitted on a desired structural fold (Fig. 1.15).

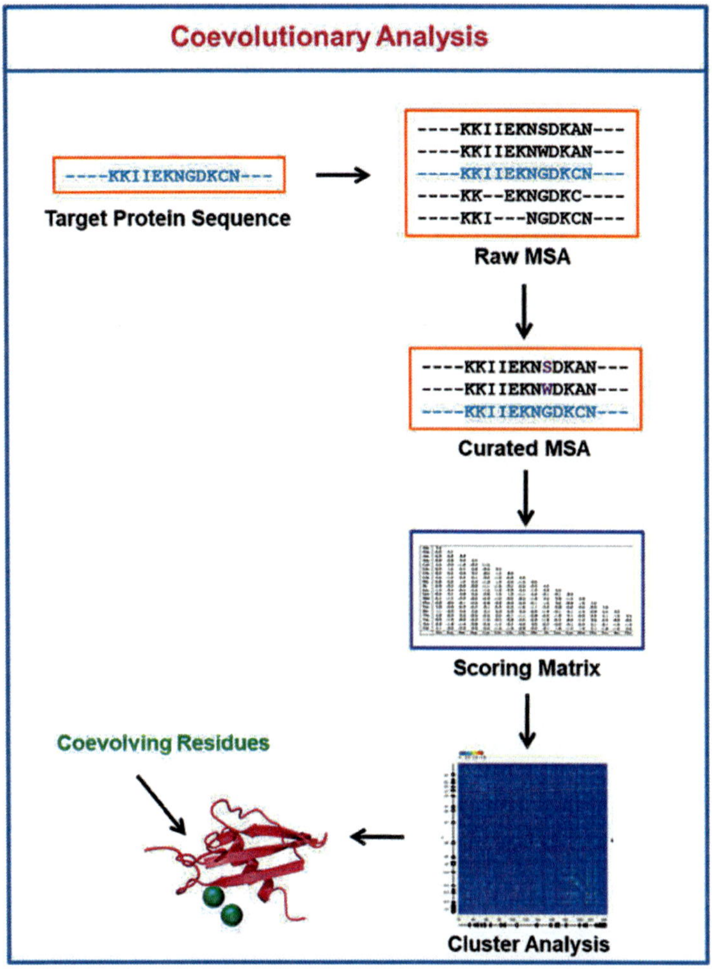

Fig. 1.14 An overview of coevolution analysis methodology to find out the coevolving residues

De novo approach to design a novel protein is divided into three major steps: (1) *Backbone building*: There are several approaches to build a backbone. One method is to combine the structural elements like helices, β-sheets via interconnecting loops with permissible torsion angles. Further, the backbone structure is aligned with the closely resembling protein structure in order to optimize the spacing between the loops and secondary structure elements (Kuhlman et al. 2002; Murzin et al. 1994; Offredi et al. 2003). (2) *Sequence design and optimization*: This step involves fitting the sequence onto a designed structural fold. Numerous methods are devised for fitting a low energy sequence onto backbone structure. One such method utilizes rotamer libraries that contain information about all the permissible conformations of different amino acids (Dunbrack Jr. 2002; Fung et al. 2008; Lassila

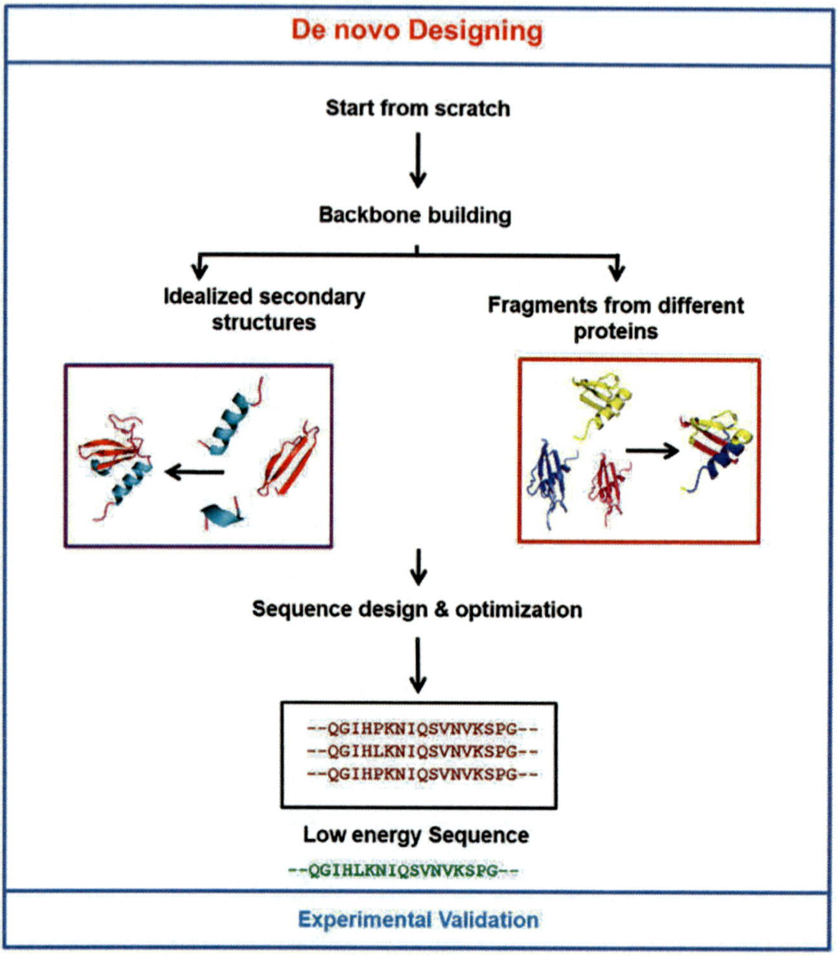

Fig. 1.15 Overview of steps involved in engineering proteins using de novo designing approach

et al. 2006; Mendes et al. 2002). (3) *Validation of selected sequence and structure pair:* Selected sequence and structure pairs are validated in silico. The sequence fitted onto a structure is examined in silico if the sequence will adopt the desired structural fold. Once the sequence has passed the in silico testing, the sequence is cloned, expressed, purified and its structure is determined either using NMR, cryo-electron microscopy or X-ray crystallography (Joh et al. 2014; Liang et al. 2009; Summa et al. 2002).

1.9.3 Combinatorial Approach

Researchers have devised a way to ease the tedious process of library screening in directed evolution strategy of designing the proteins. They made advancements in computational methods for designing proteins and wrote several algorithms, developed databases and servers to provide relaxation to experimentalists in process of protein designing. Combinatorial approach is a hybrid method of designing protein which amalgamates both computational and directed evolution strategies. In silico methods have been developed to design smart and efficient libraries which can be rapidly screened for the desired protein (Frith et al. 2008; Lane and Seelig 2014; Pantazes et al. 2007; Saraf et al. 2005).

1.10 Overview of Protein–Protein Interactions (PPIs)

While some of the protein molecules are known to perform their functions independently, there are several protein molecules which require being involved in interaction with other proteins to perform a designated biological role. This phenomenon is widely known as "protein–protein interaction," shortly abbreviated as PPI. Therefore, in recent times, research has focused more on understanding protein function in the context of their interactors, rather than the functional roles at the single protein level (Chatr-aryamontri et al. 2008). Research groups have attempted to analyze PPIs to understand the intricate network of PPIs in a cell and how the network is vital to the sustenance of cellular functions (Bahar and Jernigan 1997; Bennett et al. 1994; Betts and Sternberg 1999; Byrum et al. 2012). The growing field of proteomics and widely available data on human genome have greatly aided in this study.

Although there are several criteria on which PPIs can be classified, they are usually categorized as either transient or stable (Nooren and Thornton 2003). Stable interactions are usually linked with multi-subunit complexes. Such complexes comprise several subunits, which may be identical or distinct. In most cases, different subunits in a complex perform different biological roles. A protein complex usually serves as molecular machinery and it has been observed that the cell is abundant in these modular supramolecular complexes. Transient interactions, on the other hand, denote temporary PPIs and are governed by a set of external conditions, such as localization to other regions in the cell, phosphorylation, etc. Transient interactions can either be weak or strong, and also slow or fast. Transient PPIs have been found to be involved in several cellular roles, ranging from transport to signaling to protein folding to cell signaling and apoptosis. To truly understand the nature of PPIs, it is essential to characterize the various aspects of the interacting interfaces of the proteins—such as size, shape, complementarity, presence of buried waters, etc. Additionally, it is vital to analyze the driving forces behind two interacting proteins including hydrophobic bonding, salt bridges, van der Waals forces, etc. Several research groups have also emphasized on understanding the thermodynamic and kinetic features of several interacting proteins. All these aspects lead to a holistic analysis of interactions of proteins, and in turn, a better understanding of the cellular network.

Since PPIs are involved in several cellular functions, an anomaly in such interactions may lead to diseased conditions (Safari-Alighiarloo et al. 2014). Therefore, in recent times, identifying PPIs in the cell has turned out to be of utmost importance. The identification of PPIs in certain diseases, such as cancers, has also aided in the development of anticancer therapy that involves modulating the cancer-specific PPIs (Ivanov et al. 2013). Various experimental techniques have been established over the decades to identify PPIs. Some of these techniques are quite traditional, while a few of them are recently developed and state-of-the-art. Each of these techniques has its own set of merits and demerits. Therefore, these techniques are usually applied in combination to validate predicted PPIs. Experimental techniques, although aid in accurate identification of PPIs, are time-consuming and inefficient in their coverage of the entire proteome that comprises several thousands of protein interactions (Rao et al. 2014). Therefore, it has paved the new way for the augmentation of novel computational techniques that are highly efficient in predicting non-discovered PPIs and also in the validation of discovered ones. Apart from prediction tools, other computational methods and web resources have been established for the analysis, storage, and visualization of PPIs (Keskin et al. 2016).

1.10.1 Role of Protein Engineering in PPIs

Protein–protein interfaces play a crucial role in diverse biological actions including cell adhesion and immune responses. Biophysical insights of these interactions not only aid in understanding the mechanistic of these interactions but also enable in designing the novel PPIs with high therapeutic values (Saraf et al. 2005; Winter et al. 2012). Creating a high affinity interface between the two non-interacting protein partners is a stringent test of our understanding the physical basis of macromolecular interaction. All the protein engineering strategies, namely directed evolution, rational designing, and de novo designing either alone or in combination of each other have contributed not only in illuminating the basic properties of protein association, and their involvement in specific pathways, however, they have also aided in engineering protein pairs that can interact with each other and devoid of interaction with other endogenous cellular proteins (Karanicolas et al. 2011; Mandell and Kortemme 2009).

1.10.1.1 Redesigning the Existing PPIs Using Protein Engineering Strategies

Directed evolution techniques have been successfully employed to design novel protein interfaces that are never observed in nature earlier (Yagi et al. 2018). However, these methods can be used to redesign only one of the interacting partners at one point of time. They were unable to generate interfaces in both the interacting partners simultaneously (Binz et al. 2005). Computational algorithms have been designed to optimize the sequence of two proteins to form a stable functional complex. Karanicolas et al. have integrated rational design approach and directed evolution to design a protein complex exhibiting sub-nanomolar affinity. They designed a pair of interacting proteins named Prb and Pdar forming heterodimer with a Kd of 130 nM. Their binding affinity was further improved to 180 pM by including two-point mutations in the interface as depicted by directed evolution. Authors confirmed that the complex of proteins was formed by the designed interface using the X-ray crystal structures. This implies that the interaction can be created between the two non-interacting partners by designing complementary interaction surfaces (Karanicolas et al. 2011).

Several research groups have redesigned protein–protein or protein–peptide binding interactions. Song et al. improved the low affinity interaction between intercellular adhesion molecule-1 (ICAM-1) and integrin lymphocyte function-associated antigen-1 (LFA-1) which plays crucial immunological roles. They used various methods of rational designing for engineering ICAM-1 mutants which interact with LFA-1 with higher affinity than its cognate wild-type protein. They substituted 5 amino acids as a resultant of which hydrophobicity and packing around Glu34 of ICAM-1 was enhanced and there is an alteration in the edges of the binding interface. Glu34 is important as it is involved in coordinating with Mg^{2+} of LFA-1 I-domain, implying that rational designing can be employed to create new adhesion molecules exhibiting higher affinities that can be used for both studying a biological phenomenon and also encompass high therapeutic values (Song et al. 2006). Clark et al. took a great challenge of enhancing the affinity of high affinity protein–protein interactions that can further aid to develop therapeutic biomolecules. They exploited combination of several structural-based computational protocols to incorporate four mutations in an antibody to enhance its binding affinity to integrin VLA-1 I domain (Clark et al. 2006).

Lazar et al. redesigned the Fc/Fc gamma receptor interactions using computational programming methods and high-throughput screening. They created Fc variant which exhibits higher specificity and affinity for Fc gamma receptor. Engineered Fc regions offer a potential to generate therapeutic antibodies for antibody-based tumor therapy (Lazar et al. 2006). DeGrado and coworkers have designed a four-helix bundle protein that can specifically bind metalloporphyrin cofactor using entirely de novo approach. They employed SCADs (statistical computational design) algorithm to develop a 34-residue peptide sequence, which was then chemically synthesized. It has been observed that the peptide predominantly exists as a random coil monomer and undergoes transition to four-helix bundle upon cofactor binding. This study paves the way to sculpt selective binder for a cofactor in a de novo designed scaffold (Cochran et al. 2005).

1.10.1.2 Dissecting PPIs Using Protein Engineering Strategies

Association of two/more proteins during their interaction increases the complexity owing to the structural heterogeneity and energetic heterogeneity. This implies that different amino acids contribute differentially to the binding energetics, and some amino acids contribute quite significantly in contrast to others. It has been observed that protein interfaces are augmented in some amino acids that contribute majorly in binding energetics. Directed evolution strategy has been employed to address the minimal requirements in protein interface for the fine interaction between the two proteins.

Fellouse et al. prepared a synthetic antibody with restricted complementary determining regions (CDR) to four amino acids, namely: alanine, aspartate, serine, and tyrosine, using a phage display library. They showed that synthetic antibody binds to human VEGF with higher affinity. The crystal structure of complex of synthetic antibody and VEGF depicted that the tyrosine side chains are majorly involved in antigen recognition. In contrast, aspartate in the CDR was completely eliminated from the interaction interface. Serine and alanine residues were not involved in direct contact with antigen. However, they are playing an auxiliary role in making interactions between the antibody and VEGF by providing both spacious and conformational flexibility. The ability of tyrosine side chains to make productive contacts with the antigen is the main reason for its overabundance in natural antigen-binding sites. This finding can be employed to develop improved synthetic antibodies (Fellouse et al. 2004). Fellouse et al. further generated antibodies with antigen-binding sites restricted to only two amino acids, namely: tyrosine and serine. This antibody showed high binding affinity (Kd, 60 nM) and specificity toward its antigen. Their outcomes clearly demonstrate that these amino acids are appropriate for molecular recognition and binding (Fellouse et al. 2005). Koide et al. generated specific and tight binders for proteins like maltose-binding protein and human small ubiquitin like modifier 4 using Tyr/Ser binary interface (Koide et al. 2007). These studies convey that limited chemical diversity and high conformational complexity at the binding interface are essential for tight and specific binding of proteins and to create novel protein–protein interactions.

Investigators have exploited the selective capacities of directed evolution strategy to understand the specificity phenomenon in different protein systems. Several proteins most likely those involved in cell signaling and regulation often interact with numerous other proteins either through promiscuity, cross-reactivity or multi-specificity (Bonsor and Sundberg 2011; Clackson and Wells 1995), (Bernat et al. 2004), (Cho et al. 2010). Researchers have engineered proteins that can interact with multiple protein partners which further forms the basis of understanding specificity in PPIs. Further, energetic role of individual residues in the protein interface has been assessed by many research groups. Clackson and Wells were the first one who assessed the energetic contribution of each of the residues that participates in interaction of human growth hormone with its cognate receptor using alanine scanning strategy. They proposed that the functional epitope constitutes the amino acids making direct contacts in complex are surrounded by less important hydrophilic and partially hydrated residues (Clackson and Wells 1995). Furthermore, they observed that central hydrophobic core containing two tryptophan residues contribute to 3/fourth of the binding free energy. This shows that only subset of residues in the interface contribute more to the binding free energy and were termed as hotspots (Clackson and Wells 1995). These hotspot residues are not distributed randomly instead they are found in clusters or modules. Differences in the contribution to binding free energies were observed when these residues were mutated together and individually in the proteins. This implies that there could be a potential occurrence of cooperativity for energetic contributions of the hotspots.

Bernett et al. dissected the cooperative and additive energetics for human growth hormone with its cognate receptors using phage display libraries (Bernat et al. 2004). Pal et al. exploited combinatorial shotgun alanine scanning mutagenesis to evaluate the intramolecular cooperativity in the human growth hormone for binding its receptor (Pal et al. 2005). Additionally, the contribution of amino acids in the disordered regions of protein in binding energetics has also been assessed. However, the contribution of these amino acids is coordinated and is not dependent on single residue. Indeed, it is quite impossible that conformational changes in an amino acid significantly affect the movement of protein backbone without any movement in other neighboring residues (Cho et al. 2010).

1.11 Protein-Based Therapeutics

Proteins are known to be the functional horses in a living system. However, several environmental factors and/or mutations in the proteins can abnormally change their concentrations or conformations, which can lead to a diseased condition. To date, numerous protein-based therapeutics have been successfully designed that are being exploited to alleviate such diseases. Protein-based therapeutics exhibit several merits over small molecule drugs. They include: (1) Proteins are highly specific than small molecule drugs with respect to their actions. (2) Chances of interference of protein therapeutics with normal biological processes are minimal. (3) Protein therapeutics are well tolerated by the body than small molecule drugs. (4) Protein therapeutics are effective in cases where diseases are caused by mutations at genetic level and for which gene therapy is not available. (5) Approval of protein therapeutics by FDA for clinical use is less time-consuming than small molecule drugs.

Initially, therapeutic proteins were purified from human and animal sources, for instance α-1-proteinase inhibitor from human plasma (Dirksen et al. 1999a), pancreatic enzymes from animal pancreas (Dirksen et al. 1999b). Later, recombinant DNA technology has taken a lead for the production of proteins. Wide variety of expression systems such as bacteria, yeast, mammalian cells, transgenic plants, and animals have been exploited to express recombinant proteins (Mason et al. 2002; Wurm and Bernard 1999; Zoller 1991). The choice of using a system to express a protein depends on the production cost as well as the modifications like phosphorylation and glycosylation required in the protein for its biological function. For instance, bacterial system does not perform glycosylation, rest all the systems perform glycosylation but the pattern and extent of glycosylation are quite different. Glycosylation pattern of proteins greatly affects their biological activity, half-life, and immunogenicity. Hence, it is essential to carefully choose the expression system for the recombinant protein production.

Proteins obtained using recombinant DNA technology have several merits over proteins purified from other sources. (1) Exact human genes can be transcribed and

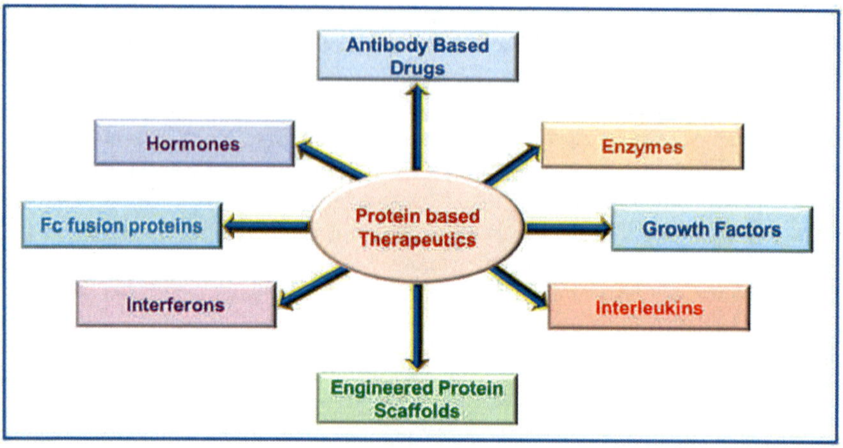

Fig. 1.16 Classification of various protein-based therapeutics on the basis of their different molecular types

translated which results in specific proteins with minimal chances of immunogenic rejection. (2) Recombinant protein production yields are high and cost-effective. (3) Recombinant DNA technology facilitates the improvement in the activity of protein and also provides a provision to inculcate new functionality in the protein (Leader et al. 2008).

Nowadays, new generation of protein therapeutics has emerged which is an outcome of an evolving protein engineering field. This allows a variation in the gene sequence of a protein to obtain desirable therapeutic potencies. Tremendous advancements have been made in engineering protein therapeutics with remarkable qualitative improvements in their therapeutic outcomes (Kimchi-Sarfaty et al. 2013). Therapeutic proteins are tailored that mainly involve three basic approaches: (1) therapeutic proteins are either chemically or genetically linked to another protein or polymer, (2) variation in the amino acid sequence of a therapeutic protein either by introducing a single or more mutations or deletions, (3) codon optimization in the recombinant proteins by introducing changes in the genetic code (Kimchi-Sarfaty et al. 2013).

Protein-based therapeutics have been allocated into differential groups depending on their molecular type. They include blood factors, anticoagulants, antibody-based drugs, enzymes, engineered protein scaffolds, hormones, growth factors, interleukins, interferons, Fc fusion proteins, bone morphogenetic proteins, etc. (Fig. 1.16). Of all these classes, antibody-based drugs are one of the fastest growing classes of therapeutic proteins. Engineered protein therapeutics exhibit enhanced specificity, stability, activity, pharmacokinetics, and pharmacodynamics (Poluri and Gulati 2017a; Poluri and Gulati 2017d). They also have reduced chances of immunogenicity and their productivity levels have also been improvised. Some of the examples of engineered therapeutic proteins are explained below.

Insulin is the first protein-based therapeutic developed by Eli Lilly at Genetech using recombinant DNA technology. Owing to the several discrepancies in the recombinant insulin, the protein engineering principles were used to engineer insulin analogs (Berenson et al. 2011a; Pandyarajan and Weiss 2012). These analogs differ from insulin as one to three amino acids have been replaced in them. These analogs exhibit properties similar to endogenous insulin but they have quick and heightened actions. The analogs with quick responses include insulin lispro, insulin aspart, and insulin gelusine. The analogs with heightened actions include insulin glargine and insulin detemir. The most imperative commercially available insulin formulation that has been engineered to provide prolonged actions is Lantus® (Vigneri et al. 2010). Supramolecular protein engineering principles were utilized to engineer zinc stapled insulin which has improved pharmacological properties and enhanced specificity in comparison with Lantus (Phillips et al. 2010). Further details about engineered insulin formulations and their utilizations have been summarized by Berenson et al. (Berenson et al. 2011b).

Enzymes are being engineered for their usage in replacement therapy and also for targeting extracellular molecules. Enzymes with an ability to degrade an amino acid are being employed to treat cancers which are auxotrophic for the particular amino acid (Jones et al. 2017; Li et al. 2020; Yu et al. 2017). For example, lymphoid tumor cells are auxotrophic for aspargine as they are devoid of asparagine synthetase activity. Oncaspar®, Enzon, a recombinant PEGylated L-asparaginase is being employed to treat leukemia. PEGylation of an enzyme enhances the pharmacokinetics of enzyme and also reduces the chances of immunogenic actions (Pasut et al. 2008). Enzyme replacement therapy is being used to replace the enzymes for which the patient is deficient. There are several genetic diseases which occur due to absence of an enzyme. These diseases are now being treated using enzyme replacement therapy (ERT). Some of the FDA-approved ERT drugs include VPRIV™ (velaglucerase alfa) (Hughes et al. 2015), Lumizyme® (alglucosidase alfa) (Keating 2012) for glycogen storage disease type II which is also known as Pompe disease; Naglazyme™ (galsulfase) for MPS VI (Maroteaux-Lamy syndrome) (Brunelli et al. 2016), Aldurazyme® (laronidase) for the treatment of MPS I (Hurler, Hurler-Scheie, or Scheie syndrome) (Jameson et al. 2019).

Scientists have geared up to engineer *antibody-based therapeutics* with improved specificity, stability, and binding affinity (He and Zhu 2015). Miklos et al. have designed a single chain variable fragment antibody which is highly thermos-resistant and enhanced antigen binding affinity by replacing 14 residues with lysine or arginine (Miklos et al. 2012). Pantezas and Maranas designed antibodies based on complementarity determining regions (CDR) by employing OptCDR (optimal complementarity determining regions). CDR is a hypervariable region where the antibody interacts with antigen. Several antibodies targeted for VEGF (vascular endothelial growth factor), capsid of hepatitis C, and fluorescein have been designed using optCDR (Pantazes and Maranas 2010).

GPCR (*G-protein coupled receptors*) plays an imperative role in the cell as they convey the signals from the extracellular environment to the effector molecules inside the cell. They aid in various biological functions such as in driving various signaling pathways, memory and sensory functions, homeostasis, migration of cells, etc. They perform all these functions by interacting with proteins, peptides, or other biological macromolecules (Kobilka 2007). Owing to their numerous functions, they are involved in various diseases. Hence, they are also targeted as therapeutics for the treatment of various diseases. Efforts are being made in development of anti-GPCR antibodies and some of them have already been formulated and are under clinical trials. Some of the anti-GPCR antibodies are summarized by Jo and Jung (Jo and Jung 2016).

Cardiovascular therapeutics have been formulated based on the proteins that aid in the formation of cardiac microvasculature. Proteins with an ability of inducing cardiomyocyte proliferation are the main targets for therapeutic designing. Numerous proteins engineered for cardiovascular therapy have been reviewed by Jay and Lee (Jay and Lee 2013).

Bleeding disorders such as hemophilia are being treated by the development of protein-based therapeutics as transfusion-based treatments have increased risk of infections. Several recombinant *clotting factors* like rFIX, rFVII etc., have been expressed and purified with improved functionalities as therapeutics for blood-related disorders. Anticoagulation activity of recombinant APC (activated protein C) was reduced by introducing mutations by site-directed mutagenesis in its two surface loops. The resultant mutants showed reduced anticoagulation activity and retained their apoptotic activity. Such APC variants are considered to be beneficial for cells with reduced bleeding risks (Mosnier et al. 2004).

Cytokines are small proteins secreted by various cell types such as epithelial cells, endothelial cells, and fibroblasts. Cytokines interact with their cognate receptors existing on the cell to accomplish their actions. Cytokines include chemokines, interleukins, interferons, colony- stimulating factors, etc. Manifold functions of cytokines are an outcome of their association in number of diseases. Hence, they have been identified as therapeutics for treating various diseases such as malignant, autoimmune, inflammatory, and other infectious diseases. Several recombinant cytokines are being employed for the treatment of diseases. For instance, recombinant interleukin-2 (Il-2) and IFN-α have been approved for treating metastatic melanoma and hairy cell leukemia, respectively (Vazquez-Lombardi et al. 2013). Inherent properties of cytokines like their pleotropic effects and short life span exert adverse effects during the disease treatment. Protein engineering strategies have been employed to overcome these pitfalls associated with cytokine-based therapeutics. Numerous strategies involving their fusion with other proteins, immune complexing, mutagenesis, and PEGylation have been used to engineer cytokines with enhanced serum half-life and specificity for their targets (Belladonna and Grohmann 2013; Boyman and Sprent 2012; Bukowski et al. 2002; Jazayeri and Carroll 2008; Perry and Jarvis 2001; Shum et al. 2018; Villarino and O'Shea 2017).

PEGylated cytokines which include PEG-IFN-α2b, PEG-IFN-α2a, PEG-G-CSF have been approved for chronic hepatitis C, chronic hepatitis B/C (Bukowski et al. 2002; Perry and Jarvis 2001), and chemotherapy-induced neutropenia (Crawford 2003), respectively. Cytokines in complex with antibodies have also been engineered for therapeutic purposes. For example, monoclonal antibody specific to IL-2 in combination with recombinant IL-2 showed an enhancement in the proliferation of CD8+ T-cells. Cytokines fused with other biological macromolecules are also being produced to further improve their functions and biophysical properties (Boyman and Sprent 2012). Cytokines fused to (1) bacterial toxins for increasing their cytotoxicity, (2) antibodies for localized delivery, (3) Fc region of antibody to enhance their half-life, (4) cytokine agonist to improve their activity. Examples of cytokines belonging to each of the category are well described by Lombardi et al. (Vazquez-Lombardi et al. 2013). Mutants of cytokines have also shown improvisation in their binding affinity, specificity and also aid in prolonging their half-life. G-CFS with enhanced thermal stability and reduced chances of protein aggregation has been engineered via the mutation of helix-breaking residues to helix-forming residues (Bishop et al. 2001). Recombinant IL-2 known as Proleukin® was subjected to cysteine to serine mutations for the free cysteine residues that were involved in creation of needless disulfide bonds resulting in protein aggregation (Marshall et al. 2003).

Numerous protein therapeutics have been formulated employing engineered *protein scaffolds*. These scaffolds offer a binding site which can be engineered specifically for the target molecule. Various parameters have to be kept in mind while choosing a protein scaffold as therapeutic. These parameters include (1) small size, (2) high solubility, (3) high chemical and thermal stability, (4) devoid of disulfide bonds and glycosylation sites. Further, these scaffolds should be easily expressible in microbial host systems. Structurally, these scaffolds should have exposed loops that can be altered in accordance with the target molecule for efficient binding. Such scaffolds can be either antibody-based scaffolds or non-antibody scaffolds (Gill and Damle 2006). Some of the scaffolds that are being currently employed for therapeutic purposes are discussed below.

Knottins are the small cysteine knot containing proteins which are also known as cysteine-knot mini proteins. They are present in toxins of scorpions, spiders, snails and also found in fungi, animals, and plants. Knottins are 30 amino acids long and structurally, they contain three antiparallel b-strands linked through loops and three intra-disulfide bridges. They are thermally, chemically, and proteolytically very stable owing to the presence of covalent linkages in their tertiary structure. Knottins are the potential therapeutic candidates owing to their smaller size, non-immunogenicity, and stability properties (Christmann et al. 1999; Reiss et al. 2006).

Affibodies are affinity proteins obtained originally from B-domain of immunoglobulin binding region of staphylococcal protein A (Uhlen et al. 1984). Chemically stable Z domain has been engineered exhibiting low affinity for Fab part and unblemished affinity for Fc part of immunoglobulins. Z domain has been originated as a result of mutations in B-domain (Nilsson et al. 1987). Affibody molecules with affinities ranging from μM to pM range specific to differential proteins such as insulin, IgE, IgM, IgA, EGFR, HER2, serum albumin have been engineered (Nygren 2008). Affibodies specific to cancer ligands serve as promising candidates for imaging tumors. Affibody specific to HER2 with 2200 fold higher affinity is being employed for visualizing HER2 expression in tumors using gamma camera (Orlova et al. 2006).

Small modular immune pharmaceuticals (*SMIPs*) are engineered proteins assembled using different parts of antibodies. They contain two domains, a target binding domain linked by a hinge region to an effector domain. Binding domain can undergo modification to provide binding site to the variety of target proteins that can be soluble proteins or receptors on cell surface. Hinge region is a flexible region acquired from immunoglobulin G1 which provides a site to combine different SMIPs. CD37-SMIP has been engineered specifically for B-cells expressing CD37 and has been suggested to function as therapeutics for treating B-cell malignancies (Zhao et al. 2007).

Tetranectins were first recognized by Borean Pharma to serve as platforms to engineer high quality antibody analogs. Tetranectins are composed of 181 residues forming two domains, namely: N-terminal trimerization domain and C-terminal binding domain. Binding domain comprises five loops for which the amino acids can be modulated according to the target molecule. Tetranectin monomers interact with each other to form trimers that are highly stable with enhanced half-life (Strino et al. 2009; Thogersen and Holldack 2006).

Adnectins are the tenth Fibronectin type II domain-based scaffolds which contain three loops that are CDR analogs of antibodies. Amino acids in these loops provide sites that can be modulated for binding the diversified target molecules. They are simpler than antibodies as they lack disulfide bonds. The first therapeutic adnectin is a PEGylated anti-angiogenic adnectin known as CT-322, which has been engineered specifically for VEGF (Lipovšek 2011).

Lipocalins are secreted non-antibody proteins, composed of 180 amino acid residues forming eight antiparallel β-sheet arranged in a funneled barrel structure with four loops at one end of the barrel. Such loops provide the sites that can be tailored to target-specific proteins. Lipocalins are being used to create anticalin scaffolds which exhibit high therapeutic potencies. Anticalins specific for VEGF and CTLA4 have been designed (Schlehuber and Skerra 2005; Skerra 2008).

Ankyrin repeat, a 33-amino acid repeat protein is organized in a β-hairpin-helix-loop-helix structure. They were first recognized in a cytoskeletal protein ankyrin and in cell cycle regulators. The β-turn and loop are composed of variable amino acids that are amenable to changes for preparing the sites for the variety of target molecules (Binz et al. 2004). Ankyrin repeat has also been employed to design DARPins (designed ankyrin repeat proteins) with potential biomedical applications (Stumpp et al. 2008). MP0112 is a DARPin with potency of inhibiting all VEGFs has been designed for diabetic macular edema treatment (Holz et al. 2014).

Avimers are the multi-domain proteins designed using in vitro exon shuffling and phage display of domains of human extracellular receptor. These multi-domain proteins are highly specific and exhibit high affinity as compared to single domain proteins. AMG220 is an avimer designed to treat Crohn's disease (Braddock 2007; Silverman et al. 2005).

Kunitz domains belong to the protease inhibitors class which is capable of reversibly inhibiting serine and trypsin proteases. These domains can be engineered to make them active against variety of proteases. DX-88 is plasma kallikrein inhibitor that has been approved to treat hereditary angioedema (Banta and Craig 2010). DX-1000 is a plasmin inhibitor that has been designed as an anticancer agent (Devy et al. 2007).

1.11.1 Protein-Based Therapeutics Targeting PPIs

As discussed in the previous sections, numerous PPIs are involved in various diseases. Henceforth, several approaches are being used to target PPIs (Fig. 1.17). These strategies include: (a) Small molecule approach; (b) Protein approach; and (c) Peptide/Peptidomimetic approach.

Small molecule approach involves designing small molecules that can specifically inhibit a PPI (Fig. 1.17a). However, this is a quite daunting task due to multifaceted nature of PPI interface. Small molecule allosteric inhibitors have also been designed that can interact with the protein at a site other than the binding site

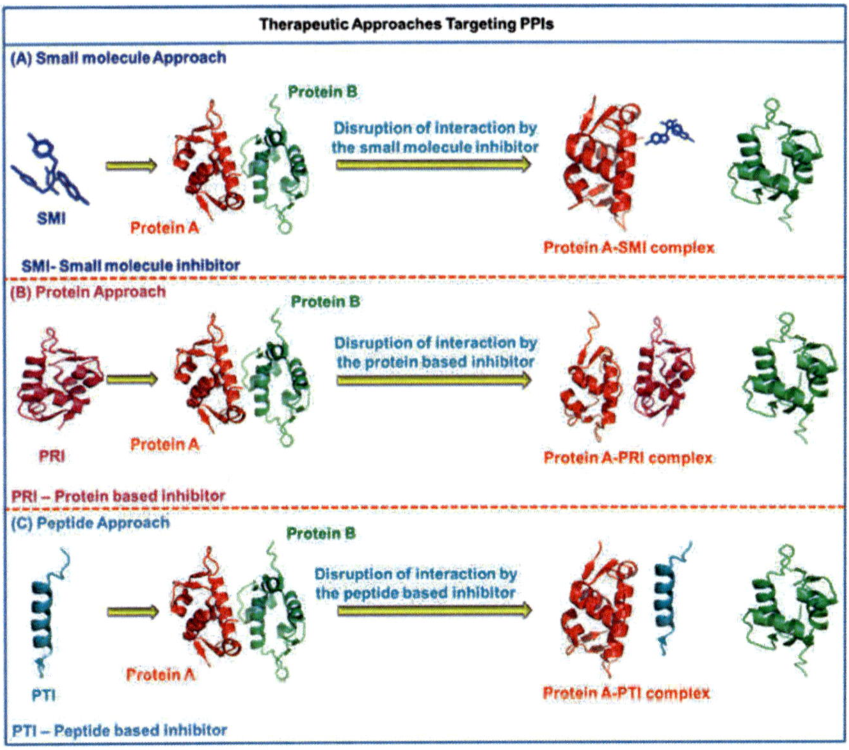

Fig. 1.17 Differential approaches to target PPIs: (**a**) small molecule approach, (**b**) protein-based approach, (**c**) peptide/peptidomimetic approach

that causes a conformational variation in the protein, which finally results in disrupting the PPI (Arkin et al. 2014).

Protein approach is used to target those PPIs that involve large interfaces. Monoclonal antibodies have been developed to compete with naturally occurring PPIs (Fig. 1.17b). Immunoglobulin domains are being engineered to form simple and stable molecules targeting PPIs. Scaffolds other than immunoglobulin domains such as DARPins (Kummer et al. 2012; Plückthun 2015), Alpharep (Guellouz et al. 2013; Urvoas et al. 2010) were also being exploited in targeting PPIs. In this approach, generally the protein-based inhibitors of PPIs directly contend with the cognate protein partner in an orthosteric manner; however, rarely they also act to utilize allosteric mechanisms (Kim et al. 2012). Primary limitation of this approach is the difficulty in the penetration of PPI inhibitors into the cell to bind to their target proteins. Hence, the applications of this approach have been limited to the extracellular targets (Bakal and Ochsenbein 2016).

In *Peptide/Peptidomimetic approach*, peptides are used as inhibitors to block PPIs (Fig. 1.17c). Inhibitory peptides have to cross the membrane for targeting the intracellular PPIs. Efforts have been made to develop cell penetrating peptides which ultimately lead to the development of intracellular peptide-based drugs (Habault and Poyet 2019; Park et al. 2019). Numerous companies and various academic labs across the globe aim to design cell permeable peptide (CPP)-based PPI inhibitors. The cell penetration ability of peptides was first identified in small neurotransmitter proteins (Dupont et al. 2015) and in viral proteins (Gump and Dowdy 2007). Properties of CPP are being employed to engineer active inhibitory peptides. The CPP sequence is being added at the N- or C-terminal points of the inhibitory peptide (Planel et al. 2010). Positively charged residues that are not involved in binding with the target are being modified to inculcate the characteristics of CPP in the inhibitory peptides (Smith et al. 2008).

Peptide-based strategies offer many advantages such as: (1) They are easily modifiable by the already available methods for protein approach such as molecular biology, protein engineering, rational/de novo designing, etc. (2) peptides being highly flexible in comparison with proteins, the computational protocols have to be specifically optimized for peptides (Donsky and Wolfson 2011; London et al. 2010; London et al. 2011; Trabuco et al. 2012). (3) They can be easily synthesized due to the development in the strategies such as solid phase chemical synthesis, and other high-throughput methods like peptide arrays to detect peptides (Dutta et al. 2010; Katz et al. 2011). There are several obstacles in the peptide strategy due to limitations of peptides like their high vulnerability to proteases in vivo.

Efforts have been made to overcome this limitation by chemical modifications of peptides. D-amino acids are being introduced in the peptides which have increased their solubility, provided protection from proteases, and also enhanced their affinity toward the target (Fasan et al. 2004). Cyclic peptides are being designed through peptide cyclization or by introduction of the helical staples (Dias et al. 2006; Fasan et al. 2004). The helical staples not only aid in penetration of peptides through cell but also provide stability to them (Bird et al. 2014; Checco et al. 2015b; Verdine and Hilinski 2012). Peptides are also being designed using non-α-amino acids with an ability to preserve their compact structure, and such structures are known as "foldamers." These foldamers exhibit 100 times higher resistance to proteolysis and also have conserved position of residues that are involved in target protein interaction (Checco et al. 2015a; Cheloha et al. 2014; Guichard and Huc 2011; Johnson and Gellman 2013; Nelli et al. 2015; Semetey et al. 2002). Non-peptide alpha helix mimetic foldamers such as aryl linked imidazolidin-2, tetraphthalamide-based and terphenyl-based foldamers have been fabricated (Buratto et al. 2014; Plante et al. 2008). These molecules have been used as inhibitors for PPIs against VEGF receptor (Haase et al. 2012), P53-mdm2 (Azzarito et al. 2012), Bax-BclXL (Ernst et al. 2003; Smith et al. 2013), small muscle myosin light chain kinase and calmodulin (Orner et al. 2001). Computational methods are being developed for the rational designing of foldamers (Lao et al. 2014; Renfrew et al. 2014). All these significant developments in the area of protein/peptide-based therapeutics to inhibit PPIs despite their interface size and compositional limitations are an extremely amazing journey for the biomedical researchers.

1.12 Conclusions

Proteins are one of the most functionally diversified entities which are a direct outcome of their high structural complexity. This structural complicacy is a consequence of the differential arrangements of the polypeptide chains that are chemically distinct owing to their varied composition and sequence of amino acids. Such heterogeneity in composition of proteins allows them to fold in unique three-dimensional structures thus inculcating unique functionalities for them. Proteins continue to evolve by mutating their amino acid residues to combat the needful environmental challenges. Some proteins perform their function alone, whereas some proteins work in conjunction with other proteins to accomplish sophisticated cellular processes. Hence, evolution in one of the protein insists an evolvement in its partner protein to maintain their functionality. Keeping in view the natural pathways of protein evolution, scientists have geared up to mimic these natural strategies in lab to engineer desired proteins. With the current advancements in computational field, researchers have amalgamated these "Mother Nature's" evolution approaches with computational means to ease the process of evolving novel proteins in the lab environment. Numerous proteins have already been tailored and are being employed in various fields including therapeutics. This has raised the protein engineering and protein-based therapeutics markets to billion dollars. However, the task of generating a fancy basket of desired proteins is quite daunting owing to the differential characteristic features of proteins and their interaction properties with other biological macromolecules especially with proteins. As proteins are one of the most complex biological macromolecules in cell and work in tandem with several other partner proteins, it is essential to decode the principles of their associations and interactions, which opened up the fascinating field of protein–protein interactions (PPIs) for the scientific community to explore. Henceforth, the following chapters in the current book focus on structural and energetic aspects of PPIs, protein complexes, experimental and computational techniques to unravel PPIs.

References

Aakre CD, Herrou J, Phung TN, Perchuk BS, Crosson S, Laub MT. (2015). Evolving new protein-protein interaction specificity through promiscuous intermediates. *Cell, 163*, 594-606.

Abecassis V, Pompon D, Truan G. (2003). Producing chimeric genes by CLERY: in vitro and in vivo recombination. *Methods Mol.Biol.*, *231*, 165-173.

Alcalde M, Zumarraga M, Polaina J, Ballesteros A, Plou FJ. (2006). Combinatorial saturation mutagenesis by in vivo overlap extension for the engineering of fungal laccases. *Combinatorial Chemistry & High Throughput Screening*, *9*, 719-727.

Arkin MR, Tang Y, Wells JA. (2014). Small-molecule inhibitors of protein-protein interactions: progressing toward the reality. *Chem.Biol.*, *21*, 1102-1114.

Arnold FH (1998) Design by Directed Evolution vol 31.

Azzarito V et al. (2012). 2-O-alkylated para-benzamide Î±-helix mimetics: the role of scaffold curvature. *Org.Biomol.Chem.*, *10*, 6469-6472.

Bahar I, Jernigan RL. (1997). Inter-residue potentials in globular proteins and the dominance of highly specific hydrophilic interactions at close separation. *Journal of molecular biology*, *266*, 195-214.

Bakal M, Ochsenbein F (2016) Targeting protein protein interactions, a wide open field for drug design vol 19.

Banner DW, Bloomer A, Petsko GA, Phillips DC, Wilson IA. (1976). Atomic coordinates for triose phosphate isomerase from chicken muscle. *Biochem.Biophys.Res.Commun.*, *72*, 146-155.

Banta E, Craig TJ. (2010). Overview of ecallantide in the treatment of hereditary angioedema types I and II. *Clinical Practice*, *7*, 565.

Batley J, Edwards D. (2009). Genome sequence data: management, storage, and visualization. *BioTechniques*, *46*, 333-334, 336.

Belladonna ML, Grohmann U. (2013). Bioengineering heterodimeric cytokines: turning promiscuous proteins into therapeutic agents. *Biotechnology & genetic engineering reviews*, *29*, 149-174.

Bello-Lopez JM, Cabrero-Martinez OA, Ibanez-Cervantes G, Hernandez-Cortez C, Pelcastre-Rodriguez LI, Gonzalez-Avila LU, Castro-Escarpulli G. (2019). Horizontal Gene Transfer and Its Association with Antibiotic Resistance in the Genus Aeromonas spp. *Microorganisms*, *7*.

Bennett M, Choe S, Eisenberg D. (1994). Domain swapping: entangling alliances between proteins. *Proceedings of the National Academy of Sciences*, *91*, 3127-3131.

Berenson DF, Weiss AR, Wan ZL, Weiss MA. (2011a). Insulin analogs for the treatment of diabetes mellitus: therapeutic applications of protein engineering. *Annals of the New York Academy of Sciences*, *1243*, E40-E54.

Berenson DF, Weiss AR, Wan ZL, Weiss MA. (2011b). Insulin analogs for the treatment of diabetes mellitus: therapeutic applications of protein engineering. *Ann.N.Y.Acad.Sci.*, *1243*, E40-E54.

Bernat B, Sun M, Dwyer M, Feldkamp M, Kossiakoff AA. (2004). Dissecting the binding energy epitope of a high-affinity variant of human growth hormone: cooperative and additive effects from combining mutations from independently selected phage display mutagenesis libraries. *Biochemistry*, *43*, 6076-6084.

Betts MJ, Sternberg MJ. (1999). An analysis of conformational changes on protein–protein association: implications for predictive docking. *Protein Engineering*, *12*, 271-283.

Bielawski JP, Yang Z. (2003a). Maximum likelihood methods for detecting adaptive evolution after gene duplication. *J Struct.Funct.Genomics*, *3*, 201-212.

Bielawski JP, Yang Z. (2003b). Maximum likelihood methods for detecting adaptive evolution after gene duplication. *Journal of structural and functional genomics*, *3*, 201-212.

Bikard D, Julie-Galau S, Cambray G, Mazel D. (2010). The synthetic integron: an in vivo genetic shuffling device. *Nucleic Acids Res.*, *38*, e153.

Binz HK et al. (2004). High-affinity binders selected from designed ankyrin repeat protein libraries. *Nature biotechnology*, *22*, 575-582.

Binz HK, Amstutz P, Pluckthun A. (2005). Engineering novel binding proteins from nonimmunoglobulin domains. *Nat.Biotechnol*, *23*, 1257-1268.

Bird GH, Irimia A, Ofek G, Kwong PD, Wilson IA, Walensky LD. (2014). Stapled HIV-1 peptides recapitulate antigenic structures and engage broadly neutralizing antibodies. *Nat.Struct.Mol. Biol.*, *21*, 1058-1067.

Bishop B, Koay DC, Sartorelli AC, Regan L. (2001). Reengineering granulocyte colony-stimulating factor for enhanced stability. *Journal of Biological Chemistry*, *276*, 33465-33470.

Bloom JD, Arnold FH. (2009). In the light of directed evolution: pathways of adaptive protein evolution. *Proc.Natl.Acad.Sci.U.S.A*, *106 Suppl 1*, 9995-10000.

Bonneau R, Baker D. (2001). Ab initio protein structure prediction: progress and prospects. *Annu. Rev.Biophys.Biomol.Struct.*, *30*, 173-189.

Bonsor DA, Sundberg EJ. (2011). Dissecting protein-protein interactions using directed evolution. *Biochemistry*, *50*, 2394-2402.

Boyman O, Sprent J. (2012). The role of interleukin-2 during homeostasis and activation of the immune system. *Nature Reviews Immunology*, *12*, 180-190.

Braddock M. (2007). 11th Annual inflammatory and immune diseases drug discovery and development summit: 12–13 March 2007, San Francisco, USA. *Expert Opinion on Investigational Drugs*, *16*, 909-917.

Brunelli MJ, Atallah ÁN, da Silva EM. (2016). Enzyme replacement therapy with galsulfase for mucopolysaccharidosis type VI. *Cochrane Database of Systematic Reviews.*

Bryngelson JD, Onuchic JN, Socci ND, Wolynes PG. (1995). Funnels, pathways, and the energy landscape of protein folding: a synthesis. *Proteins*, *21*, 167-195.

Bukowski RM, Tendler C, Cutler D, Rose E, Laughlin MM, Statkevich P. (2002). Treating cancer with PEG Intron: pharmacokinetic profile and dosing guidelines for an improved interferon-alpha-2b formulation. *Cancer: Interdisciplinary International Journal of the American Cancer Society*, *95*, 389-396.

Buratto J et al. (2014). Structure of a complex formed by a protein and a helical aromatic oligoamide foldamer at 2.1 Ã… resolution. *Angew.Chem.Int.Ed Engl.*, *53*, 883-887.

Byrum S, Smart SK, Larson S, Tackett AJ (2012) Analysis of stable and transient protein–protein interactions. In: Chromatin Remodeling. Springer, pp. 143-152

Cadwell RC, Joyce GF. (1992). Randomization of genes by PCR mutagenesis. *Genome research*, *2*, 28-33.

Chakraborty S, Mohan PK, Hosur RV. (2012). Residual structure and dynamics in DMSO-d6 denatured Dynein Light Chain protein. *Biochimie*, *94*, 231-241.

Chatr-aryamontri A, Ceol A, Licata L, Cesareni G. (2008). Protein interactions: integration leads to belief. *Trends in biochemical sciences*, *33*, 241-241.

Checco JW et al. (2015a). Targeting diverse protein-protein interaction interfaces with Î±/Î^{2}-peptides derived from the Z-domain scaffold. *Proc.Natl.Acad.Sci.U.S.A*, *112*, 4552-4557.

Checco JW et al. (2015b). Î±/Î^{2}-Peptide Foldamers Targeting Intracellular Protein-Protein Interactions with Activity in Living Cells. *J Am.Chem.Soc.*, *137*, 11365-11375.

Cheloha RW, Maeda A, Dean T, Gardella TJ, Gellman SH. (2014). Backbone modification of a polypeptide drug alters duration of action in vivo. *Nat.Biotechnol*, *32*, 653-655.

Chen K, Arnold FH. (1993). Tuning the activity of an enzyme for unusual environments: sequential random mutagenesis of subtilisin E for catalysis in dimethylformamide. *Proceedings of the National Academy of Sciences*, *90*, 5618-5622.

Chen L, Wang F, Wang X, Liu Y-G. (2013). Robust one-tube Ω-PCR strategy accelerates precise sequence modification of plasmids for functional genomics. *Plant and cell physiology*, *54*, 634-642.

Chen Z, Zeng AP. (2016). Protein engineering approaches to chemical biotechnology. *Current opinion in biotechnology*, *42*, 198-205.

Cho S et al. (2010). Assessing energetic contributions to binding from a disordered region in a protein-protein interaction. *Biochemistry*, *49*, 9256-9268.

Choi SH, Kim HS, Lee EY. (2009). Comparative homology modeling-inspired protein engineering for improvement of catalytic activity of Mugil cephalus epoxide hydrolase. *Biotechnology letters*, *31*, 1617-1624.

Choi Y, Hua C, Sentman CL, Ackerman ME, Bailey-Kellogg C Antibody humanization by structure-based computational protein design. In: MAbs, 2015. vol 6. Taylor & Francis, pp 1045-1057

Christmann A, Walter K, Wentzel A, Krätzner R, Kolmar H. (1999). The cystine knot of a squash-type protease inhibitor as a structural scaffold for Escherichia coli cell surface display of conformationally constrained peptides. *Protein engineering*, *12*, 797-806.

Clackson T, Wells JA. (1995). A hot spot of binding energy in a hormone-receptor interface. *Science*, *267*, 383-386.

Clark LA et al. (2006). Affinity enhancement of an in vivo matured therapeutic antibody using structure-based computational design. *Protein Sci.*, *15*, 949-960.

Cobb RE, Chao R, Zhao H. (2013a). Directed Evolution: Past, Present and Future. *AIChE J*, *59*, 1432-1440.

Cobb RE, Si T, Zhao H. (2012). Directed evolution: an evolving and enabling synthetic biology tool. *Curr.Opin.Chem.Biol.*, *16*, 285-291.

Cobb RE, Sun N, Zhao H. (2013b). Directed evolution as a powerful synthetic biology tool. *Methods*, *60*, 81-90.

Cochran FV, Wu SP, Wang W, Nanda V, Saven JG, Therien MJ, DeGrado WF. (2005). Computational de novo design and characterization of a four-helix bundle protein that selectively binds a nonbiological cofactor. *J Am.Chem.Soc.*, *127*, 1346-1347.

Coco WM (2003) RACHITT. In: Directed Evolution Library Creation. Springer, pp. 111-127

Coco WM et al. (2002). Growth factor engineering by degenerate homoduplex gene family recombination. *Nature biotechnology*, *20*, 1246-1250.

Cowan SW, Newcomer ME, Jones TA. (1990). Crystallographic refinement of human serum retinol binding protein at 2A resolution. *Proteins*, *8*, 44-61.

Crameri A, Raillard S-A, Bermudez E, Stemmer WP. (1998). DNA shuffling of a family of genes from diverse species accelerates directed evolution. *Nature*, *391*, 288-291.

Crawford J Once-per-cycle pegfilgrastim (Neulasta) for the management of chemotherapy-induced neutropenia. In: Seminars in oncology, 2003. Elsevier, pp 24-30

Crick F. (1970). Central dogma of molecular biology. *Nature*, *227*, 561-563.

Dalby PA. (2011). Strategy and success for the directed evolution of enzymes. *Current opinion in structural biology*, *21*, 473-480.

David R, Korenberg MJ, Hunter IW. (2000). 3D-1D threading methods for protein fold recognition. *Pharmacogenomics*, *1*, 445-455.

Dennig A, Shivange AV, Marienhagen J, Schwaneberg U. (2011). OmniChange: the sequence independent method for simultaneous site-saturation of five codons. *PloS one*, *6*, e26222.

Deshler JO. (1992). A simple method for randomly mutating cloned DNA fragments by using chemical mutagens and the polymerase chain reaction. *Genetic Analysis: Biomolecular Engineering*, *9*, 103-106.

Devy L et al. (2007). PEGylated DX-1000: pharmacokinetics and antineoplastic activity of a specific plasmin inhibitor. *Neoplasia*, *9*, 927-937.

Dias RL, Fasan R, Moehle K, Renard A, Obrecht D, Robinson JA. (2006). Protein ligand design: from phage display to synthetic protein epitope mimetics in human antibody Fc-binding peptidomimetics. *J Am.Chem.Soc.*, *128*, 2726-2732.

Dirksen A et al. (1999a). A randomized clinical trial of α1-antitrypsin augmentation therapy. *American journal of respiratory and critical care medicine*, *160*, 1468-1472.

Dirksen A et al. (1999b). A randomized clinical trial of alpha(1)-antitrypsin augmentation therapy. *Am.J Respir.Crit Care Med.*, *160*, 1468-1472.

Donsky E, Wolfson HJ. (2011). PepCrawler: a fast RRT-based algorithm for high-resolution refinement and binding affinity estimation of peptide inhibitors. *Bioinformatics.*, *27*, 2836-2842.

Dunbrack Jr RL. (2002). Rotamer libraries in the 21st century. *Current opinion in structural biology*, *12*, 431-440.

Dupont E, Prochiantz A, Joliot A. (2015). Penetratin Story: An Overview. *Methods Mol.Biol.*, *1324*, 29-37.

Dutta S, GullÃ¡ S, Chen TS, Fire E, Grant RA, Keating AE. (2010). Determinants of BH3 binding specificity for Mcl-1 versus Bcl-xL. *J Mol.Biol.*, *398*, 747-762.

Edelheit O, Hanukoglu A, Hanukoglu I. (2009). Simple and efficient site-directed mutagenesis using two single-primer reactions in parallel to generate mutants for protein structure-function studies. *BMC biotechnology*, *9*, 1-8.

Edison AS. (2001). Linus Pauling and the planar peptide bond. *Nat.Struct.Biol.*, *8*, 201-202.

Eirin-Lopez JM, Rebordinos L, Rooney AP, Rozas J. (2012). The birth-and-death evolution of multigene families revisited. *Genome dynamics*, *7*, 170-196.

Engler C, Gruetzner R, Kandzia R, Marillonnet S. (2009). Golden gate shuffling: a one-pot DNA shuffling method based on type IIs restriction enzymes. *PloS one*, *4*, e5553.

Ernst JT, Becerril J, Park HS, Yin H, Hamilton AD. (2003). Design and application of an alpha-helix-mimetic scaffold based on an oligoamide-foldamer strategy: antagonism of the Bak BH3/Bcl-xL complex. *Angew.Chem.Int.Ed Engl.*, *42*, 535-539.

Esvelt KM, Carlson JC, Liu DR. (2011). A system for the continuous directed evolution of biomolecules. *Nature*, *472*, 499-503.

Faber MS, Whitehead TA. (2019). Data-driven engineering of protein therapeutics. *Current opinion in biotechnology*, *60*, 104-110.

Fasan R, Dias RL, Moehle K, Zerbe O, Vrijbloed JW, Obrecht D, Robinson JA. (2004). Using a beta-hairpin to mimic an alpha-helix: cyclic peptidomimetic inhibitors of the p53-HDM2 protein-protein interaction. *Angew.Chem.Int.Ed Engl.*, *43*, 2109-2112.

Fasman GD (1990) Prediction of Protein Structure and the Principles of Protein Conformation vol 2nd Edition.

Fellouse FA, Li B, Compaan DM, Peden AA, Hymowitz SG, Sidhu SS. (2005). Molecular recognition by a binary code. *J Mol.Biol.*, *348*, 1153-1162.

Fellouse FA, Wiesmann C, Sidhu SS. (2004). Synthetic antibodies from a four-amino-acid code: a dominant role for tyrosine in antigen recognition. *Proc.Natl.Acad.Sci.U.S.A*, *101*, 12467-12472.

Finney-Manchester SP, Maheshri N. (2013). Harnessing mutagenic homologous recombination for targeted mutagenesis in vivo by TaGTEAM. *Nucleic acids research*, *41*, e99-e99.

Firnberg E, Ostermeier M. (2012). PFunkel: efficient, expansive, user-defined mutagenesis. *PloS one*, *7*, e52031.

Frith MC, Saunders NF, Kobe B, Bailey TL. (2008). Discovering sequence motifs with arbitrary insertions and deletions. *PLoS Comput Biol*, *4*, e1000071.

Fujii R, Kitaoka M, Hayashi K (2014) Error-prone rolling circle amplification greatly simplifies random mutagenesis. In: Directed Evolution Library Creation. Springer, pp. 23-29

Fujii W, Kano K, Sugiura K, Naito K. (2013). Repeatable construction method for engineered zinc finger nuclease based on overlap extension PCR and TA-cloning. *PloS one*, *8*, e59801.

Fung HK, Welsh WJ, Floudas CA. (2008). Computational de novo peptide and protein design: rigid templates versus flexible templates. *Industrial & engineering chemistry research*, *47*, 993-1001.

Gangele K, Gulati K, Joshi N, Kumar D, Poluri KM. (2020). Molecular insights into the differential structure-dynamics-stability features of interleukin-8 orthologs: Implications to functional specificity. *International journal of biological macromolecules*, *164*, 3221-3234.

Gangele K, Jamsandekar M, Mishra A, Poluri KM. (2019). Unraveling the evolutionary origin of ELR motif using fish CXC chemokine CXCL8. *Fish & shellfish immunology*, *93*, 17-27.

Gaytan P, Contreras-Zambrano C, Ortiz-Alvarado M, Morales-Pablos A, Yanez J. (2009). TrimerDimer: an oligonucleotide-based saturation mutagenesis approach that removes redundant and stop codons. *Nucleic acids research*, *37*, e125-e125.

Gibbs MD, Nevalainen KH, Bergquist PL. (2001). Degenerate oligonucleotide gene shuffling (DOGS): a method for enhancing the frequency of recombination with family shuffling. *Gene*, *271*, 13-20.

Gill DS, Damle NK. (2006). Biopharmaceutical drug discovery using novel protein scaffolds. *Current opinion in biotechnology*, *17*, 653-658.

Gonzalez-Perez D, Molina-Espeja P, Garcia-Ruiz E, Alcalde M. (2014). M utagenic O rganized R ecombination P rocess by H omologous In Vivo G rouping (MORPHING) for Directed Enzyme Evolution. *PLoS One*, *9*, e90919.

Goodsell DS. (1991). Inside a living cell. *Trends Biochem.Sci.*, *16*, 203-206.

Gratz A, Jose J. (2008). Protein domain library generation by overlap extension (PDLGO): a tool for enzyme engineering. *Analytical biochemistry*, *378*, 171-176.

Graur D, Li W. (2000). Fundamentals of molecular evolution, 2nd edn. Sinauer Ass. *Inc, Sunderland, xiv*, *481*.

Greener A, Callahan M, Jerpseth B (1996) An efficient random mutagenesis technique using an E. coli mutator strain. In: In vitro mutagenesis protocols. Springer, pp 375-385

Guellouz A et al. (2013). Selection of specific protein binders for pre-defined targets from an optimized library of artificial helicoidal repeat proteins (alphaRep). *PLoS.One.*, *8*, e71512.

Guichard G, Huc I. (2011). Synthetic foldamers. *Chem.Commun.(Camb.)*, *47*, 5933-5941.

Gulati K, Jamsandekar M, Poluri KM. (2017). Mechanistic insights into molecular evolution of species-specific differential glycosaminoglycan binding surfaces in growth-related oncogene chemokines. *Royal Society open science*, *4*, 171059.

Gulati K, Poluri K. (2019). Role of Engineered Proteins as Therapeutic Formulations. *Pharmaceutical Biocatalysis: Fundamentals, Enzyme Inhibitors, and Enzymes in Health and Diseases*, 159.

Gulati K, Poluri KM. (2016). An Overview of Computational and Experimental Methods for Designing Novel Proteins. *Recent patents on biotechnology*, *10*, 235-263.

Gump JM, Dowdy SF. (2007). TAT transduction: the molecular mechanism and therapeutic prospects. *Trends Mol.Med.*, *13*, 443-448.

Gupta RS. (1998). Protein phylogenies and signature sequences: A reappraisal of evolutionary relationships among archaebacteria, eubacteria, and eukaryotes. *Microbiology and molecular biology reviews: MMBR*, *62*, 1435-1491.

Haase HS, Peterson-Kaufman KJ, Lan Levengood SK, Checco JW, Murphy WL, Gellman SH. (2012). Extending foldamer design beyond Î±-helix mimicry: Î±/Î²-peptide inhibitors of vascular endothelial growth factor signaling. *J Am.Chem.Soc.*, *134*, 7652-7655.

Habault J, Poyet J-L. (2019). Recent advances in cell penetrating peptide-based anticancer therapies. *Molecules*, *24*, 927.

Hadfield A, Kryger G, Ouyang J, Petsko GA, Ringe D, Viola R. (1999). Structure of aspartate-beta-semialdehyde dehydrogenase from Escherichia coli, a key enzyme in the aspartate family of amino acid biosynthesis. *J Mol.Biol.*, *289*, 991-1002.

He L, Zhu J. (2015). Computational tools for epitope vaccine design and evaluation. *Current opinion in virology*, *11*, 103-112.

Heine A et al. (1998). An antibody exo Diels-Alderase inhibitor complex at 1.95 angstrom resolution. *Science*, *279*, 1934-1940.

Hidalgo A, Schließmann A, Bornscheuer UT (2014) One-pot Simple methodology for CAssette Randomization and Recombination for focused directed evolution (OSCARR). In: Directed Evolution Library Creation. Springer, pp. 207-212

Holz FG, Schmitz-Valckenberg S, Fleckenstein M. (2014). Recent developments in the treatment of age-related macular degeneration. *The Journal of clinical investigation*, *124*, 1430-1438.

Hughes DA et al. (2015). Velaglucerase alfa (VPRIV) enzyme replacement therapy in patients with Gaucher disease: Long-term data from phase III clinical trials. *American journal of hematology*, *90*, 584-591.

Hull J. (2012). Cystic fibrosis transmembrane conductance regulator dysfunction and its treatment. *Journal of the Royal Society of Medicine*, *105*, 2-8.

Ivanov AA, Khuri FR, Fu H. (2013). Targeting protein–protein interactions as an anticancer strategy. *Trends in pharmacological sciences*, *34*, 393-400.

Jameson E, Jones S, Remmington T. (2019). Enzyme replacement therapy with laronidase (Aldurazyme®) for treating mucopolysaccharidosis type I. *Cochrane Database of Systematic Reviews*.

Jay SM, Lee RT. (2013). Protein engineering for cardiovascular therapeutics: untapped potential for cardiac repair. *Circulation research*, *113*, 933-943.

Jazayeri JA, Carroll GJ. (2008). Fc-based cytokines: prospects for engineering superior therapeutics. *BioDrugs*, *22*, 11-26.

Jo M, Jung ST. (2016). Engineering therapeutic antibodies targeting G-protein–coupled receptors. *Experimental & molecular medicine*, *48*, e207-e207.

Joh NH et al. (2014). De novo design of a transmembrane Zn2+-transporting four-helix bundle. *Science*, *346*, 1520-1524.

Johnson LM, Gellman SH. (2013). Î±-Helix mimicry with Î±/Î²-peptides. *Methods Enzymol.*, *523*, 407-429.

Jones BJ, Lim HY, Huang J, Kazlauskas RJ. (2017). Comparison of Five Protein Engineering Strategies for Stabilizing an alpha/beta-Hydrolase. *Biochemistry*, *56*, 6521-6532.

Joshi N, Nagar N, Gulati K, Gangele K, Mishra A, Kumar D, Poluri KM. (2020). Dissecting the differential structural and dynamics features of CCL2 chemokine orthologs. *International journal of biological macromolecules*, *156*, 239-251.

Juo ZS, Chiu TK, Leiberman PM, Baikalov I, Berk AJ, Dickerson RE. (1996). How proteins recognize the TATA box. *J Mol.Biol.*, *261*, 239-254.

Karanicolas J et al. (2011). A de novo protein binding pair by computational design and directed evolution. *Mol.Cell*, *42*, 250-260.

Katz C, Levy-Beladev L, Rotem-Bamberger S, Rito T, Rudiger SG, Friedler A. (2011). Studying protein-protein interactions using peptide arrays. *Chem.Soc.Rev.*, *40*, 2131-2145.

Keating GM. (2012). Agalsidase Alfa. *BioDrugs*, *26*, 335-354.

Keskin O, Tuncbag N, Gursoy A. (2016). Predicting protein–protein interactions from the molecular to the proteome level. *Chemical reviews*, *116*, 4884-4909.

Khan FI, Wei D-Q, Gu K-R, Hassan MI, Tabrez S. (2016). Current updates on computer aided protein modeling and designing. *International journal of biological macromolecules*, *85*, 48-62.

Kim B, Eggel A, Tarchevskaya SS, Vogel M, Prinz H, Jardetzky TS. (2012). Accelerated disassembly of IgE-receptor complexes by a disruptive macromolecular inhibitor. *Nature*, *491*, 613-617.

Kimchi-Sarfaty C, Schiller T, Hamasaki-Katagiri N, Khan MA, Yanover C, Sauna ZE. (2013). Building better drugs: developing and regulating engineered therapeutic proteins. *Trends Pharmacol.Sci.*, *34*, 534-548.

Kipnis Y, Dellus-Gur E, Tawfik DS. (2012). TRINS: a method for gene modification by randomized tandem repeat insertions. *Protein Engineering, Design & Selection*, *25*, 437-444.

Kitamura K, Kinoshita Y, Narasaki S, Nemoto N, Husimi Y, Nishigaki K. (2002). Construction of block-shuffled libraries of DNA for evolutionary protein engineering: Y-ligation-based block shuffling. *Protein Engineering*, *15*, 843-853.

Kobilka BK. (2007). G protein coupled receptor structure and activation. *Biochimica et Biophysica Acta (BBA)-Biomembranes*, *1768*, 794-807.

Koide A, Gilbreth RN, Esaki K, Tereshko V, Koide S. (2007). High-affinity single-domain binding proteins with a binary-code interface. *Proc.Natl.Acad.Sci.U.S.A*, *104*, 6632-6637.

Kolkman JA, Stemmer WP. (2001). Directed evolution of proteins by exon shuffling. *Nature biotechnology*, *19*, 423-428.

Koonin EV, Wolf YI, Karev GP. (2002). The structure of the protein universe and genome evolution. *Nature*, *420*, 218-223.

Krishna Mohan P, Mukherjee S, Chary KV. (2008). Differential native state ruggedness of the two Ca2+-binding domains in a Ca2+ sensor protein. *Proteins: Structure, Function, and Bioinformatics, 70*, 1147-1153.

Kuhlman B, O'Neill JW, Kim DE, Zhang KY, Baker D. (2002). Accurate computer-based design of a new backbone conformation in the second turn of protein L. *Journal of molecular biology, 315*, 471-477.

Kummer L et al. (2012). Structural and functional analysis of phosphorylation-specific binders of the kinase ERK from designed ankyrin repeat protein libraries. *Proc.Natl.Acad.Sci.U.S.A, 109*, E2248-E2257.

Kuriyan J, Eisenberg D. (2007). The origin of protein interactions and allostery in colocalization. *Nature, 450*, 983-990.

Kyte J, Doolittle RF. (1982). A simple method for displaying the hydropathic character of a protein. *J Mol.Biol., 157*, 105-132.

Labrou NE. (2010). Random mutagenesis methods for in vitro directed enzyme evolution. *Current Protein and Peptide Science, 11*, 91-100.

Lane MD, Seelig B. (2014). Advances in the directed evolution of proteins. *Current opinion in chemical biology, 22*, 129-136.

Lao BB, Drew K, Guarracino DA, Brewer TF, Heindel DW, Bonneau R, Arora PS. (2014). Rational design of topographical helix mimics as potent inhibitors of protein-protein interactions. *J Am. Chem.Soc., 136*, 7877-7888.

Lassila JK, Privett HK, Allen BD, Mayo SL. (2006). Combinatorial methods for small-molecule placement in computational enzyme design. *Proceedings of the National Academy of Sciences, 103*, 16710-16715.

Lazar GA et al. (2006). Engineered antibody Fc variants with enhanced effector function. *Proc. Natl.Acad.Sci.U.S.A, 103*, 4005-4010.

Leader B, Baca QJ, Golan DE. (2008). Protein therapeutics: a summary and pharmacological classification. *Nat.Rev.Drug Discov., 7*, 21-39.

Lederer F, Glatigny A, Bethge PH, Bellamy HD, Matthew FS. (1981). Improvement of the 2.5 A resolution model of cytochrome b562 by redetermining the primary structure and using molecular graphics. *J Mol.Biol., 148*, 427-448.

Lee HC, Lai K, Lorenc MT, Imelfort M, Duran C, Edwards D. (2012). Bioinformatics tools and databases for analysis of next-generation sequence data. *Briefings in functional genomics, 11*, 12-24.

Leon O, Roth M. (2000). Zinc fingers: DNA binding and protein-protein interactions. *Biol.Res., 33*, 21-30.

Levinthal C (1968) Are There Pathways for Protein Folding vol 65.

Li YF, Zhou W, Blankenship RE, Allen JP. (1997). Crystal structure of the bacteriochlorophyll a protein from Chlorobium tepidum. *J Mol.Biol., 271*, 456-471.

Li Z, Jiang Y, Guengerich FP, Ma L, Li S, Zhang W. (2020). Engineering cytochrome P450 enzyme systems for biomedical and biotechnological applications. *The Journal of biological chemistry, 295*, 833-849.

Liang H et al. (2009). De novo design of a βαβ motif. *Angewandte Chemie, 121*, 3351-3353.

Lipovšek D. (2011). Adnectins: engineered target-binding protein therapeutics. *Protein Engineering, Design & Selection, 24*, 3-9.

London N, Movshovitz-Attias D, Schueler-Furman O. (2010). The structural basis of peptide-protein binding strategies. *Structure., 18*, 188-199.

London N, Raveh B, Cohen E, Fathi G, Schueler-Furman O. (2011). Rosetta FlexPepDock web server--high resolution modeling of peptide-protein interactions. *Nucleic Acids Res., 39*, W249-W253.

Lovell SC, Robertson DL. (2010). An integrated view of molecular coevolution in protein–protein interactions. Molecular biology and evolution, 27, 2567-2575.

Lutz S. (2010). Beyond directed evolution—semi-rational protein engineering and design. *Current opinion in biotechnology, 21*, 734-743.

Lutz S, Iamurri SM. (2018). Protein Engineering: Past, Present, and Future. *Methods in molecular biology*, *1685*, 1-12.
Mandell DJ, Kortemme T. (2009). Computer-aided design of functional protein interactions. *Nat. Chem.Biol.*, *5*, 797-807.
Marshall SA, Lazar GA, Chirino AJ, Desjarlais JR. (2003). Rational design and engineering of therapeutic proteins. *Drug discovery today*, *8*, 212-221.
Mason HS, Warzecha H, Mor T, Arntzen CJ. (2002). Edible plant vaccines: applications for prophylactic and therapeutic molecular medicine. *Trends Mol.Med.*, *8*, 324-329.
Matsumura I, Rowe LA. (2005). Whole plasmid mutagenic PCR for directed protein evolution. *Biomolecular engineering*, *22*, 73-79.
Matsuo Y, Yamazaki T. (1989). Nucleotide variation and divergence in the histone multigene family in Drosophila melanogaster. *Genetics*, *122*, 87-97.
Mendes J, Guerois R, Serrano L. (2002). Energy estimation in protein design. *Current opinion in structural biology*, *12*, 441-446.
Merrick WC, Cavallius J, Kinzy TG, Zoll WL (1993) Evolution of the EF-Tu Family. In: The Translational Apparatus. Springer, pp. 669-678
Miklos AE et al. (2012). Structure-based design of supercharged, highly thermoresistant antibodies. *Chemistry & biology*, *19*, 449-455.
Mohan PK, Barve M, Chatterjee A, Ghosh-Roy A, Hosur RV. (2008). NMR comparison of the native energy landscapes of DLC8 dimer and monomer. *Biophysical chemistry*, *134*, 10-19.
Mohan PK, Chakraborty S, Hosur RV. (2010). Hierarchy of local structural and dynamics perturbations due to subdenaturing urea in the native state ensemble of DLC8 dimer. *Biophysical chemistry*, *153*, 17-26.
Mohan PK, Hosur RV. (2009). Structure-function-folding relationships and native energy landscape of dynein light chain protein: nuclear magnetic resonance insights. *Journal of biosciences*, *34*, 465-479.
Mohan PK, Joshi MV, Hosur RV. (2009). Hierarchy in guanidine unfolding of DLC8 dimer: regulatory functional implications. *Biochimie*, *91*, 401-407.
Mohan U, Banerjee UC. (2008). Molecular evolution of a defined DNA sequence with accumulation of mutations in a single round by a dual approach to random chemical mutagenesis (DuARCheM). *Chembiochem*, *9*, 2238-2243.
Moore JC, Arnold FH. (1996). Directed evolution of a para-nitrobenzyl esterase for aqueous-organic solvents. *Nature biotechnology*, *14*, 458-467.
Morgan GJ. (1998). Emile Zuckerkandl, Linus Pauling, and the molecular evolutionary clock, 1959-1965. *Journal of the history of biology*, *31*, 155-178.
Mosnier LO, Gale AJ, Yegneswaran S, Griffin JH. (2004). Activated protein C variants with normal cytoprotective but reduced anticoagulant activity. *Blood*, *104*, 1740-1744.
Moss GP (2020) Recommendations of the Nomenclature Committee of the International Union of Biochemistry and Molecular Biology on the Nomenclature and Classification of Enzymes by the Reactions they Catalyse, https://www.qmul.ac.uk/sbcs/iubmb/enzyme/.
Murakami H, Hohsaka T, Sisido M. (2002). Random insertion and deletion of arbitrary number of bases for codon-based random mutation of DNAs. *nature biotechnology*, *20*, 76-81.
Murzin AG, Lesk AM, Chothia C. (1994). Principles determining the structure of β-sheet barrels in proteins I. A theoretical analysis. *Journal of molecular biology*, *236*, 1369-1381.
Nei M, Rooney AP. (2005). Concerted and birth-and-death evolution of multigene families. *Annu. Rev.Genet.*, *39*, 121-152.
Nelli YR et al. (2015). Isosteric substitutions of urea to thiourea and selenourea in aliphatic oligourea foldamers: site specific perturbation of the helix geometry. *Chemistry.*, *21*, 2870-2880.
Nelson DL, Cox (2013) Lehninger Principles of biochemistry vol 6th Edition.
Ness JE et al. (2002). Synthetic shuffling expands functional protein diversity by allowing amino acids to recombine independently. *Nature biotechnology*, *20*, 1251-1255.

Ng PC, Henikoff S. (2001). Predicting deleterious amino acid substitutions. *Genome research, 11*, 863-874.

Nilsson B et al. (1987). A synthetic IgG-binding domain based on staphylococcal protein A. *Protein Engineering, Design and Selection, 1*, 107-113.

Nooren IM, Thornton JM. (2003). Diversity of protein–protein interactions. *The EMBO journal, 22*, 3486-3492.

Nygren PÅ. (2008). Alternative binding proteins: Affibody binding proteins developed from a small three-helix bundle scaffold. *The FEBS journal, 275*, 2668-2676.

Offredi F et al. (2003). De novo backbone and sequence design of an idealized α/β-barrel protein: evidence of stable tertiary structure. *Journal of molecular biology, 325*, 163-174.

Ohta T. (1991). Multigene families and the evolution of complexity. *Journal of molecular evolution, 33*, 34-41.

Onuchic JN, Luthey-Schulten Z, Wolynes PG. (1997). Theory of protein folding: the energy landscape perspective. *Annu.Rev.Phys Chem., 48*, 545-600.

Orlova A et al. (2006). Tumor imaging using a picomolar affinity HER2 binding affibody molecule. *Cancer research, 66*, 4339-4348.

Orner BP, Ernst JT, Hamilton AD. (2001). Toward proteomimetics: terphenyl derivatives as structural and functional mimics of extended regions of an alpha-helix. *J Am.Chem.Soc., 123*, 5382-5383.

Ostermeier M, Nixon AE, Benkovic SJ. (1999). Incremental truncation as a strategy in the engineering of novel biocatalysts. *Bioorganic & medicinal chemistry, 7*, 2139-2144.

Ota T, Nei M. (1994). Divergent evolution and evolution by the birth-and-death process in the immunoglobulin VH gene family. *Molecular biology and evolution, 11*, 469-482.

Ovchinnikov S, Kamisetty H, Baker D. (2014). Robust and accurate prediction of residue-residue interactions across protein interfaces using evolutionary information. *Elife., 3*, e02030.

Pal G, Ultsch MH, Clark KP, Currell B, Kossiakoff AA, Sidhu SS. (2005). Intramolecular cooperativity in a protein binding site assessed by combinatorial shotgun scanning mutagenesis. *Journal of molecular biology, 347*, 489-494.

Pandyarajan V, Weiss MA. (2012). Design of non-standard insulin analogs for the treatment of diabetes mellitus. *Current diabetes reports, 12*, 697-704.

Pantazes R, Maranas CD. (2010). OptCDR: a general computational method for the design of antibody complementarity determining regions for targeted epitope binding. *Protein Engineering, Design & Selection, 23*, 849-858.

Pantazes RJ, Saraf MC, Maranas CD. (2007). Optimal protein library design using recombination or point mutations based on sequence-based scoring functions. *Protein Engineering, Design & Selection, 20*, 361-373.

Park SE, Sajid MI, Parang K, Tiwari RK. (2019). Cyclic Cell-Penetrating Peptides as Efficient Intracellular Drug Delivery Tools. *Molecular pharmaceutics, 16*, 3727-3743.

Pasut G, Sergi M, Veronese FM. (2008). Anti-cancer PEG-enzymes: 30 years old, but still a current approach. *Adv.Drug Deliv.Rev., 60*, 69-78.

Patrick WM, Firth AE, Blackburn JM. (2003). User-friendly algorithms for estimating completeness and diversity in randomized protein-encoding libraries. *Protein engineering, 16*, 451-457.

Patwardhan RP, Lee C, Litvin O, Young DL, Pe'er D, Shendure J. (2009). High-resolution analysis of DNA regulatory elements by synthetic saturation mutagenesis. *Nature biotechnology, 27*, 1173-1175.

Perry CM, Jarvis B. (2001). Peginterferon-α-2a (40kD). *Drugs, 61*, 2263-2288.

Petsko GA, Ringe D (2004) Protein Structure and Function.

Phillips NB et al. (2010). Supramolecular protein engineering: design of zinc-stapled insulin hexamers as a long acting depot. *J Biol.Chem., 285*, 11755-11759.

Planel S, Salomon A, Jalinot P, Feige JJ, Cherradi N. (2010). A novel concept in antiangiogenic and antitumoral therapy: multitarget destabilization of short-lived mRNAs by the zinc finger protein ZFP36L1. *Oncogene, 29*, 5989-6003.

Plante J, Campbell F, Malkova B, Kilner C, Warriner SL, Wilson AJ. (2008). Synthesis of functionalised aromatic oligamide rods. *Org.Biomol.Chem.*, *6*, 138-146.

Plückthun A. (2015). Designed ankyrin repeat proteins (DARPins): binding proteins for research, diagnostics, and therapy. *Annual review of pharmacology and toxicology*, *55*, 489-511.

Poluri KM, Gulati K (2016) Protein engineering techniques: Gateways to synthetic protein universe. Springer,

Poluri KM, Gulati K (2017a) Biotechnological and Biomedical Applications of Protein Engineering Methods. In: Protein Engineering Techniques. Springer, pp. 103-134

Poluri KM, Gulati K (2017b) Expanding the synthetic protein universe by guided evolutionary concepts. In: Protein Engineering Techniques. Springer, pp. 27-59

Poluri KM, Gulati K (2017c) Generating the Fancy Protein Basket with De Novo and Combinatorial Approaches. In: Protein Engineering Techniques. Springer, pp. 85-102

Poluri KM, Gulati K (2017d) Protein engineering techniques: gateways to syhthetic protein universe vol DOI https://doi.org/10.1007/978-981-10-2732-1.

Poluri KM, Gulati K (2017e) Rational Designing of Novel Proteins Through Computational Approaches. In: Protein Engineering Techniques. Springer, pp. 61-83

Poluri KM, Gulati K (2017f) World of Proteins: Structure-Function Relationships and Engineering Techniques. In: Protein engineering techniques. Springer, pp. 1-25

Qi X, Guo Q, Wei Y, Xu H, Huang R. (2012). Enhancement of pH stability and activity of glycerol dehydratase from Klebsiella pneumoniae by rational design. *Biotechnology letters*, *34*, 339-346.

Qu G, Zhu T, Jiang Y, Wu B, Sun Z. (2019). [Protein engineering: from directed evolution to computational design]. *Sheng wu gong cheng xue bao = Chinese journal of biotechnology*, *35*, 1843-1856.

Rao VS, Srinivas K, Sujini G, Kumar G. (2014). Protein-protein interaction detection: methods and analysis. *International journal of proteomics*, *2014*.

Reidhaar-Olson JF, Sauer RT. (1988). Combinatorial cassette mutagenesis as a probe of the informational content of protein sequences. *Science*, *241*, 53-57.

Reiss S et al. (2006). Inhibition of platelet aggregation by grafting RGD and KGD sequences on the structural scaffold of small disulfide-rich proteins. *Platelets*, *17*, 153-157.

Renfrew PD, Craven TW, Butterfoss GL, Kirshenbaum K, Bonneau R. (2014). A rotamer library to enable modeling and design of peptoid foldamers. *J Am.Chem.Soc.*, *136*, 8772-8782.

Rokas A, Williams BL, King N, Carroll SB. (2003). Genome-scale approaches to resolving incongruence in molecular phylogenies. *Nature*, *425*, 798-804.

Rooney AP, Ward TJ. (2005). Evolution of a large ribosomal RNA multigene family in filamentous fungi: birth and death of a concerted evolution paradigm. *Proceedings of the National Academy of Sciences of the United States of America*, *102*, 5084-5089.

Safari-Alighiarloo N, Taghizadeh M, Rezaei-Tavirani M, Goliaei B, Peyvandi AA. (2014). Protein-protein interaction networks (PPI) and complex diseases. *Gastroenterology and Hepatology from bed to bench*, *7*, 17.

Saraf MC, Gupta A, Maranas CD. (2005). Design of combinatorial protein libraries of optimal size. *Proteins: Structure, function, and bioinformatics*, *60*, 769-777.

Saven JG. (2011). Computational protein design: engineering molecular diversity, nonnatural enzymes, nonbiological cofactor complexes, and membrane proteins. *Current opinion in chemical biology*, *15*, 452-457.

Schlehuber S, Skerra A. (2005). Lipocalins in drug discovery: from natural ligand-binding proteins to 'anticalins'. *Drug Discovery Today*, *10*, 23-33.

Schroeder HW, Jr., Cavacini L. (2010). Structure and function of immunoglobulins. *J Allergy Clin. Immunol.*, *125*, S41-S52.

Semetey V, Rognan D, Hemmerlin C, Graff R, Briand JP, Marraud M, Guichard G. (2002). Stable helical secondary structure in short-chain N,N'-linked oligoureas bearing proteinogenic side chains. *Angew.Chem.Int.Ed Engl.*, *41*, 1893-1895.

Shao Z, Zhao H, Giver L, Arnold FH. (1998). Random-priming in vitro recombination: an effective tool for directed evolution. *Nucleic Acids Research*, *26*, 681-683.

Sharma M, Kumar D, Poluri KM. (2016). Elucidating the pH-dependent structural transition of T7 bacteriophage endolysin. *Biochemistry, 55*, 4614-4625.

Sharma M, Kumar D, Poluri KM. (2018). Unraveling the differential structural stability and dynamics features of T7 endolysin partially folded conformations. *Biochimica et Biophysica Acta (BBA)-General Subjects, 1862*, 924-935.

Shum T, Kruse RL, Rooney CM. (2018). Strategies for enhancing adoptive T-cell immunotherapy against solid tumors using engineered cytokine signaling and other modalities. *Expert opinion on biological therapy, 18*, 653-664.

Sieber V, Martinez CA, Arnold FH. (2001). Libraries of hybrid proteins from distantly related sequences. *nature biotechnology, 19*, 456-460.

Silverman J et al. (2005). Multivalent avimer proteins evolved by exon shuffling of a family of human receptor domains. *Nature biotechnology, 23*, 1556-1561.

Silverman RB (2000) The organic chemistry of enzyme-catalyzed reactions vol 2nd edition.

Singh RK, Lee JK, Selvaraj C, Singh R, Li J, Kim SY, Kalia VC. (2018). Protein Engineering Approaches in the Post-Genomic Era. *Current protein & peptide science, 19*, 5-15.

Sinha R, Shukla P. (2019). Current Trends in Protein Engineering: Updates and Progress. *Current protein & peptide science, 20*, 398-407.

Skerker JM, Perchuk BS, Siryaporn A, Lubin EA, Ashenberg O, Goulian M, Laub MT. (2008). Rewiring the specificity of two-component signal transduction systems. *Cell, 133*, 1043-1054.

Skerra A. (2008). Alternative binding proteins: anticalins–harnessing the structural plasticity of the lipocalin ligand pocket to engineer novel binding activities. *The FEBS journal, 275*, 2677-2683.

Smith BA, Daniels DS, Coplin AE, Jordan GE, McGregor LM, Schepartz A. (2008). Minimally cationic cell-permeable miniature proteins via alpha-helical arginine display. *J Am.Chem.Soc., 130*, 2948-2949.

Smith BJ, Lee EF, Checco JW, Evangelista M, Gellman SH, Fairlie WD. (2013). Structure-guided rational design of Î±/Î²-peptide foldamers with high affinity for BCL-2 family prosurvival proteins. *Chembiochem., 14*, 1564-1572.

Smith GP. (1974). Unequal crossover and the evolution of multigene families. *Cold Spring Harb. Symp.Quant.Biol., 38*, 507-513.

Song G, Lazar GA, Kortemme T, Shimaoka M, Desjarlais JR, Baker D, Springer TA. (2006). Rational design of intercellular adhesion molecule-1 (ICAM-1) variants for antagonizing integrin lymphocyte function-associated antigen-1-dependent adhesion. *J Biol.Chem., 281*, 5042-5049.

Steipe B, Schiller B, Plückthun A, Steinbacher S (1994) Sequence statistics reliably predict stabilizing mutations in a protein domain. Elsevier,

Stemmer WP. (1994a). DNA shuffling by random fragmentation and reassembly: in vitro recombination for molecular evolution. *Proceedings of the National Academy of Sciences, 91*, 10747-10751.

Stemmer WP. (1994b). Rapid evolution of a protein in vitro by DNA shuffling. *Nature, 370*, 389-391.

Strino F, Lii J-H, Gabius H-J, Nyholm P-G. (2009). Conformational analysis of thioglycoside derivatives of histo-blood group ABH antigens using an ab initio-derived reparameterization of MM4: implications for design of non-hydrolysable mimetics. *Journal of computer-aided molecular design, 23*, 845.

Stumpp MT, Binz HK, Amstutz P. (2008). DARPins: a new generation of protein therapeutics. *Drug discovery today, 13*, 695-701.

Sullivan BJ, Durani V, Magliery TJ. (2011). Triosephosphate isomerase by consensus design: dramatic differences in physical properties and activity of related variants. *Journal of molecular biology, 413*, 195-208.

Summa CM, Rosenblatt MM, Hong J-K, Lear JD, DeGrado WF. (2002). Computational de novo design, and characterization of an A2B2 diiron protein. *Journal of molecular biology, 321*, 923-938.

Tan S, Tan HT, Chung MC. (2008). Membrane proteins and membrane proteomics. *Proteomics.*, *8*, 3924-3932.

Tee KL, Wong TS. (2013). Polishing the craft of genetic diversity creation in directed evolution. *Biotechnology advances*, *31*, 1707-1721.

Thogersen C, Holldack J. (2006). A tetranectin-based platform for protein engineering. *Innovations Pharmac Technol*, 27-30.

Tiwari MK, Singh R, Singh RK, Kim I-W, Lee J-K. (2012). Computational approaches for rational design of proteins with novel functionalities. *Computational and Structural Biotechnology Journal*, *2*, e201204002.

Trabuco LG, Lise S, Petsalaki E, Russell RB. (2012). PepSite: prediction of peptide-binding sites from protein surfaces. *Nucleic Acids Res.*, *40*, W423-W427.

Uhlen M, Guss B, Nilsson B, Gatenbeck S, Philipson L, Lindberg M. (1984). Complete sequence of the staphylococcal gene encoding protein A. A gene evolved through multiple duplications. *Journal of Biological Chemistry*, *259*, 1695-1702.

Urvoas A et al. (2010). Design, production and molecular structure of a new family of artificial alpha-helicoidal repeat proteins (Î±Rep) based on thermostable HEAT-like repeats. *J Mol.Biol.*, *404*, 307-327.

Vazquez-Lombardi R, Roome B, Christ D. (2013). Molecular engineering of therapeutic cytokines. *Antibodies*, *2*, 426-451.

Verdine GL, Hilinski GJ. (2012). Stapled peptides for intracellular drug targets. *Methods Enzymol.*, *503*, 3-33.

Vigneri R, Squatrito S, Sciacca L. (2010). Insulin and its analogs: actions via insulin and IGF receptors. *Acta Diabetol.*, *47*, 271-278.

Villarino A, O'Shea JJ. (2017). Extraordinary effects of unnatural pairings. *eLife*, *6*.

Voet D, Voet JG (2011) Biochemistry vol 4th edition.

Vojtechovsky J, Chu K, Berendzen J, Sweet RM, Schlichting I. (1999). Crystal structures of myoglobin-ligand complexes at near-atomic resolution. *Biophys.J*, *77*, 2153-2174.

Wang F, Zeltwanger S, Hu S, Hwang TC. (2000). Deletion of phenylalanine 508 causes attenuated phosphorylation-dependent activation of CFTR chloride channels. *The Journal of physiology*, *524*, 637-648.

Wang Q et al. (2010). Prospecting metagenomic enzyme subfamily genes for DNA family shuffling by a novel PCR-based approach. *Journal of Biological Chemistry*, *285*, 41509-41516.

Watanabe K, Ohkuri T, Yokobori S-i, Yamagishi A. (2006). Designing thermostable proteins: ancestral mutants of 3-isopropylmalate dehydrogenase designed by using a phylogenetic tree. *Journal of molecular biology*, *355*, 664-674.

Wilding M, Hong N, Spence M, Buckle AM, Jackson CJ. (2019). Protein engineering: the potential of remote mutations. *Biochemical Society transactions*, *47*, 701-711.

Winter A, Higueruelo AP, Marsh M, Sigurdardottir A, Pitt WR, Blundell TL. (2012). Biophysical and computational fragment-based approaches to targeting protein-protein interactions: applications in structure-guided drug discovery. *Quarterly reviews of biophysics*, *45*, 383-426.

Wong TS, Tee KL, Hauer B, Schwaneberg U. (2004). Sequence saturation mutagenesis (SeSaM): a novel method for directed evolution. *Nucleic Acids Research*, *32*, e26-e26.

Wurm F, Bernard A. (1999). Large-scale transient expression in mammalian cells for recombinant protein production. *Curr.Opin.Biotechnol*, *10*, 156-159.

Xiang Z. (2006). Advances in homology protein structure modeling. *Current Protein and Peptide Science*, *7*, 217-227.

Xu D, Xu Y. (2004). Protein databases on the internet. *Current protocols in molecular biology*, *68*, 19.14. 11-19.14. 15.

Yagi S, Akanuma S, Yamagishi A. (2018). Creation of artificial protein-protein interactions using alpha-helices as interfaces. *Biophysical reviews*, *10*, 411-420.

You L, Arnold F. (1996). Directed evolution of subtilisin E in Bacillus subtilis to enhance total activity in aqueous dimethylformamide. *Protein Engineering, Design and Selection*, *9*, 77-83.

Yu H, Yan Y, Zhang C, Dalby PA. (2017). Two strategies to engineer flexible loops for improved enzyme thermostability. *Scientific reports*, *7*, 41212.

Zeldovich KB, Shakhnovich EI. (2008). Understanding protein evolution: from protein physics to Darwinian selection. *Annu.Rev.Phys.Chem.*, *59*, 105-127.

Zhang Y. (2007). Template-based modeling and free modeling by I-TASSER in CASP7. Proteins: Structure, Function, and Bioinformatics, 69, 108-117.

Zhao X et al. (2007). Targeting CD37-positive lymphoid malignancies with a novel engineered small modular immunopharmaceutical. *Blood, The Journal of the American Society of Hematology*, *110*, 2569-2577.

Zoller MJ. (1991). New molecular biology methods for protein engineering. *Curr.Opin.Biotechnol*, *2*, 526-531.

Zuckerkandl E, Pauling L. (1965a). Molecules as documents of evolutionary history. *J Theor.Biol.*, *8*, 357-366.

Zuckerkandl E, Pauling L. (1965b). Molecules as documents of evolutionary history. *Journal of theoretical biology*, *8*, 357-366.

Structural Aspects of Protein–Protein Interactions

2

Abstract

Many biological functions in the cell are mediated when a protein interacts with one or more proteins in a process known as protein–protein interaction (PPI). These PPIs can be classified based on six distinct criteria such as contact criteria, interaction lifetime, affinity, nature of interacting partners, identity of formed complex, and presence/absence of artifacts. This chapter delineates how PPIs are functionally significant for the sustenance of life and discusses all the parameters that are utilized to analyze a particular PPI, particularly the interface between the two interacting proteins. Further, it discusses the models that have been proposed to describe PPIs and also traces the pathways down which PPIs have evolved. The chapter concludes with interactions that are mediated either by particular subunits, domains and/or motifs.

Keywords

Protein–protein interactions · Interaction lifetime · Metabolic pathways · Protein interactome · Misordered proteins · Conformational ensemble

2.1 Introduction

Genome sequencing of several organisms over the years led to the finding that organism complexity at the phenotype level cannot be described solely by the gene numbers in an organism. Therefore, the post-genomic era attempted to explain it by putting forward the proof for novel mechanisms like RNA splicing, post-translational modifications, cellular localization, and tissue complexity. For instance, it was determined that more than 90% of human genes have alternately spliced

K. M. Poluri et al., *Protein-Protein Interactions*,
https://doi.org/10.1007/978-981-16-1594-8_2

isoforms of mRNA, and thereby accounting for around nine different transcripts per protein coding gene (Flicek et al. 2014; Wang et al. 2008). Intensive research in the fields of biophysics, biochemistry, cell biology, molecular biology, and structural biology gave rise to an ample amount of information about the structural and functional properties of several individual proteins. Yet, the compendium of tens of thousands of proteins in the cell failed to explain an organism's phenotypical intricacy, that was partly explained by the occurrence of proteins physically interacting with other proteins in the cell to regulate their function (Berggård et al. 2007). This phenomenon, known as protein–protein interaction (PPI), was known to the world even before the advent of the proteomics era. However, the vastness and the intricacy of the network formed by the myriad of protein–protein interactions (PPIs) were yet to be determined.

Currently, it has been believed that each protein in a eukaryotic cell exhibits six to eight interacting protein partners on an average (Tong et al. 2004). Usually, PPIs can be categorized on the basis of six grounds such as (a) interaction lifetime, (b) contact criteria, (c) affinity, (d) identity of formed complex, (e) nature of interacting partners, and (f) presence/absence of artifacts. PPIs are indispensable to biological processes and in turn the sustenance of living organisms. The roles of some PPIs are well-established. For instance, the formation of antibody−antigen, protease−inhibitor, or hormone−receptor complexes involves PPIs (Clackson et al. 1998; Granelli-Piperno and Reich 1978; Sundberg 2009). Several other cellular processes such as communication between cells, replication, transcription, membrane transport, catalysis of metabolic reactions, regulation (activation or repression) of other proteins, alteration of enzymes' kinetic parameters, etc. are also performed by the aid of PPIs. Events like cell growth and transformation that involve signaling are also regulated by the virtue of competitions and versatility of protein–protein interactions (Couzens et al. 2013; Romano et al. 2014). In a nutshell, PPIs help in the construction of signaling and metabolic pathways in the cell to maintain homeostasis. If the interactions are altered, these pathways often become dysfunctional and lead to several diseased conditions (De Las Rivas and Fontanillo 2012).

The entire map of protein interactions occurring in an organism is known as the interactome. The complicated molecular relationships in the living cell can be unraveled by mapping all such interactions. It can aid systems biologists in furnishing deeper insights into the mechanisms of several diseases (Ideker and Sharan 2008; Jiang et al. 2013; Tourette et al. 2014; Vidal et al. 2011).

Owing to the great importance attributed to PPIs, the study of such interactions has received critical attention from the scientific community. In order to have better insights into the functions of PPIs, it is essential that the structural aspects of PPIs are unraveled. Therefore, researchers have focused on elucidating the characteristic parameters of PPIs, especially the interface properties such as size, residue composition, secondary structure, hydrophobicity, hydrogen bonding, packing, crevices, and surface complementarity, etc. (Jones and Thornton 1996). In addition to these parameters, some research groups have also analyzed the water molecules present in the protein interfaces. Further, several structure-based studies have attempted to explain the plasticity of PPIs, the property by virtue of which one protein gains the capability to share interactions with several proteins (Sundberg and Mariuzza 2000). The plasticity has been largely deemed to be dependent on the conformational variations associated with the binding events of proteins. Several models have cropped up to explain the mechanism behind protein interactions and how the binding accommodates such conformational changes (Goh et al. 2004). The current chapter provides a detailed account of all the fundamental aspects of PPIs, with special emphasis on their structural features. Moreover, it throws light onto PPI evolution and some unique PPIs that utilize motifs to mediate binding.

2.2 Protein–Protein Interactions and their Criteria

It is essential to have a clear idea regarding how PPIs are defined. Ordinarily, they are defined as physical contacts between proteins in cells or in vivo. However, it should be noted that the mere presence of a "functional contact" between two proteins should not be confused with the phenomenon of direct physical interaction between them (Chatr-Aryamontri et al. 2008; Mackay et al. 2007). For instance, any protein residing in the basal transcription apparatus or in the ribosome shares function-oriented contacts with other protein partners present in the particular complex. However, that does not ensure that all the proteins in that specific complex interact (Cowell and Hurst 1996; Jeronimo et al. 2007). Similarly, other kind of functional links between biomolecules in the cell such as DNA, proteins, metabolites should not be deemed as physical interactions among proteins. A variety of techniques are dedicated for the investigation of such functional relationships or links. For instance, transcriptome expression profiling aids to elucidate gene co-expression (Prieto et al. 2008). Similarly, double mutant synthetic lethality is employed to investigate genetic interactions (Mani et al. 2008). Proteins are also extensively involved in interactions with other cellular entities such as DNA, RNA, cofactor, or ligand. Although such interactions are important to gain better insights into the interactome, data obtained from the study of these interactions should be kept segregated from that of PPI data. Despite the fact that definitions of PPI focus on the "physical contact" between proteins, it should be ensured that such a contact is specific and significant. In other words, proteins bumping into each other simply by the virtue of chance should not be included under PPIs. Further, PPIs should not include interactions that are indulged in universal functions such as protein folding, protein degradation, etc.

For instance, all the proteins make contact with entities like ribosome and most of them "touch" the degradation machinery or chaperones at some point in their lifetime (Freilich et al. 2018). Chaperones are very well-known for their imperative role in protein homeostasis, which is also known as proteostasis. The proteostasis refers to the process of maintaining the folding of proteins, their quality, and turnover number. Now, the question arises that how chaperones maintain the entire cellular and organism level proteome. The answer to this question is that chaperones engage in PPIs both with each other and also with their client proteins. These interactions allow the chaperones to form complexes with their clients and test them for their mis-folding. These complexes also allow the linkage of proteins to other cellular processes such as trafficking and degradation. The network of PPIs driven by chaperones also enables them to "hand off" a client protein from one chaperone to other and also aid them in deciding the quality of their client proteins. Hence, it is essential to get better insights into the PPIs to determine the potential PPI drug targets in the chaperone network for rebalancing the process of proteostasis (Freilich et al. 2018).

However, from a PPI point of view, such generalized interactions should be filtered out. To summarize, the definition of PPIs should take two criteria into consideration (De Las Rivas and Fontanillo 2010). First, the interaction should be intended or deterministic and not accidental, that is, the interaction should not take place as a consequence of other biomolecular forces or events. Second, the interactions should not be generic in nature, i.e. the interactions should be intended for functions that are completely distinct from basic cellular functions such as protein synthesis or protein degradation.

Another essential element that should be kept in mind during defining PPIs is their biological ambience. All the possible PPIs would not happen in any cell at any point of time. Instead, it is dependent on the expression of proteins that are restricted to only specific developmental phases or certain cell cycle stages in particular tissues. Therefore, factors like cell type, protein modifications, cell cycle stage, presence/absence of other cofactors/binding partners, developmental stage, environmental conditions, etc. play an imperative role in determining the real occurrence of PPIs.

2.3 Classification of Protein–Protein Interactions

PPIs are diverse in nature, owing to which they are ascribed with different functions. Defining PPIs on several grounds can furnish better insights into the characteristic features of PPIs. Therefore, researchers have focused on classifying PPIs into several categories (Fig. 2.1). PPIs can be categorized on the following criteria: (a) direct or indirect (b) transient or permanent (c) obligate or non-obligate (d) homo-oligomeric or hetero-oligomeric (e) disordered or ordered (f) biological or crystal complex and (g) domain–domain or domain–peptide interactions. However, the segregation of PPIs into one of these classes is not simple or straightforward owing to the existence of a continuum (Nooren and Thornton 2003). Which class a particular PPI would fall in is highly dependent on the physiological state of the cell. For instance, an interaction that behaves transiently in vivo might become permanent under certain cell culture conditions. There is paucity of data on protein folding as well as dynamics of protein association at different physiological conditions. In such a scenario, available parameters of structural characterization fail to distinguish PPIs into one of the above-mentioned categories. Nevertheless, classification might be suggested both by proteins' function and their subcellular locations. For instance, proteins that constitute a signaling pathway would be, in all possibility, involved in

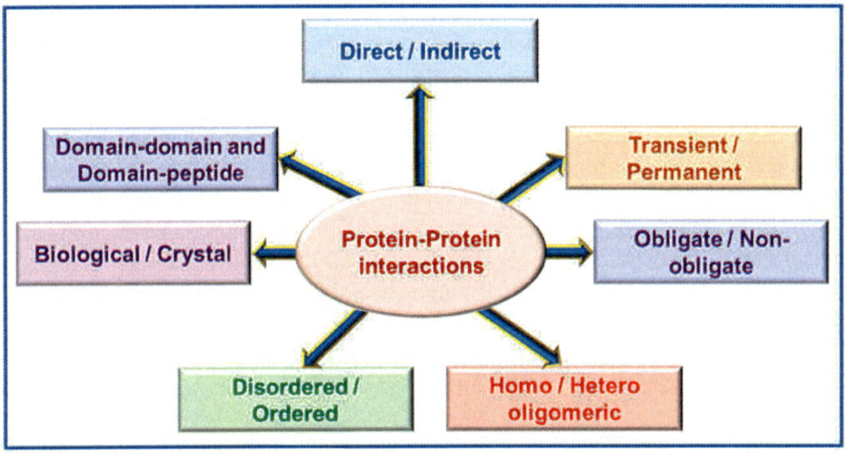

Fig. 2.1 Schematic of classification of PPIs

transient interactions since well-organized signaling is dependent on the quick association and dissociation of the signaling proteins.

2.3.1 Direct or Indirect PPIs

PPIs involve molecular docking between proteins in a living organism in in vivo conditions. If the two proteins have their molecular interfaces in contact with each other, the interaction is known as "direct." In contrast, if the two proteins are physically restrained at a certain distance, but their interaction is mediated by other intermediate-acting protein partners in a complex, the interaction is deemed to be "indirect" (Peng et al. 2017).

2.3.2 Transient or Permanent PPIs

PPIs can be classified into transient or permanent based on factors like stability and lifetime (Acuner Ozbabacan et al. 2011). The notion that PPIs denote the physical association of proteins does not imply that the contacts are permanent or static. This is because the machinery of the cell constantly experiences the processes of turnover and reassembly. Some proteins undergo a very stable association and permanently form a complex following the interaction (Jones and Thornton 1996). For instance, in macromolecular protein complexes such as ATP synthase, which comprises eight different proteins in mammals, the constituting PPIs are permanent. Similarly, mammalian cytochrome oxidase (COX) consisting of 13 proteins involves permanent PPIs (Kadenbach and Hüttemann 2015). The proteins constituting such complexes are often referred to as the "subunits" of these complexes (De Las

Rivas and Fontanillo 2010). For example, 13 proteins in COX imply the presence of 13 subunits in COX. These subunits are encoded differentially, in which ten subunits are encoded by nuclear DNA and rest of the three subunits are encoded by mitochondrial DNA. The nuclear encoded subunits are indulged in proton translocation and oxygen consumption. Core of COX comprises three mitochondrial encoded subunits which possess high catalytic activity. This catalytic activity is suppressed by the binding of ten nuclear encoded subunits (Kadenbach and Hüttemann 2015). The rings of GroELs or the subunits of ribosomes also encompass interactions that are permanent (Xu et al. 1997; Yusupov et al. 2001). Similarly, complexes assembled from proteins such as RNA polymerase and hemoglobin involve stable interactions (Byrum et al. 2012).

On the other hand, there are several PPIs that associate and dissociate on a temporary basis in vivo (Ngounou Wetie et al. 2013; Wang et al. 2014). They are less stable and have a shorter lifetime than permanent PPIs. Proteins indulging in such interactions are intended to perform transient functions such as regulation of gene expression (activation or repression) by uniting with other protein molecules such as activators and transcription factors. Similarly, binding interactions constituting chaperone-protein, hormone-receptor, protein-protease, or consecutive components in a signaling pathway involve transient PPIs. Transient interactions are usually established with the aid of non-covalent bonds and other weaker forces such as hydrogen bonds, hydrophobic interactions, ionic bonds, and van der Waals interactions (Westermarck et al. 2013). Transience or the temporary association–dissociation cycles of interactions enable the cell to readily respond to alterations in external stimuli and relay response signals as required. For instance, Ras protein which is involved in critical regulatory networks of the cell indulge in transient interactions with its effector protein partners such as PI3K, Raf, etc. to act as a signaling switch (Plowman and Hancock 2005). The stability of a particular PPI is a function of the binding free energy. Generally, interacting partners involved in forming permanent complexes have higher possibilities of being co-expressed and co-localized than those constituting transient interactions (Keskin et al. 2016).

2.3.3 Obligate or Non-Obligate PPIs

Based on the stability and affinity of proteins in monomeric states and in complex, PPIs can be categorized into non-obligate and obligate interactions. If the proteins that form the complex are unstable as monomers in vivo, then they set up obligate interactions in the complex. On the other hand, if the constituting protein components of a complex can exist independently, then the interactions present in the complex are termed as non-obligate (Levy and Teichmann 2013). Proteins involved in obligate interactions fold and interact simultaneously to form steady complexes. This is because the proteins individually cannot exist as folded, stable structures, but attain stability only on complex formation. Therefore, protein components mediating obligate interactions are known as two-state folders. On the other hand, components constituting non-obligate interactions are referred to as three-state folders. This is because the individual proteins are first folded. In the second step, the folded proteins associate with each other to aid in complex formation. It is difficult to demarcate between obligate/non-obligate and transient/permanent complexes. The classifications amalgamate at the boundaries such that most of the non-obligate PPIs are usually transient in nature (Janin et al. 2008), whereas obligate PPIs are compulsorily permanent ones (Nooren and Thornton 2003). For instance, supramolecular complexes in the cell and other stable cellular machineries contain obligate and permanent interactions. On the other hand, interactions mediated by antigen–antibody and thrombin–rodniin inhibitor are examples of non-obligate, permanent interactions (DeLisi 2013; Nooren and Thornton 2003). As the constituents of such protein complexes are not co-localized initially, they need to be stabilized independently (Nooren and Thornton 2003). Non-obligate, transient behavior is observed in bovine G protein which is a hetero-trimer. In the trimeric complex, the interaction between Gbg and Ga is transient (Ji et al. 1998; Terakita et al. 2002).

2.3.4 Homo-Oligomeric or Hetero-Oligomeric PPI Complexes

This is the simple and straightforward basis for the categorization of PPIs. If the complex constituting proteins are identical, that is, the interactions are mediated between identical protein chains, the complex is known as a homo-oligomer. On the other hand, if the interactions embrace non-identical protein chains, then the formed complex is referred to as hetero-oligomer. Identical units of homo-oligomers are usually arranged in a symmetric fashion thereby providing scaffold for stable macromolecules (Goodsell and Olson 2000). The organization of such homologous protein units can either utilize an isologous style or heterologous style (Monod et al. 1965). An isologous interaction is mediated by the same surfaces on both the interacting partners and further oligomerization is possible only with the utilization of a different interface. On the other hand, a heterologous association utilizes dissimilar interfaces that in the absence of a cyclic/closed symmetry would lead to a state of infinite aggregation. Majority of the homodimers exist as oligomers and it is not possible to segregate them into independent, correctly folded monomers. For instance, in F-type ATP synthase, which is a multi-subunit complex converting energy stowed in electrochemical gradients of protons to synthesize ATP, the C-ring (or the C-subunit assembly) is used to transduce electrochemical energy to mechanical energy and vice versa. The C-ring consists of a cyclic, highly symmetric structure and is a well-known homo-oligomer (Pogoryelov et al. 2010). Other homo-oligomeric complexes include obligate homodimers such as P22 Arc repressor and non-obligate homodimers such as sperm lysin. Unlike homo-oligomers that offer a stable scaffold, owing to their symmetry, the hetero-oligomers have variable stability. Examples of hetero-oligomers include obligate heterodimers such as human cathepsin D and non-obligate heterodimers such as RhoA and RhoGAP signaling machinery complex (Nooren and Thornton 2003).

2.3.5 Disordered and/or Ordered PPI Complexes

PPIs mediated by disordered proteins are not a well-defined class of PPIs. Intrinsically disordered proteins or IDPs have unstructured regions whose corresponding amino acid residues fail to form a folded, stable structure. The region of disorder might correspond to either a large region of the protein or a small portion (Dunker et al. 2001). Such proteins are abundantly found in the cells of eukaryotes. For instance, histone protein tails and proteins controlling cell cycle and signaling have been reported to comprise IDPs. The disordered proteins have the ability to bind to several proteins via the formation of a compatible conformation with the interacting protein partners. The interactions established by such proteins are, however, transient in nature (Dunker et al. 1998). Disordered segments that are large in size undergo simultaneous folding and binding with their partners, whereas shorter disordered segments that behave like linkers function in the assembly of the cell's macromolecular complexes (van der Lee et al. 2014). The fact that the interactions are mediated by weak binding affinities is energetically coherent. It offsets the high entropic cost of the stabilization of a single conformation from the dynamic pool that the IDPs sample (Dyson and Wright 2005).

In some instances, however, favorable interactions are fine-tuned that furnish significantly strong affinities (Ferreon et al. 2013). When an IDP binds to a partner, the IDP folds upon it, and the formed complex shares a greater surface area compared to a PPI comprising of similar-sized folded proteins. This large buried surface area, in turn, maximizes the specificity of the interaction. Further, having a great degree of conformational freedom, the IDPs can adopt various conformations and interact with multiple partners (Hsu et al. 2013). This versatility in interactions thus makes IDPs important components of cell signaling pathways and attributes them the functions of hub proteins and scaffold proteins. The diversity of conformations of IDPs also increases the repertoire of interactions without the need of expression of several proteins (Berlow et al. 2015). Upon binding to interacting protein partners, IDPs transit from a disordered state to an ordered state and form a complex that still remains flexible in the bound conformation. The flexibility provides greater access to enzymes and potential interacting partners. The access to enzymes also makes it logical why multiple post-translational modification (PTM) sites are present in the disordered domains. Such modification often forms the central element of the cell's metabolic regulation. PTM not only regulates (activates/deactivates) the protein, but also modifies the IDP's affinity of interaction with other partners (Varadi et al. 2015).

2.3.6 Biological or Crystal PPI Complexes

Protein structures that are determined by X-ray crystallography are stored in the Protein Databank (PDB). However, they are often found to contain contacts that are non-biological in nature. These contacts are deemed to be experimental artifacts (Janin 1997). Such non-biological interactions are often similar to those mediated by biological contacts; however, it is essential to distinguish between the two classes to accurately analyze properties related to protein binding (Keskin et al. 2004). The feature that discriminates the most among the biological and crystal contacts is the interface area size. The second distinguishing feature is the conservation rate. Biological interfaces usually have larger contacts and greater degree of conservation. Features of interface size and conservation combined can distinguish biological contacts with 98.3% accuracy (Valdar and Thornton 2001). However, there do exist a few exceptions where very large interfaces are also mediated entirely by non-biological or crystal contacts. The third distinguishing feature is the composition of amino acid. When the amino acids comprising the interface are akin to those comprising the rest of the protein surface, then the interface is known as non-biological. Other properties that distinguish between non-biological and biological contacts are hydrophobicity, free energy changes, and interacting forces across the interface, such as salt bridges and hydrogen bonds.

2.3.7 Domain–Domain and Domain–Peptide Interactions

Protein interactions can be categorized into domain–domain or domain–peptide interactions depending on the nature of folds (Aloy and Russell 2006). It is possible to narrow down the segments or domain of proteins that mediate the interaction. By performing experiments with smaller constructs and recurrent "domain signatures" in pairs of proteins that are known to interact, it is possible to pinpoint the interacting regions of the protein partners. If the same domains are identified multiple times in the interacting pair, then the interaction, in all probability, is mediated by those domains. There are few fundamental differences between domain–domain and domain–peptide interactions. Usually, domain–peptide interactions are less tight than domain–domain interactions. Also, domain–peptide interactions are usually transient in nature, as a result of which they are known as transient-peptide-mediated interactions. In such contacts, the interacting domain does not change its conformation when binding with specific peptide motifs. Consequently, they can change their interacting partners easily (London et al. 2010). The constancy of conformation on binding to peptides also minimizes the entropic cost associated with binding. Peptides comprise interfaces that display better packing than interfaces of protein–protein interactions. Peptide interfaces also contain more hydrogen bonds that usually involve the backbone of the peptide. Further, the interface comprises "hotspot" residues that possess the greatest contribution to the binding energy (London et al. 2010). Lastly, peptides have the tendency to bind to the broadly available pockets on the surface of the protein.

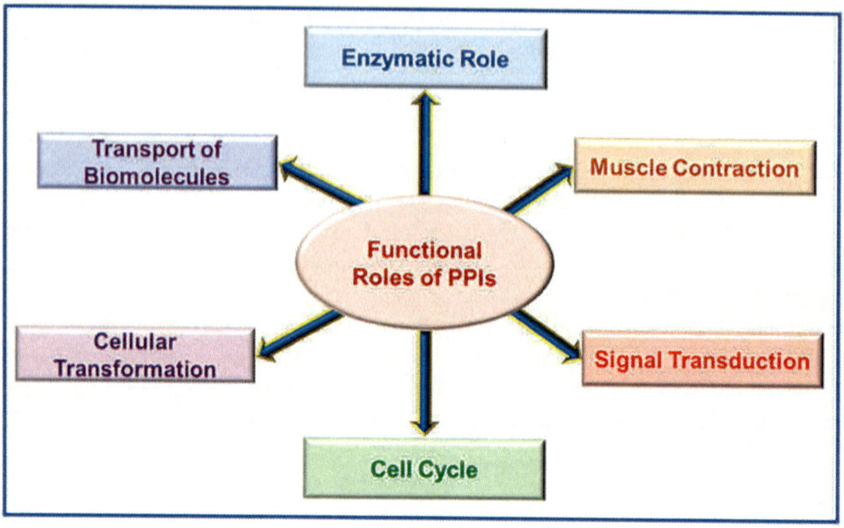

Fig. 2.2 Schematic illustration of the various functional roles of PPIs

2.4 Functional Relevance of PPIs and the Interactome

PPIs play a central role in several biological activities such as muscle contraction, cell signaling, metabolism, cell cycle, regulation of other biomolecules, etc. (Fig. 2.2). Aberrance in protein interactions (such as loss of PPI) may thus lead to several diseased conditions. In recent years, the total map of protein interaction networks (i.e., the interactome) is being extensively studied to get better understanding of the functional involvement of PPIs in human health and diseases (Jubb et al. 2017; Lan et al. 2015; Safari-Alighiarloo et al. 2014).

2.4.1 Role of PPIs in Enzyme Function

The importance of physical association between proteins has been unraveled in several biochemical pathways that are central to metabolic reactions. When a number of metabolic enzymes were discovered and purified, it has been observed that most of them had molecular weights in the range of mega-Daltons. This molecular weight range, being very large for singular proteins, led researchers to speculate that enzymes were actually molecular complexes where several individual proteins worked in consortium. It took around 10 years to resolve the multi-protein enzyme complex into its constituent elements. Once the individual constituent proteins were recognized, each of them was linked with overall reaction's individual catalytic steps (Koike et al. 1960). Re-association experiments led to the reconstitution of the complex from the individual constituent proteins, thereby proving that the interactions among the various proteins were non-covalent in nature (Hayakawa et al. 1966).

Over the years, with the aid of experiments such as gel electrophoresis, the compositions of several enzymes were unraveled and it was determined that most of the proteins that serve as enzymes in the metabolic reactions were composed of interacting subunits (Frieden 1971). Research on the mechanism of enzyme subunit regulation led researchers to conclude that oligomers (i.e., enzymes comprising of more than one subunit) are better suited to participate in metabolic reactions as they have certain desired properties associated with symmetry that are not applicable to monomers. PPIs are thus acknowledged as central players in the dynamic adaptation process of enzyme activity that acts in accordance with the metabolic status of the cell (Monod et al. 1965).

Further, interactions that are mediated between subunits of multi-subunit proteins are critical for allosteric proteins to function. For instance, the hemoglobin protein–protein interface experiences minor conformational changes that, in turn, are responsible for cooperative binding of oxygen (Stites 1997). Additionally, non-enzymatic proteins were found to interact with enzymes. Such interactions were heterologous in nature and were intended for the modification of catalytic parameters. For instance, the enzyme activity of aldolase is drastically enhanced via its interaction with F-actin protein (Arnold and Pette 1970).

Further, in several metabolic reactions, protein molecules that act as electron carriers associate with enzymatic proteins such as reductases. These proteins receive electrons and then dissociate from the enzymes. Following this, they associate with enzymes such as oxidases which act as acceptors of the electrons. In such processes, the highly specific association between the proteins certifies an efficient transfer of electrons in metabolic reactions and thus provides functional advantage. Indeed, such interactions are regular in mitochondrial oxidative phosphorylation processes (Hanukoglu 1996).

2.4.2 Role of PPIs in Muscle Contraction

Myosin filament proteins act as "molecular motors." They convert chemical energy into mechanical energy by hydrolyzing ATP. This generates force and aids in movement. The greatest instance of such movement is muscle contraction. The fact that muscle contraction requires actin–myosin interaction was first revealed when it was observed that the conformation of myosin changes on ATP addition (Dainty et al. 1944). Later, it was further determined that Myosin B, which was one of the previously discovered forms, was in reality a complex of myosin and actin. Both these proteins physically interacted with each other to yield the complex of actomyosin (Guba 1943; Jakus and Hall 1947). The physical association between the two proteins was also validated using experimental techniques such as immunoprecipitation (IP) (Kesztyüs et al. 1949). Additionally, it was demonstrated that the association of actin with myosin induces the ATPase activity of the latter. Therefore, actin–myosin interaction proved to be a key element for the enzymatic activity and the process of contraction of muscles (Banga and Szent-Györgyi 1942).

A working model was set up for the molecular mechanism underlying muscle contraction. The model states that the muscle contraction cycle begins with myosin, strongly bound to actin in the absence of ATP. Once ATP binds, myosin dissociates from actin. ATP hydrolysis then triggers conformational alteration in myosin that dislocates its head group. The dislocation leads to the binding of the head of myosin to a different location on the actin filament and the release of the hydrolysis products (i.e., ADP and inorganic phosphate). Once the myosin head acquires back its original conformation, the sliding of the actin filament is driven. Actin–myosin interactions have also been demonstrated to be responsible for different types of movements that involve non-muscle cells. They have also been depicted in a variety of cellular biological activities such as cell division and crawling movements of cells across a surface (Cooper and Hausman 2000).

2.4.3 Role of PPIs in Signal Transduction

Extracellular signals regulate the activity of the living cells. These signals are propagated in the interior of the cells by the interactions between different proteins that act as signaling molecules. Early studies revealed that cyclic AMP (cAMP) is produced in the cells following hormonal stimulation. cAMP, in turn, is responsible for activating phosphorylase, while phosphorylase is directly regulated by phosphorylase kinase, the latter does not bind to cAMP. This conjecture led researchers to search for a kinase that was cAMP-dependent. The purified protein was identified as protein kinase A (PKA). In accordance with the hypothesis, PKA was found to activate phosphorylase kinase with the aid of phosphorylation (Braun and Gingras 2012). This laid out the critical importance of regulated, stable PPIs in signaling pathways.

Among the classical signaling pathways, the cascades mediated by heterotrimeric G proteins are one of the best understood (Mayer 2006; Tripathi and Poluri 2020). For instance, in the p-adrenergic pathway, a receptor that gets stimulated by an agonist activates several molecules of a heterotrimeric G protein. Each of the activated protein molecules, in turn, can stimulate a molecule of the enzymatic protein adenylyl cyclase. The end product of the enzymatic reaction cAMP triggers several molecules of PKA that finally phosphorylates many other proteins involved in that signaling pathway. Although the cascade apparently seems simple, it comprises some critical PPIs: (1) interaction of the heterotrimeric G-protein with the ligand receptor, (2) interaction of the alpha-subunit of G-protein with the cyclase enzyme, (3) interaction of the P and γ-subunits of G-protein with p-adrenergic receptor kinase (P-ARK), etc.

Further investigation on PKA also elucidated the role of PPIs in cell signaling regulation. PKA is activated following a wide range of external signals (Taylor et al. 1990). PKA mediates such diverse signals, often producing completely opposite physiological effects. This led researchers to perform a more detailed analysis to unravel how the regulation is mediated. It was determined that different stimuli activate different subpopulations of PKA in a selective manner (Smith and Scott 2006). For instance, a tight binding between the type II-B PKA to microtubules was demonstrated via its physical association with a microtubule-associated protein (Bregman et al. 1989). Such interactions with A-kinase anchoring proteins (AKAPs), proved by genetic experiments, were found to be critical for the selective induction of PKA signals and the production of different physiological outcomes (Rosenmund et al. 1994). Modules mediated by AKAP–PKA interactions provided the first instance of spatial regulation by PPIs. It laid out the concept of compartmentalization and the roles PPIs play in maintaining such spatial insulation. This concept was later extrapolated to other signaling cascades such as Raf-Ras-MAPK signaling (Mor and Philips 2006). The role of PPIs in regulating signal transduction was also observed in *E. coli,* where the binding affinity of gene-regulatory catabolite activator protein was found to be connected to a monomer–dimer equilibrium of the protein (Brown and Crothers 1989).

2.4.4 Role of PPIs in Cell Cycle

Maturation-promoting factor, also called as mitosis-promoting factor and abbreviated as MPF, is the factor that drives the meiotic and mitotic phases of the cell cycle. Following the purification of the factor, it was found that the factor is actually a complex of two proteins—cyclin and cyclin-dependent kinase (CDK). The kinase is activated via physical interaction with the cyclin in a cell-cycle dependent manner (Braun and Gingras 2012). Experimental approaches utilizing several model systems unraveled the basic mechanisms underlying the eukaryotic cell cycle and concluded that the PPI mediated by cyclin-CDK is central to the entire procedure. MPF induces the cell to move into cell cycle's M phase (mitotic phase) from G_2 phase by phosphorylation of several cell-cycle related proteins. MPF itself is activated by its interaction with a phosphatase protein that removes a phosphate group. The phosphate moiety inhibits the activity of MPF, prior to activation by the phosphatase. MPF acts on a host of protein targets and all the resulting PPIs are extremely crucial for cell cycle progression. MPF interacts with condensins, which aid in chromatin condensation during prophase. MPF is involved in interaction with several microtubule-associated proteins that contribute to formation of the mitotic spindle. MPF also interacts with lamins to aid in the process of nuclear envelope degradation. MPF is also reported to interact with histone proteins H_1 and H_3 (Cooper and Hausman 2000).

2.4.5 Role of PPIs in Cellular Transformation

Cell cycle's de-regulation and cell transformation are the leading factors behind the aberrant growth of cells that in turn leads to cancer. Protein interactions also play an imperative role in such cellular transformation processes. This can be demonstrated by the fact that hijacking of the host's protein interactions by viral proteins or the loss/gain of critical PPIs via mutations can cause the cell to transform. For example, when Neu receptor is mutated such that it prefers dimerization, the corresponding *neu* gene is said to have been converted into an oncogene. The preference for dimerization in the mutated protein in turn activates its tyrosine kinase domain (Sternberg and Gullick 1989).

Several critical cellular PPIs are targeted by both tumor mutations and viral proteins. For instance, both tumor and virus replication target p53 and Rb proteins as pre-requisites (Ou et al. 2011). One unusual DNA virus, Polyoma (Py), however does not employ its early region proteins to deactivate p53 via direct physical association (Dilworth 1990). Instead, its protein PyST interacts with the cellular protein phosphatase 2A (PP2A) complex to preclude the downstream stabilization of p53 by ARF protein. Interestingly, PP2A subunit mutations have also been observed in human tumors (Wang et al. 1998), implicating that protein interactions mediated by PP2A also play key roles in cellular transformation and cancer.

Vital cell-regulating proteins are not always directly subject to mutations in cancer. Instead, mutations that affect interactions downstream or upstream of that pathway de-regulate the functions of such critical proteins. For instance, the pro-apoptotic proteins, Bak and Bax do not undergo direct mutation in human cancer. Instead, mutation (translocation) in the BCL2 gene makes it express constitutively in follicular lymphoma. Consequently, Bak and Bax bind to BCL2 and lose their functional roles in the protective process of cell suicide (Kirkin et al. 2004). Similarly, when Rb is mutated or Cyclin D is amplified, E2F protein is released from Rb-mediated PPIs that regulate cell growth by acting in a repressive manner (Dick and Rubin 2013).

Host immune response modulation mediated by PPIs has also been demonstrated to play a crucial role in cellular transformation. Tumor mutations and viral proteins have been reported to converge on several cellular target proteins to moderate the host immune response. For instance, adenovirus encodes proteins from its E3 region that have the capacity to modify the immune response. This is best exemplified by E3-19K protein that interacts with the major histocompatibility class I complex, thereby retain it in the endoplasmic reticulum (ER) and block the presentation of viral antigen (Burgert and Kvist 1985). Such mechanisms of escape of the immune surveillance often lead to carcinogenic conditions. On a similar note, adenovirus E3-14.7 K protein binds to tumor necrosis factor α and inhibits caspase 8 activation and apoptosis (Chen et al. 1998; Horton et al. 1991). Further, interaction between E3-14.5 K (RIDβ) and E3-10.4 K (RIDα) induces TRAIL and FAS receptor degradation (Tollefson et al. 1998, 2001). This, in turn, shuts off the extrinsic cell death pathway and often leads to conditions such as squamous cell carcinoma.

How viral proteins hijack the host's PPI networks has long intrigued the researchers. Performing a number of biophysical studies and structural studies, scientists have put forward two general mechanisms of hijack: (a) acting as molecular mimics and (b) taking advantage of promiscuity of IDPs (Ou et al. 2011). In the former strategy, the viral proteins mimic the cognate interactions that are naturally present in the host. The viral proteins achieve this by outcompeting the endogenous

PPIs either through greater expression levels and/or enhanced affinity of binding. For instance, viral SV40 ST competes with endogenous B56γ regulatory subunit to bind with the heterotrimeric PP2A complex. This, in turn, re-directs the function of PP2A and activates Akt instead (Arroyo and Hahn 2005).

The second hijack mechanism was proposed by utilizing bioinformatics analyses of "hub proteins," that is, host proteins that have multiple interacting protein partners. The studies implicated that the intrinsic disordered regions of proteins enable their interaction with many different partners (Dunker et al. 2005). Such disordered regions in proteins demonstrate disparate conformations existing in rapid equilibrium states. However, they adopt one single conformation after binding to an interactor. The permit of such conformational samplings and the potential utility of the same region to bind with number of partners bring to a great extent promiscuity in such protein interactions. The binding of intrinsic disordered regions with interacting partners is highly specific, which is due to the reason that the disordered region has the capacity to fold into a structure that is complementary to the interactor's binding interface. Such interactions, however, still maintain low affinity owing to the entropic cost of the folding process (Wright and Dyson 1999).

On a similar note, viral proteins essentially act as hub proteins, having the ability to bind with several cellular proteins. Therefore, it seems plausible that viral proteins utilize a similar kind of optimized interacting strategy to seize interactions mediated by host hub proteins (Pelka et al. 2008). One of the best examples of a viral protein that employs an intrinsically disordered region for interaction is the conserved region 1 (CR1) domain of adenovirus E1A. It binds with the TAZ2 (transcriptional adaptor zinc finger-2) domain of host CBP. Experimental analysis, using NMR, has elucidated that the otherwise disordered CR1 region, upon binding to TAZ2, leads to the formation of two helices that fit into two binding grooves of TAZ2. p53 has also been reported to interact with CBP using its disordered region (Ferreon et al. 2009). However, having a lesser interaction area as compared to CR1-TAZ2 interaction, p53-TAZ2 interaction has a lower binding affinity. Consequently, E1A outcompetes the interaction of p53 with TAZ2. CR1 is demonstrated to compensate for the entropic cost of folding by exhibiting an interface that is more complementary with TAZ2 interface than p53. Indeed, E1A contains a second CR1 region that also interacts with TAZ2 (Stein et al. 1990) and further enhances the specificity of the interaction. E1A-CBP has been reported to cause cellular transformation by guiding CBP to promoter regions of genes that are responsible for cell division and growth. Once directed to the promoters, the proteins act as transcriptional activators of these genes.

2.4.6 Other Roles of PPIs

Some of the major functions of PPIs have been discussed in the above sections. Apart from those, PPIs are also involved in a variety of other functions that are crucial for the sustenance of the cell (Aboulwafa et al. 2019; Mattaj and Englmeier 1998; Seedorf et al. 1999). For example, PPIs play crucial role in transport of biological macromolecules (Mattaj and Englmeier 1998; Seedorf et al. 1999). Importin is a protein that carries/transports other cargo proteins into the nucleus. It performs this function via interaction with specific recognition sequences on the cargo protein, known as nuclear localization signal (NLS). Importin itself comprises two subunits: α and β. Importin-β family members can either directly associate with cargo proteins to transport them across the nucleus or they interact with Importin-α to form heterodimers. When present in the form of a heterodimer, Importin-β interacts with the pore complex (a complex of several interacting proteins), while Importin-α interacts with the NLS of the cargo protein, thereby acting as an adaptor. The cargo-importin complex docks at the nuclear pore complex and then translocates through the aqueous pore channel. The trimer formed by NLS-Importin α- Importin β is dissociated in the nucleus, after its association with Ran GTP. The two importin proteins are then recycled (transferred) back to the cytoplasm for the next trip of transport (Mattaj and Englmeier 1998). Exportin belongs to the same protein family and plays an essential role in the export of cargo proteins from the nucleus to the cytoplasm. Together, Importin and Exportin constantly shuttle cargo proteins between the interior of the nucleus and the cytoplasm (Seedorf et al. 1999).

Apart from transport of cargo proteins, PPIs are also involved in mediating the transport of other metabolites such as monosaccharides. For instance, the protein complex of phosphotransferase system (PTS) in *Escherichia coli* is involved in the transport of sugar. It uptakes sugar from the surrounding medium and simultaneously phosphorylates them. Finally, it releases the resulting sugar phosphates in the cell cytoplasm. This prokaryotic transport system comprises two energy coupling proteins: Enzyme I (EI) and Enzyme II (EII) complex. The EII itself is a complex of three proteins: IIA, IIB, and IIC. There is however an exception; the EII complex of the mannose family comprises four protein domains: IIA, IIB, IIC, and IID. Here, the extra domain of IID is employed for the activity of the IIC transporter (Aboulwafa et al. 2019).

Elucidating the other subtle aspects of protein interactions is also essential. Oligomeric proteins that are mediated by PPIs often engender binding sites—for instance, antibodies. Further, protein interactions are vital for maintaining quaternary structures of proteins, although direct functional roles are not attributed to such structures, it has been observed that monomeric proteins are more likely to degrade or unfold than their oligomeric versions (Ingram 1957; Watowich et al. 1993).

The impact of anomalous PPIs in various diseased conditions has also been illustrated in several instances. For example, the oligomerization of mutant hemoglobin in sickle cell anemia is a well-characterized phenomenon depicting the effect of an unwanted PPI (Ingram 1957; Watowich et al. 1993). The specificity of PPIs is lost during the process of protein aggregation. In many diseases such as Alzheimer's, the spongiform encephalopathies, familial amyloidotic polyneuropathy, familial Mediterranean fever, etc., amyloid fibrils comprising of aggregated proteins are observed inside tissues (Fredrickson et al. 1989). Deposition of amyloid fibrils is not only held as the causative agent of diseases such as Alzheimer's, but is also held responsible for lethal conditions such as renal failure (Rosenberg 1993). Owing to the number of functions that are associated with PPIs, it has been monitored that the loss of interactions among the proteins can have unpleasant outcomes. For instance, when α-1-antitrypsin undergoes oxidation at a specific methionine, it loses its interaction with several proteases including elastase. This consequently leads to elastase activation, which in turn leads to the development of diseases such as adult respiratory distress syndrome and emphysema. Such oxidative disruption of the protease-inhibitor interaction is usually observed in smokers' lungs (Abrams et al. 1981; Gadek et al. 1979; Johnson and Travis 1979).

Apart from the intrinsic significance of PPIs with regard to the living cell, PPIs are also important as a subject of research. This is because they aid researchers to answer fundamental questions regarding protein folding phenomenon. The various interacting forces that are responsible for protein folding such as hydrophobicity, van der Waals interactions, hydrogen bonding, electrostatic interactions, etc. are also accountable for PPIs. Paucity of models and the great extent of conformational flexibility of denatured protein states have acted as barrier in resolving the protein folding problem. Quantifying the accurate magnitudes of the above-listed forces however has proven vital in the theoretical calculations of the stability of proteins. Studying PPIs furnishes the requisite quantifications because they involve well-defined backbone conformations of the interacting partners and thereby exclude the problem of conformational uncertainty (Stites 1997).

A second aspect of the protein folding problem has also been answered by extrapolating information from PPI studies. Crystal structures of proteins with high molecular weights have demonstrated that beyond a threshold molecular weight, proteins, even comprising of single polypeptide chains, are divided into independent subunits (or domains). Since the independent folding units or domains mediate interactions among them in a way similar to PPIs, the studies on inter-protein interactions are deemed to be essential to understand intra-protein interactions (Jaenicke 1991).

2.4.7 PPI Interactome and Its Functional Relevance

Although the exact number of PPIs taking place in the human body is unknown, they are estimated to range from hundred thousand to about a million (Venkatesan et al. 2009). Data on protein interactions aids in assigning the biological functions to the unknown proteins that are revealed by genome sequencing (Oliver 2000). This aspect led researchers to inquire all the biological processes that the encoded proteins are involved in rather than focusing on all proteins involved in a particular biological function (Fischer 2010). Available data on PPI constitutes only a small part of the large pool of human PPIs. Nevertheless, the interactome or the network of such interactions is useful in discerning molecular links existing between proteins and various diseased conditions (Menche et al. 2015) (Fig. 2.3).

Analyses derived from global interaction mapping experiments have led to the conclusion that many biological properties such as mRNA expression regulation are interrelated with physical interactions between proteins (Ge et al. 2001). Other correlations of protein interactions include their structural information. In several

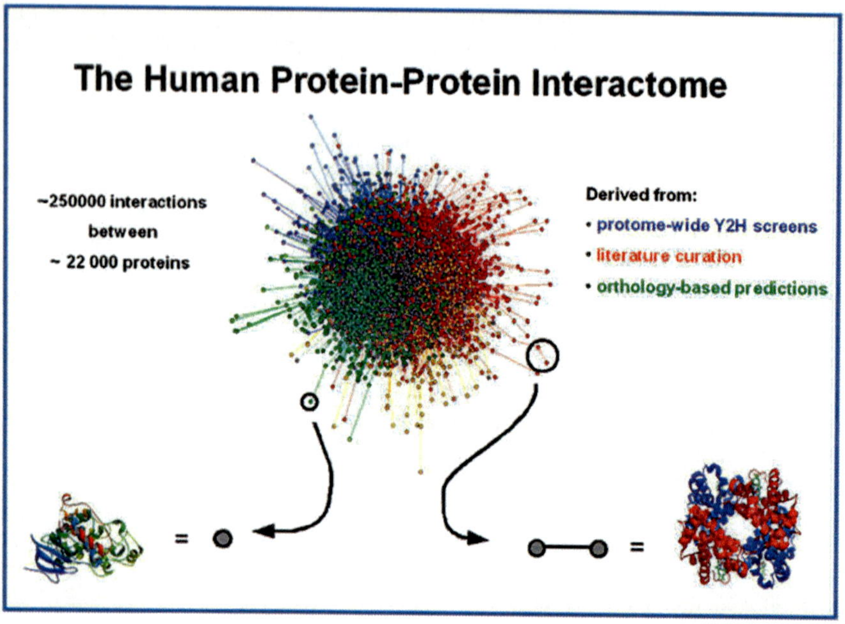

Fig. 2.3 Graphical representation of current human protein–protein interactome (Adapted from Chaurasia et al. (Chaurasia and Futschik 2010))

instances, structural data has been exploited to unravel protein interactions utilizing bioinformatics (Jansen et al. 2003; Jensen et al. 2009). In a nutshell, analyzing the interactome by employing statistical and computational tools has furnished significant insights into how the biological systems are organized on a large scale, and also how genotype-to-phenotype relationships are mediated (Vidal et al. 2011). The topology of protein interaction networks denotes proteins having one or few interactions as "edges" and proteins having a very large number of interaction partners as "hubs." Detailed study of the interactome led to a very fascinating observation: In both humans and plants, pathogenic proteins (bacterial, viral, and fungal proteins) target hub proteins (Keskin et al. 2016).

Even with regard to the human body, proteins that are associated with similar diseases have higher likelihood to physically interact thereby validating the presence of distinctive disease-specific functional modules (Goh et al. 2007; Lim et al. 2006). The fact that the interactome is functionally relevant is further obvious from the successful studies that have defined novel disease and cancer-related genes by analyzing the network (Lage et al. 2007; Pujana et al. 2007). Protein interaction networks also have the potential to aid in analyzing data on genetic diseases. Further, the integration of computational data on protein interactions with transcriptional data in tumors has helped in predicting cancer prognosis (Taylor et al. 2009).

It is however very difficult to solve this complete interactome owing to several technical aspects. First, several protein interactions are transient in nature and dissociate quickly once the protein partners interact (Nooren and Thornton 2003). Second, some proteins require to be post-translationally modified for their interaction with partner proteins. Third, many proteins are spatially heterogeneous, that is, they are localized in different compartments of the cell and are not transported to other locations. In such a scenario, proteins are precluded from interacting with one another, even though their interaction is possible from a physicochemical point of view. Fourth, owing to tissue specificity, proteins are differentially expressed in distinct cell types. Therefore, different cell types have disparate interactomes.

Systems biology era therefore attempts to reveal new interactions and unravel the principles of interaction via the construction of protein interaction networks. It is believed that a collaborative effort between experimental and computational methods can help in unraveling the complete interactome.

2.5 Geometric Analysis of Interaction Interface

Proteins interact with each other via their interfaces. To date, many structures of protein–protein complexes are available. The data available in Protein Data Bank (PDB) is chemical in nature, including atomic positions and geometric data. Therefore, tools that may convert the information into terms of biological and/or physical relevance have been developed. Particularly, three geometric tools are applicable to define the interface shared by two macromolecules, namely A and B in a complex AB. They include: (a) distance, (b) buried surface, and (c) alpha-complex.

2.5.1 Distance

Chemical groups or atoms x of A and y of B form part of the A:B interface if they fulfill the criteria $d_{xy} < d_0$, where d_0 depends on the atomic or group radii r_x and r_y, and on a cut-off value r_0 in the range 0.5–2 Å:

$$d_{xy} < d_0 = r_x + r_y + r_0 \tag{2.1}$$

This definition, although simple, depends on an arbitrary cut-off (Janin 2010).

2.5.2 Buried Surface

The A:B interface consists of the entire solvent accessible surface of A or B that does not fall under the accessible surface of AB (Chothia and Janin 1975) (Fig. 2.4). Calculating the buried surface area (BSA) is one of the most convenient ways to measure the interface size. It can be calculated as:

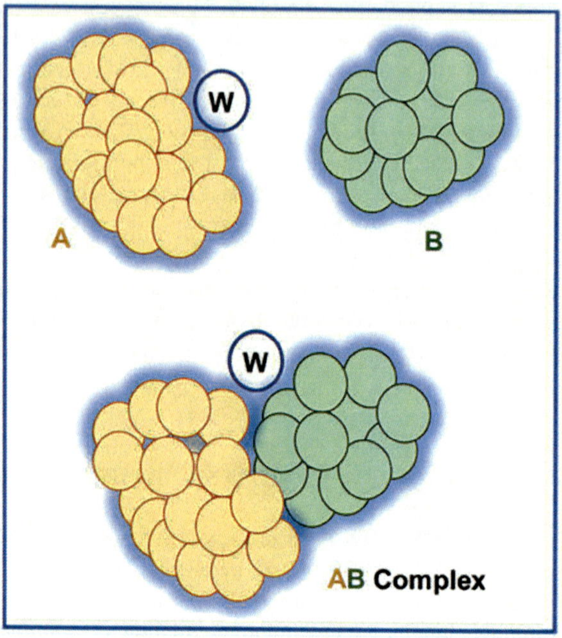

Fig. 2.4 Illustration showing the buried and solvent accessible protein surfaces. The solvent accessible surface is defined by center of the water probe (W). The A:B interface contains A (top) and B (top) solvent accessible surface points that are no longer accessible in the AB complex (bottom)

$$BSA = ASA_A + ASA_B - ASA_{AB} \tag{2.2}$$

where ASA_A, ASA_B, and ASA_{AB} correspond to the accessible surface areas (ASA) of A, B, and the AB complex. BSA contributing residues and atoms form part of the interface. Moreover, it has been observed that the BSA is proportional to the number of such residues.

2.5.3 Alpha-Complex

Alpha-complex is defined on the basis of a geometric construction known as the Voronoi diagram (Edelsbrunner and Mücke 1994). This Voronoi diagram was first utilized for proteins by Richards (Richards 1977). The diagram attributes each atom with a complex polyhedron known as the Voronoi cell. This cell comprises all the space points that are closer to that atom compared to any other atom. The different sizes of the chemical moieties are accounted via the replacement of the Euclidean distance by the power distance p(x):

$$p(x) = d^2 - r^2 \tag{2.3}$$

where r corresponds to the radius of the sphere representing the atom and d corresponds to distance of point x to its center.

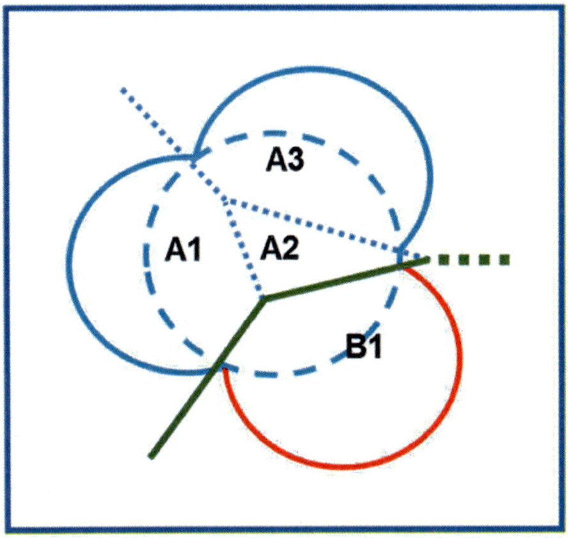

Fig. 2.5 Depicting the Voronoi interface in two dimensions: The atoms of A and B are represented by blue and red circles; their radii are increased by the radius of the probe; the dotted blue lines are "monocolor" Voronoi facets that are shared by atoms of A, the green lines are "bicolor" facets shared by both A and B. The two facets drawn as complete lines are included in the A:B interface; they are part of the alpha-complex, but the B1:A3 facet (green dashes) is not

The Voronoi cell of an atom consists of all points x having a power distance to that atom lesser than to any other atom (Aurenhammer 1987; Gellatly and Finney 1982). The radical plane that comprises the interaction of the spheres also consists of the facets of the Voronoi cell.

The Voronoi diagram, or in other words, the power diagram, provides the regular definition of contacts (Fig. 2.5). According to this concept, if the Voronoi cells of two atoms share a facet, then the corresponding two atoms are in contact. The interface area may be subsequently calculated by adding up the areas of the shared facets. Atoms lying on the molecular surface are problematic since they always comprise facets that are unbounded in nature. The alpha-complex provides a sophisticated solution to this problem (Edelsbrunner and Mücke 1994). It is constructed similar to a power diagram, but it restricts each atom's Voronoi cell to its associated sphere, also seeking interactions between the regions that are restricted.

Therefore, an alpha-complex does not consist of a facet between two atoms if there is no interaction between the associated spheres, or if the intersection region does not include the facet; i.e., the facet lies outside the intersection. Such a definition of an interface may still comprise some facets that are unbounded. They might either be removed via an iterative retraction process performed (Headd et al. 2007) or via testing suitable geometric criteria as performed by Cazals et al. (Cazals et al. 2006). In the Cazals procedure, Voronoi interface is described by approximating the shape of the buried molecular surface. The Voronoi interface area is calculated by utilizing the areas of the shared facets, has a linear correlation with the BSA. However, it was observed that for a set of protein interaction interfaces, the Cazals' method furnished a correlation ($R^2 = 0.98$) that was way better than the Ban protocol ($R^2 = 0.85$) (Chakrabarti and Janin 2002). One notable result of the Cazals' method is the fact that Voronoi interface, as compared to the buried surface, usually consists of significantly more atoms, specifically those coming from the main chain. This is because approximately 13% of the atoms that have shared facets do not contribute toward the BSA since they are inaccessible to the solvent.

2.6 Characteristic Features of Interaction Interface

There are several types of protein complexes with regard to physicochemical properties, binding specificity, affinity, and other structural aspects of the interfaces. For instance, oligomeric proteins that comprise several identical subunits and lack a folded state in the monomeric form are naturally different from heterodimeric proteins such as antigen-antibody or protease-inhibitor that have independently stable subunits. Even homo-dimeric proteins that are made from two identical monomeric subunits have distinctly different interface properties from hetero-dimers that comprise interactions between two diverse proteins. Nevertheless, numerous protein–protein complexes analyzed by several groups have revealed the presence of some broad structural similarities.

Interface regions need to be studied extensively to understand and predict the preferential binding patterns of proteins. There are several physicochemical properties of the interface that are vital for understanding the nature of interactions between protein molecules (Fig. 2.6). For instance, interface characteristics such as buried surface area by the interacting partners, buried water molecules, non-polar

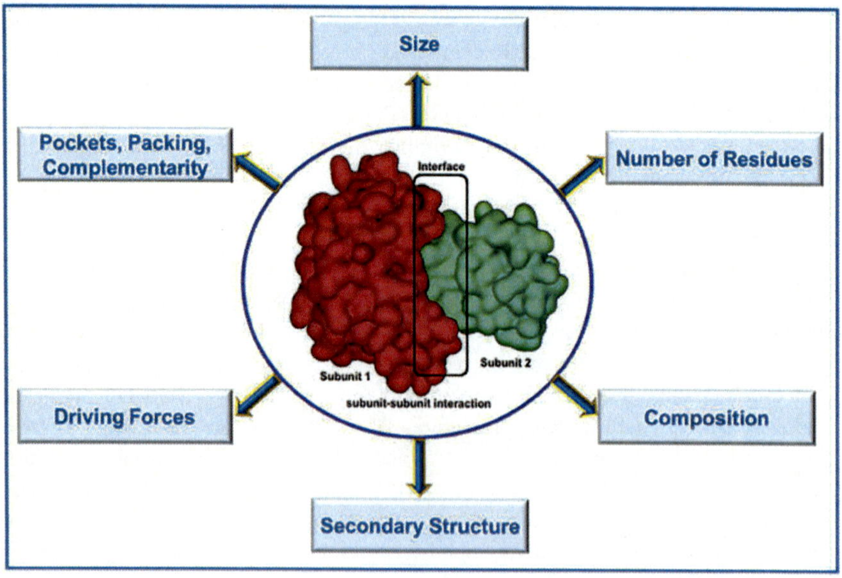

Fig. 2.6 Schematic depicting the characteristic features of PPI interface

fraction, interacting forces including salt bridges and hydrogen bonds, charge distribution, interface composition, hotspot residues, residue conservation, interface residue flexibility, interaction strength, interface complementarity, and types of secondary structures present, etc. are important parameters to analyze the nature of interface interactions (Jones and Thornton 1996; Keskin et al. 2008).

2.6.1 Interface Size

The part of *interface size* that is considered to constitute the protein–protein interface varies from one study to another. The definitions of the interface are usually based on the proximity of the two interacting proteins or whether the accessibility of one protein partner to the solvent is precluded by the other. Protein interaction interfaces are usually classified into rim regions and core regions. While core regions closely resemble the protein interior, the rims are more solvent-exposed and bear similarity to the surface of the protein with regard to residue frequency (Bahadur et al. 2003). Further, the regions involved in binding have less flexibility compared to the remaining surface regions of the proteins.

The interface surface areas that are involved in PPIs have been easily determined with the aid of crystal structures. For instance, an initial study determined that the interface size in each subunit in dimeric proteins varies from 670 to 4890 $Å^2$ (Janin et al. 1988). Later on, an examination of several non-homologous dimers determined that protein interaction interfaces have a greater range of surface areas, varying from 368 to 4761 $Å^2$ (Jones and Thornton 1997). They concluded that the interface regions were more than 6.5–29.4% of the accessible surface areas of the individual constituting monomer units. This range was slightly greater than the range (6.6–23.3%) suggested by Argos et al. (Argos 1988). The relatively small discrepancy in the interface areas of these dimeric structures infers that major deviations from the suggested range are highly improbable.

Researchers have also compared the BSA involved in homodimers and non-obligate interactions of two heterogeneous proteins. The authors found out that for the heterogeneous interactions, most of the interfaces had a BSA less than 2000 $Å^2$. On the other hand, it was seen that for homodimers, majority of the interfaces were large, approximately twice the BSA observed in the heterogeneous protein interactions (Janin 2010).

Examining multi-subunit oligomeric proteins demonstrated that their corresponding interface regions were greater than those involved in dimeric proteins (Janin et al. 1988). On a similar note, the percentage of surface area of the constituting monomer units was higher for oligomers than for dimers. For instance, monomers forming tetrameric complexes were observed to bury 20.9% of their surface areas upon oligomer formation (Argos 1988). On the other hand, dimeric interactions such as antibody-antigen or protease-inhibitor complexes that had a total interface region of 1600 ± 350 $Å^2$ involved $600 - 1000$ $Å^2$ of BSA for each of their individual protein partners. This corresponded to 5–20% of BSA of the constituting interactors (Davies et al. 1990; Janin and Chothia 1990).

In an another study, Miller et al. examined the solvent-accessible surface of several oligomeric proteins and determined that there is a high correlation between the interface parameter and molecular weight of the protein complex (Miller et al. 1987). This was validated by a later study that was performed involving even a greater set of protein complexes (Miller et al. 1989). The findings of such studies implied that the surface area of a subunit that is buried in an interaction interface increases with the rise in number of subunits involved in the complex formation, provided all the subunits are of identical molecular weight. Such a direct correlation between surface area and molecular weight, although intuitive, but appeared surprising to the researchers. This empirical relationship also enabled them to roughly predict the involved surface area in PPIs using the known molecular weights of the interactors.

Extensive study on the protein complexes has also unraveled the fact that the individual monomers in an oligomer can have unstable tertiary structures, as the burying of large surface areas during oligomer formation exposes a great extent of hydrophobic surfaces on the isolated monomers (Jaenicke 1991). Studies involving thermodynamic data and numerous protein heterodimers also observed a direct relationship between BSA and binding affinity. Protein complexes that comprised of a BSA less than 2000 $Å^2$ had an inverse relation between the free energy per unit BSA and total BSA. For instance, for a total BSA of 694 $Å^2$, the free energy per unit area was approximately 13 cal mol^{-1} $Å^{-2}$. On the other hand, for protein interactions having total BSA beyond 2000 $Å^2$, the free energy per unit area was only about 4 cal mol^{-1} $Å^{-2}$. The results implied that the surface energy density is lesser for larger complexes than for smaller ones. The authors also delineated that although the percentage of hydrophobic surface area depends on the total BSA, it is approximately 60%. It is speculated that the imposition of evolutionary limits on an interface's total hydrophobic surface area not only precludes non-specific interactions, but also blocks aggregation that might occur on exposure of the interfaces of unbound proteins (Chen et al. 2013).

Indeed, researchers have demonstrated that most of the surface of a protein is capable of indulging in PPIs from a geometric viewpoint. The authors conducted the study utilizing Half Interfaces (HI) that is defined as the protein's surface region involved in interaction with another protein. They observed that for both geometric and known HIs, the fraction of interfacial amino acids to surface amino acids correlates with the number of available templates and solved structures, respectively. The study established that the overlapping HI regions in a protein usually comprised of six to ten amino acid residues that were more conserved as compared to remaining HIs. Indeed, this study presumed that these HI regions had very low free energies of association that allowed a protein to bind with several partners, while allowing to preserve the most favored region for interaction (Tonddast-Navaei and Skolnick 2015).

Studies that involved receptor-mediated protein interactions and globular complexes demonstrated that such PPIs had larger interfaces than those involved in oncogenic and transient interactions (Kar et al. 2009). For instance, interaction between proteins Stefin B and Papain is non-obligate in nature and thus involves a smaller interface area in comparison with that of the obligate methylmalonyl–CoA mutase interaction (Stubbs et al. 1990).

2.6.2 Number of Interface Amino Acids

The requirement to bury at least 600 $Å^2$ area in order to mediate a PPI seems daunting as this necessitates a side area of 25.5 Å, that is greater than majority of the flat surfaces generally found on a small protein. However, a study by Argos et al. determined that for protein with smaller molecular weights, association of a higher percentage of their water-accessible surface area aids in mediating a stable interaction utilizing a minimum area (Argos 1988). For example, the solvent accessible surface areas of the side chains of phenylalanine, methionine, and tryptophan were determined to be 135 $Å^2$, 115 $Å^2$, and 180 $Å^2$, respectively. This implied that the burial of just one or two hydrophobic amino acids could theoretically make up for most of the BSA involved in forming a dimer.

However, in reality more residues contribute to the average interface. In one of the earliest surveys on protein interaction interfaces, a contact residue was defined as a residue within a distance of 0.5 Å from an atom across the interface (Janin and Chothia 1990). This study examined four antibody–antigen complexes and 15 protease–inhibitor complexes and determined that the average number of interface residues was 34 $\pm$ 7 (17 of each monomer). This number indicated a more or less realistic measure of the interface size. In recent times, the ASA is calculated and utilized to mark the residues that are indulged in the interface. The ASA of an amino acid residue in the monomeric state and that in the complex state is estimated. Then their difference is calculated to know if the concerned amino acid is an interface residue. If the difference is larger than the threshold value of 1 $Å^2$, then the particular residue is assigned a role in the interface formation (Nooren and Thornton 2003). On the other hand, two residues each arising from one chain of the interacting complex are regarded as contacting when the distance between any atom of those residues is less than the threshold value of 4.5 Å (Gao and Skolnick 2010).

2.6.3 Amino Acid Composition at the Interface

In general, it has been observed that solvent accessible regions of proteins comprise hydrophilic residues, whereas the buried regions are richer in hydrophobic residues. Residues present in the protein core have solvent accessibilities that are lesser than those for residues present at the interface. On the other hand, non-interface residues exhibit high solvent accessibilities than the interface residues. Therefore, the distribution of residues in these three groups depends not only on the functions that these regions serve, but also on the variance in their accessibilities to the solvent. In order to examine if certain residues possess particular preference for an interface region, one must segregate the effect imparted by solvent accessibility. This can be performed via the comparison of PPI interfaces with a group of amino acid residues that have similar accessibilities to the solvent.

Yan et al. analyzed the amino acid compositions of interface and non-interface regions of proteins by separating out the effect of solvent accessibility (Yan et al. 2008). They concluded that in the interface region, hydrophobic residues are more preferable than hydrophilic residues. The presence of hydrophobic residues in the interfaces is very critical for stabilizing the protein complexes. However, the authors did not observe this trend when the solvent accessibility parameter was not taken out of consideration. The comparative study was performed by placing the residues on the hydropathy index according to their increasing order of hydrophobicity. The study demonstrated that protein cores comprise the largest fraction of hydrophobic amino acids including phenylalanine, cysteine, leucine, valine, methionine, etc., whereas the non-interface surfaces comprise the least hydrophobic fraction. Further, their study revealed that hydrophobic residues show highly positive propensities for both heteromeric and homomeric types of interfaces, whereas hydrophilic residues such as glutamine, glutamate, aspargine, aspartate, lysine have negative valued propensities. Between homomeric and heteromeric interfaces, hydrophobic residues have higher propensities for the former type of interfaces as predicted earlier (Bahadur et al. 2003; Jones and Thornton 1996).

Other studies have also marked the presence of hydrophobic residues in the interface. Young et al. analyzed 38 protein complexes and found that for 25 of them, about one-third of the interface region comprised of the greatest hydrophobic cluster (Young et al. 1994). The data obtained by Korn and Burnett also produced similar results (Korn and Burnett 1991). By determining the hydropathy profiles of interface surface, protein interior and non-interface surface of two antibody–protein complexes, and 40 multi-subunit proteins, they concluded that interfaces were more hydrophobic compared to the exterior. Although there were some hydrophilic exceptions at the interface, the authors noticed a complementarity in the interface hydropathy. While hydrophobic patches in one protein were compared with hydrophobic regions in the partner protein across the interface, hydrophilic regions were also found comparable.

Janin et al. analyzed oligomeric protein interfaces to reveal that the residue composition of the PPI interface was intermediate between that of the protein interior and exterior. However, it had more resemblance to the interior as compared the solvent exposed regions (Janin et al. 1988). However, they made a remarkable observation that in contrast to the protein interior, where charged residues are infrequent, the interface had a high number of buried charged residues. Particularly, they found that arginines were disproportionately present in the interface and accounted for 9.9% of the interface surface area. Jones and Thornton also comprehensively analyzed interface compositions of dimeric proteins and demonstrated that among the polar residues, arginine and asparagine had preference to be located in the interface region (Jones and Thornton 1995). On the other hand, methionine and proline were the favored non-polar residues in the interface. They also observed that among the aromatic amino acids, tyrosine was preferred in the interfaces, while tryptophan and phenylalanine were not. Padlan examined the complementarity-determining region (CDR) of seven crystal Fab structures. He also demonstrated the presence of aromatic residues, especially tyrosine, in the binding pocket compared to the rest of the molecule (Padlan 1990). Results obtained by Mian et al. were in partial agreement with Padlan, which determined that tyrosine, asparagine, tryptophan, and serine were likely to be found in the interface (Mian et al. 1991).

Although there are some disparities in the results obtained by different research groups, there are more similar trends like the incorporation of more number of non-polar amino acids in the interface region in comparison with the rest of the protein. Some of the similarities and differences regarding the special preference of interfacial amino acids can be rationalized. For example, it has been speculated by Janin et al. that the capability of the guanidinium moiety to donate several hydrogen bonds might be accountable for its over-representation of arginine residue statistically (Janin et al. 1988). The second most preferred type of residues include aromatic residues like tyrosine. Although their percentages as interfacial residues vary more than that of arginine, the extent of their preference depends largely on the interface type (Jones and Thornton 1996).

The preference is clearer in case of antibody-protein complexes than for homodimers. One plausible explanation is that majority of the homodimers spend most of their lifetime in the dimerized forms, whereas both proteinaceous antigen and antibody require native folding as individual monomers. Therefore, it is highly reasonable to make assumption that the two types of protein complexes vary in the nature of their interfaces and their subsequent need of different amino acid side chains. Similarly, there are differences in the interface composition which depends on the PPI type. For instance, transient interfaces are less hydrophobic than permanent complexes and less conserved than obligate complexes (Mintseris and Weng 2005). Similar to the presence of aromatics and arginine, the presence of methionine in the interface region can also be rationalized. For instance, calmodulin that has several protein interactors has eight exposed methionine residues in its binding site. Gellman has proposed that the flexibility rendered by methionine's side chain along with sulfur's high polarizability aids in the interaction of the protein with various partners (Gellman 1991). Similar to aromatic residues, methionine has a low hydrophobic transfer free energy compared to its aliphatic side chain sizes (Fauchere 1983).

The presence of certain amino acids in the interface can also be rationalized from the energetics viewpoint. Some interfacial amino acid residues have significant contribution to the interaction and are therefore more critical energetically (Bogan and Thorn 1998). These amino acid residues can be experimentally determined by alanine scanning mutagenesis whereby the suspected residue is mutated to alanine and the binding energy variation is monitored (Clackson and Wells 1995). Hotspots are usually clustered in close proximity and guarded from solvent (Keskin et al. 2005). Chen et al. determined that with an increase in buried surface area (BSA), the fraction of BSA taken up by the hotspots decreases until it is approximately 2000 Å^2. Beyond this threshold point, a small fraction of the total buried surface area is taken up by the hotspots. Conversely, a larger fraction is occupied by the hotspots for smaller interfaces (Chen et al. 2013). For example, phenylalanine (F42), tyrosine (Y45), and glutamic acid (E62) are the hotspot residues found in the interface between IL-2 and its receptor (Tuncbag et al. 2010).

Sometimes, residues at the interface are also post-translationally modified to aid in protein binding. For instance, in weak transient homo-oligomers and hetero-oligomers, there is significant presence of phosphosites at the interface (Nishi et al. 2011). These phosphosites are conserved with respect to the rest of the protein and are usually present in the hotspot residues in hetero-oligomers.

2.6.4 Structural Motifs and Secondary Structural Elements at the Interfaces

The structural analysis of PPI interfaces is incomplete without the delineation of secondary structures and motifs present in the interfaces. The study of a large number of proteins has elucidated that helix-loop-helix and leucine zippers are the two common structural motifs present in PPI interfaces (Lin et al. 1995; Slingsby et al. 1993; Smith-Gill 1991). Since this implied that proteins might interact in a restricted number of ways, a more exhaustive study on protein interaction interfaces was carried out by Tsai et al. (Tsai et al. 1996). The authors examined around 1600 interfaces from the Protein Data Bank (PDB) belonging to around 350 families. These families were not dependent on the folds and sequences of the participating proteins, but on the interface architecture. The authors observed the presence of similar interface interactions between dissimilarly folded proteins and vice versa and concluded that large number of motifs are possible for PPI interfaces.

In order to understand if some particular types of secondary structures had preferential domination at the interface interactions, Argos studied the interfaces of oligomeric proteins (Argos 1988). The author demonstrated that the interfacial distribution of secondary structures resembled the protein exterior more closely as compared to that of the interior, although the secondary structural differences between the exterior and the interior were not very significant. Moreover, Argos revealed that majority of the interfaces did not have long lengths of sheet or helix at interaction sites. On the other hand, a contradictory result was generated by Miller who showed that helix—helix, sheet–sheet, and/or helix–sheet interactions were quite common interface interactions in oligomeric proteins (Miller et al. 1989). The differences in the results obtained by Argos and Miller can be described by the fact that, although Miller demonstrated most of the interfaces to be having helix-mediated interactions, Argos argued that these interactions did not involve majority of the interface surface area or residues.

Guharoy and Chakrabarti performed a detailed structural study of both homodimeric and heterodimeric PPI interfaces. Authors elucidated that the former type is mainly comprised of the regular secondary structures such as strands and helices, while the latter usually comprised of non-regular structures such as loops and turns. They also revealed that while homodimers had prominent helix-mediated interactions and longer helical segments, the interfacial helices in heterocomplexes increased in length with the increase in interface area. Further, they noticed a critical difference between homodimers and heterodimers with respect to the helices' relative orientation. The helices had a preference for aligned packing in homodimers, while they had a tendency to pack at 90° in heterodimers (Guharoy and Chakrabarti 2007).

2.6.5 Driving Forces of Interaction at the Interface

It has been reported that protein–protein complexes form in solution via an entropy-driven process (Creighton 1993). Indeed, it has been proposed that when hydrophobic patches are buried, a large gain in entropy is yielded that in turn enables the protein interactions and stabilizes the formed complexes. Yan et al. characterized protein interfaces in detail to determine that the stability of PPIs is greatly dependent on electrostatic interactions, hydrogen bonds, hydrophobic interactions, and salt bridges (Yan et al. 2008). The authors have also noted that despite being infrequent in the interface, disulfide bonds have a considerable contribution to the stability and rigidity of smaller protein complexes (Glaser et al. 2001). Further, the type of driving forces at the interface depends on the PPI type. For example, obligate complexes are dominated by hydrophobic interactions, whereas hydrogen bonds and salt bridges are more favored in transient complexes. The authors found an abundance of aromatic residues at the interface and reasoned that their presence supported the consideration that hydrophobic interactions are the major stabilizing forces of PPIs. This is because the aromatic residues have the ability to mediate strong hydrophobic interactions between their bulky and hydrophobic moieties. Apart from hydrophobic interactions, other studies have shown that interactions between charged residues also contribute to the PPIs stabilization (Sheinerman et al. 2000). Bahar and Jernigan have also elucidated the predominant importance of interactions between hydrophilic residue pairs that are at close distances, apart from the long-distance acting hydrophobic interactions (Bahar and Jernigan 1997).

Janin et al. demonstrated that the number of hydrogen bonds that are formed between subunits is proportional to the subunit interface area (Janin et al. 1988). Authors found that one hydrogen bond is formed on average for 200 Å^2 of the surface area of the interface. Furthermore, the authors have also reported exceptional interfaces where no or very few hydrogen bonds were present. In a study by Janin and Clothia, it was found that the nature of hydrogen bonds depends on the nature of protein complexes (Janin and Chothia 1990). For example, majority of the hydrogen bonds involved side chain atoms for antibody–protein antigen interactions, whereas protease–inhibitor interactions comprised of hydrogen bonds majorly involving main chain atoms. They also revealed that while several charged groups mediated hydrogen bonding, salt bridges were rarely formed. Another study by Jones and Thorton showed that around 56% of their studied PPIs contained no salt bridge interactions. Moreover, they revealed that there was no correspondence between the number of salt bridges and interface size (Jones and Thornton 1996).

2.6.6 Transient Pockets, Packing, Cavities, and Shape Complementarity

Protein surface is not flat, but rather engraved with crevices, pockets, and indentations (Dundas et al. 2006). When the protein is not involved in binding with its partners, these surface imperfections may be occupied by water (Clodfelter et al. 2006). However, when the protein interacts with its partner, the water may or may not be replaced by the interactor. If the pocket remains unfilled by the interacting protein, the pocket is known as "unfilled pocket." On the other hand, if a pocket disappears on binding, it is known as a "complemented pocket." Such a pocket denotes regions that are tightly fit (Li et al. 2004). The hotspot residues that have characteristics of tight packing are thus favored in complemented pockets rather than unfilled pockets. However, an interesting observation has been made with regard to complemented pockets. It has been seen that these pockets usually pre-exist binding. Li et al. studied 18 protein complexes having complemented pockets and showed that in 16 of them, the pockets pre-existed in the unbound forms (Li et al. 2004). The root-mean-square deviations (RMSD) of the pocket-lining atoms between the unbound and the bound states were very small, implying that the pockets were the constitutive features of the native state of proteins.

Eyrisch et al. have utilized molecular dynamic simulations to reveal that several transient pockets open up on the surfaces of proteins at their interaction interfaces. These pockets are fluid-like and thus open and close swiftly. The authors found the pocket volumes were comparable to those of ligand binding pockets. By performing a second set of simulations, the authors have validated that most of these transient pockets were very much reproducible (Eyrisch and Helms 2007). Another interesting study investigated the factors of the native conformational dynamics of proteins that was responsible for the creation of these transient pockets. The findings of the study implicated aspects like backbone rearrangement and solvent polarity to influence the formation of such pockets at protein interfaces (Eyrisch and Helms 2009).

Janin and Clothia analyzed interface packing of antibody–antigen complexes and protease–inhibitor complexes and concluded that the packing was as dense as that of crystalline amino acids with very few apparent cavities (Janin and Chothia 1990). Walls and Sternberg also analyzed three antibody–lysozyme structures to reveal that their interface packing densities were identical to that of the global packing (Walls and Sternberg 1992). Yan et al. have shown that the presence of aromatic rings in the interface has a role in packing. This is because tighter packing and better fit is usually induced if two aromatic rings are arranged in parallel (Yan et al. 2008). However, tight packing does not implicate that the cavities are not present at the interface. A study based on the interfacial regions between the antibody domains CL and CH1 demonstrated the presence of a large cavity (Padlan 1990). The author proposed that this cavity was responsible for reorienting central residues of the interface without reorienting the entire interface. This was in turn required because each of the five CH1 chains (α, γ, δ, ε, and μ) had to bear an ability to interact with the two distinct CL chains (λ and κ).

The complementarity between the shapes of the two binding sites is an important parameter to consider during PPI interface analysis. The gap volume index (GV index) which is a measure of the gap volume between the interacting interfaces, normalized with the interface area, throws light about interface complementarity. Smaller the GV index, better the complementarity. For instance, the interface complementarity of Stefin B/Papain complex (GV index = 2.12 Å) is lower than methylmalonyl-CoA mutase complex (GV index = 1.65 Å). The parameter of shape complementarity is not of much importance for the interactions that are transient in nature (Keskin et al. 2016). The shape complementarity depends on the type of protein complexes formed post-interaction. Lawrence and Colman showed that dimeric proteins had the greatest shape complementarity, while antibody–protein complexes had the worst complementarity. The complementarity of protease–inhibitor complexes was in between the complementarities of these two complexes (Lawrence and Colman 1993). Pawowski et al. studied the factors that are responsible for the specificity and quaternary structural diversity of EF-hand calcium-binding protein family. Authors have revealed that hydrophobic interface location on the protein surface and the masking of other potential sites for interaction via additional sequence fragments controlled the molecular interactions. In simple terms, they stated that when shape complementarity is dictated by the topology of interface, it may have considerable contribution to the specificity of the interaction (Pawłowski et al. 1996).

2.7 Water Molecules at the PPI Interface

Water molecules are known to be indulged in facilitating interactions between proteins. These water molecules are present at the interface, and the majority of them play the role of forming hydrogen bonds with either of the interacting partners. The interface residues of one protein may either share direct interactions with the partner protein or they may be involved in indirect water-mediated interactions. Interactions may also be two-fold indirect in case there are two water molecules involved. Low-affinity homologous protein complexes are usually found to comprise several doubly-indirect interactions (Lisova et al. 2014). Mutagenesis experiments that have changed a tyrosine to a phenylalanine residue have demonstrated that water-mediated interactions have substantial contribution to the interaction energy (England et al. 1997). Interface water molecules may also play a subtle role as "nanoscale buffers" by acting as hydrogen donors or acceptors. Simply via re-orientation, these water molecules may swap their roles as acceptors to roles as donors. The consequent buffering helps in mediating a "wet" interaction and in turn the interface integrity (Ahmed et al. 2011).

Ahmed et al. analyzed the data on 4741 water molecules extracted from a set of 179 high-resolution X-ray crystal structures of PPI complexes. Almost 21% of these water molecules were found to be involved in bridging, that is, they interacted with both the protein partners, while 26% did not show any interaction with either of the partners. On the other hand, the majority of the water molecules (53%) were erected to be involved in interaction with only one of the proteins. The authors observed this trend irrespective of the resolution of the diffraction data. It was further established that the interactions involving residue backbones that account for 21.5% of total interactions were in line for all the classes, and that the first class was usually more dominated by interactions that were polar residue-mediated. The authors further demonstrated that the bridging waters rendered an average stabilization energy of −0.46 kcal/mol to the PPI. On analyzing the water molecules that shared favorable interactions with either of the proteins, the authors have concluded that the phenomenon was well conserved. Around 42% of such waters had a solvent-accessible surface area lesser than 10 $Å^2$ and were therefore greatly buried. Moreover, 69% of these water molecules, being present in hydrophobic environments as "hydrophobic bubbles" and served important biological roles in PPI mediation (Ahmed et al. 2011).

2.8 Interaction-Associated Conformational Changes and Proposed PPI Models

Descriptions of numerous molecular motions are stored in the database of Macromolecular Movements (Echols et al. 2003). The conformational movements described in this database may arise due to alteration of pH levels, temperature, or other environmental conditions. However, the majority of these conformational changes largely arise due to association between proteins. Different models such as (a) lock and key model; (b) induced fit model; (c) pre-existing equilibrium/ conformational selection model have been put up to explain the mechanism of binding between two proteins (Fig. 2.7).

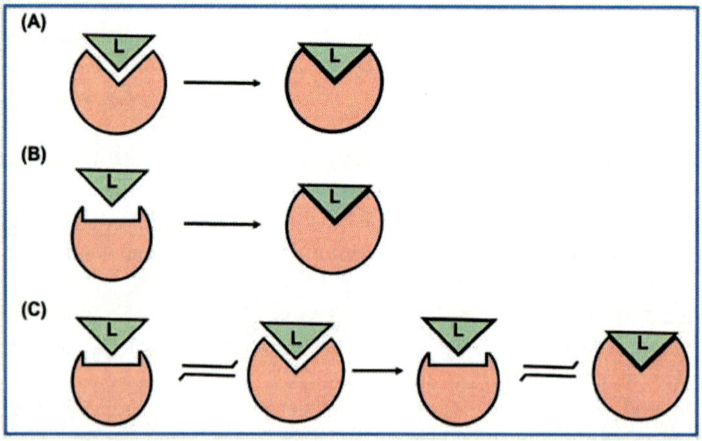

Fig. 2.7 Schematic depicting differential models of binding between two proteins: (**a**) Lock and key model, (**b**) induced fit model, and (**c**) pre-existing equilibrium model. L represents a ligand molecule

2.8.1 Lock and Key Model

One of the earliest concepts, known as the "lock and key" concept, was proposed in the year 1894 by Emil Fischer (Behr 2008). According to this model, one of the interacting proteins has an indentation or cavity that its cognate protein partner perfectly fits into. Consequently, the interaction between these two proteins is mediated without any change in the conformation. However, scientists later observed that the PPIs usually involve structural changes that are greater than local-scale rearrangements. Therefore, this classical model failed to explain protein–protein binding (Goh et al. 2004) (Fig. 2.7a).

2.8.2 Induced Fit Model

A second concept, which is known as the "induced fit" model, was first given by Daniel Koshland in 1958 (Koshland Jr. 1958). According to this model, one assumes that there is a certain degree of flexibility in the protein's active site to house its partner. In other words, this model proposes that the binding site of a protein is not as rigid as a lock and a key, but is rather analogous to a hand and a glove. This model states that ligand (or the interactor protein) induces a conformational change in the protein's binding site and subsequently shifts it toward an active state (Fig. 2.7b). Therefore, a geometric fit between the two proteins is ensured only following the structural changes induced by preliminary interaction and/or recognition.

A study based on electron cryo-microscopy has illustrated how induced fit model works for the interaction between actin and myosin. Holmes et al. have elucidated that the opening of the nucleotide-binding pocket of myosin is combined structurally to the closing of the actin-binding cleft on myosin (Holmes et al. 2003). The binding is initiated via a weak stereospecific interaction in an open conformation that brings the myosin's lower domain in contact with the actin filament. This stage then gradually progresses to strong binding, marked by the closure of the myosin domain's cleft (Rayment et al. 1993). This is followed by the rotation of the upper myosin domain, and the contact of the cardiomyopathy loop with the actin surface, which in turn enhanced the total interaction surface area by two folds. Consequently, a displacement of 6–9 A° was created in the switch 1 element, which comprises the nucleotide-binding pocket. The displacement movement is thus an indicative of the fact that strong actin binding is combined with the opening of the nucleotide-binding pocket. Other studies based on this PPI have suggested that strong binding of actin to the cross-bridge of myosin is the event most probably responsible for the closure of the actin-binding cleft (Yengo et al. 2002). Betts and Sternberg studied the conformational changes underlying PPIs for 39 complexes (Betts and Sternberg 1999). They concluded that although several PPIs follow the mechanism of induced fit, there are still many other protein complexes that have large accompanying structural changes and thus cannot be explained by induced fit mechanism alone (Chen et al. 2003; Goh et al. 2004). Bosshard has stated that only in the case of a very strong match between the interacting sites, the initial complex has enough longevity and strength for induced fit to take place within a reasonable period of time frame (Bosshard 2001).

2.8.3 Pre-Existing Equilibrium/Conformational Selection Model

In recent years, a third concept known as the "pre-existing equilibrium/conformational selection" model has emerged to explain association events between proteins (Ma et al. 1999; Tsai et al. 1999) (Fig. 2.7c). This model has been devised on the theories of protein folding that are based on the funnel-like free energy landscape concept of proteins (Dill and Chan 1997; Karplus 1997). This model postulates that a protein's native state may be an ensemble of similar conformations in equilibrium instead of a single conformation. Among all the conformations in the ensemble, the ones that are the most suited usually associate with the binder, in turn, shifting the equilibrium toward the formation of a complex. Corresponding to the binders, there are several minima near the native state in the energy landscape. The ensemble of near-native conformations increases if the protein is more flexible and lowers the energy barrier between the binders. On association of the protein with the binder, global structural changes might take place that are not just restricted to side chain rearrangements near the interaction site, but rather movements of the entire domains (Tobi and Bahar 2005).

Volkman et al. studied the motions of NtrC protein both in its inactive unphosphorylated state and active phosphorylated state. The results from NMR relaxation experiments indicated the exchange of the protein between its active and inactive conformations. The authors delineated that both conformations are populated in the unphosphorylated protein; however, there is shift of the equilibrium toward the active protein species upon phosphorylation. These results therefore demonstrated that the mechanism underlying activation was essentially the dynamic shift of population between two conformations that were pre-existing (Volkman et al. 2001).

Betts et al. analyzed the conformational changes associated with PPIs by taking into account structures of 39 pairs of complexes and their unbound counterparts. By averaging the conformational changes over the non-interface and interface regions and over individual residues, the authors evaluated that the conformational movements majorly involved the main and the side chains, whereas large interfacial changes were associated with the formation of the complex. On the other hand, changes of the non-interface residues were owing to flexibility and disorder. The authors concluded that although the general underlying mechanism followed the induced fit model, systems that have limited conformational change can be treated to a first approximation by the classical lock and key model (Betts and Sternberg 1999). Studies on conformational movements underlying protein–protein association have greatly aided in the progression of efficient docking algorithms that consider conformational changes in their refinement steps. Mechanistic models have been assigned to several PPIs based on their available experimental data (Goh et al. 2004). Some of these PPIs and their corresponding models have been summarized in Table 2.1.

Table 2.1 Some of the proteins along with their binding partners which form protein complexes employing differential interaction models. Table adapted from Goh et al. (Goh et al. 2004)

Protein	Binding partner	Models[a]
Antibody Spe7	TrxShear3	Equilibrium
$G\alpha_i$ subunit	RGS14	Induced
NtrC	P	Equilibrium
Importin-β	SREPB-2	Induced
Leukemia inhibitory factor	GP130	Induced
GPCR kinase 2	$G\beta 1\gamma 2$	Induced
Her2	Herceptin fab	Induced
GroEL	GroES	Allosteric
Proto-oncogene Cbl	ZAP-70 kinase	Induced
CDK2	Cyclin A	Induced
$PPAR\gamma$	Steroid receptor coactivator-1	Equilibrium
XRCC4	DNA-ligase IV	Induced
CheY	CheA	Induced
GpIb-IX-V	Thrombin	Equilibrium
Myosin	Actin	Induced
FcαR1	Fc	Induced
FliS	FliC	Induced
Lir-1	Hla-A2	Induced
Btub	Colicin E3	Induced
Erythropoietin	Erythropoietin receptor	Induced
F_1-ATPase	IF_1	Induced
Calmodulin	Calmodulin-binding domain of skeletal light chain myosin kinase	Allosteric
Mms2	Ubc13	Induced
RAN	RAN-binding protein 2	Induced

[a]This classification relies on the experimental data available (Goh et al. 2004).

2.9 Structural Plasticity

There are certain proteins that can associate with many proteins employing surface residues that essentially belong to the same group. It is therefore essential to understand the characteristic properties that retain the uniqueness of these consensus binding sites. Recent studies based on structural data have demonstrated that plasticity plays two roles in PPIs. First, it allows the accommodation of mutations in protein interfaces as they coevolve. Second, it enables a single molecular surface to interact with several, structurally different interactors. The conformational changes underlying PPI give rise to this flexibility of proteins. The degree of flexibility, in turn, appears to enhance with the quantity of surface area buried in the protein complex (Lo Conte et al. 1999). Recently advancements in nuclear magnetic resonance (NMR) techniques enable the analysis of intramolecular motions over different timescales at the residue level (Cavanagh and Akke 2000). A number of studies based on these NMR techniques have demonstrated a significant overlap in the amino acid residues that are important in PPIs and those that have great structural flexibility (Barile and Pellecchia 2014; Gell and Mackay 2008; O'Connell et al. 2009; Zhao et al. 2020). For instance, Burz et al. developed a high-throughput method known as in-cell NMR to probe the structural alterations that accompany PPIs. The method involves the sequential expression of two or more targeted proteins within a single cell in a time-dependent manner and monitoring the interactions between the targeted proteins using in-cell NMR spectroscopy. The spectra will entail the entire interactions between the proteins by defining the structural details of all the interacting surfaces in the proteins at atomic resolution (Burz et al. 2006). Chemical shift perturbation is a simple NMR method to probe the binding of proteins (Jaiswal et al. 2019; Williamson 2013). Recently, Somlyay developed a novel NMR-based methodology which combines ^{19}F NMR spectroscopy tagging with PRE (paramagnetic relaxation enhancement) to map PPIs. The method also provides detailed atomic level information about the conformational changes in the proteins upon the formation of protein complex (Somlyay et al. 2020). Zhao et al. have also recently determined PPIs by employing Xe-129 NMR. They utilized this methodology owing to its better detection sensitivity and selectivity over the low intrinsic background signals. This in turn is due to the hyperpolarized chemical exchange saturation transfer (hyper-CEST) of ^{129}Xe (Zhao et al. 2020). Such recent technological advancements in the NMR field have provided wealth of information in era of structural and dynamic changes induced in the proteins upon interaction with their binding protein partners at atomic resolution.

This trend was particularly observed for residues of proteins that had multiple interacting partners (Feher and Cavanagh 1999; Kay et al. 1996). Although it is evident that proteins having the ability to bind multiple interactors at a single site share several molecular features observed in other PPIs, they follow a differential set of rules when it comes to structural plasticity.

The Fc fragment of immunoglobulin G demonstrates enormous structural plasticity as it has several well-known interacting proteins such as protein G, protein A, neonatal Fc receptor, and rheumatoid factor (Sundberg and Mariuzza 2000). Remarkably, a common region between the CH_2 and CH_3 domains of the Fc fragment acts as the binding site for all the four proteins. Wells et al. performed patch analysis of this region from crystal structures to explain the high convergence of this binding site (DeLano et al. 2000). The authors observed that the consensus interaction site was part of an even greater region that had hydrophobic and highly accessible binding surfaces. The authors concluded that the consensus site was able to interact with several proteins owing to its greater conformational flexibility as was evident from the distinct structural movements in that site induced by each of the different interactors. Further, the authors observed that most of the conformational change, mediated by two methionine residues, rearranged the binding site in different ways. For instance, the structural changes formed a pocket during interaction with protein G, whereas they flattened out the binding site while interacting with protein A or rheumatoid factor. Other residue side chains in the binding site also

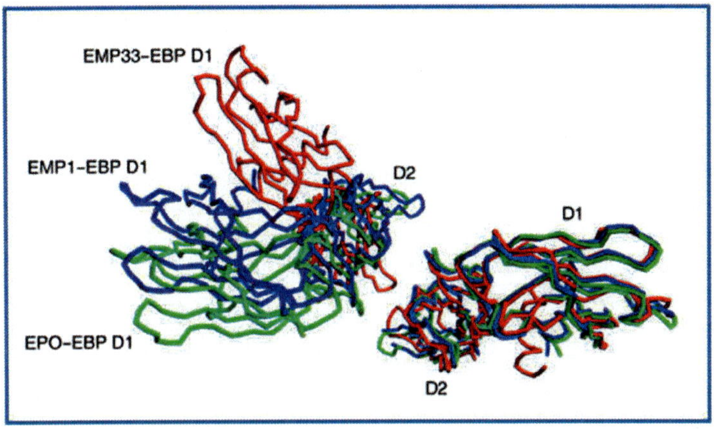

Fig. 2.8 Relative juxtapositions of the dimer pairs of erythropoietin (EPO) binding protein (EBP) when bound by the EPO (green) natural agonist, EMP1 (blue), a weak peptide agonist, and EMP33 (red), an antagonist peptide. The D1 domains of each EBP dimer pair's second EBP molecule are located approximately 15° and 60°, away from the axis along the two EBP D1 domains of the EPO–EBP complex when bound by EMP1 and EMP33, respectively (Adapted from (Sundberg and Mariuzza 2000))

adopt distinct rotamer positions to interact with different proteins. Furthermore, plasticity of the site is rendered by the inherent flexibility of the hinge region between the CH_2 and CH_3 domains.

Similarly, a study remodeled the interface between hGH and its receptor via deletion of a critical residue in the receptor. This deletion created a large cavity and the research group then attempted to select a hGH pentamutant that would fill the cavity and restore the binding. It has been observed that the 5 hGH mutations pertaining to the interaction interface were able to rescue the interaction with the receptor via the induction of conformational changes at the global scale (Atwell et al. 1997). Livnah and colleagues elucidated how structural plasticity in erythropoietin receptor (EPOR) leads to different signaling events when bound with either agonist or antagonist. EPOR is known to dimerize in the presence of both natural and synthetic erythropoietin (EPO), ultimately leading to its activation. The authors studied the crystal structure of EPOR bound to an antagonist and observed that although the dimerization still occurred, the orientation between the receptor molecules was varied relative to when bound with the agonist (Fig. 2.8) (Livnah et al. 1998).

Structural plasticity has also been widely reported in the recognition domains of immune system components such as T-cell receptors (TCRs) and antibodies. This plasticity is indispensable owing to the limited number of immune moieties compared to the great variety of foreign antigens. Antibodies have six complementarity determining region (CDR) loops extending from the variable light (V_L) and heavy (V_H) chains and TCRs have differential α- and β-domains (Vα and Vβ). Both CDRs and the variable domains can re-arrange conformationally to give rise to the enormous plasticity.

Despite the fact that several biologically relevant molecules have common binding sites for all their interactors, there are many proteins that bind to different interactors using distinct surfaces. For example, MHC class I molecules interact with natural killer (NK) cell receptors, TCRs, and CD8 utilizing different sites (Sundberg and Mariuzza 2000). In a nutshell, multi-partner proteins can either contain multiple binding sites, that they utilize to interact with all their partners simultaneously, or they can utilize the same binding site at different times to share interactions that are mutually exclusive. For the second class of multi-partner proteins, the interactions involved are usually transient in nature and less essential (Keskin and Nussinov 2007; Kim et al. 2006; Tuncbag et al. 2008). Hub proteins are classical multi-partner proteins and they have been categorized into singlish-interface hubs and multi-interface hubs. p53 and Mdm2 are two well-known multi-interface proteins that have been studied extensively (Tuncbag et al. 2009).

2.10 Evolution of PPIs

The physiological environment and biological needs govern how the affinity and structure of a PPI is tuned. Therefore, the evolution of PPIs is presumably out of the requirement to optimize efficacy at the functional level. However, this does not necessarily denote strong interactions. Often, selection pressure in the form of cellular functioning may pave the way for novel interactions that are weak and transient in nature. On the other hand, PPI evolution can yield complexes that are obligate in nature. This simply reflects the requirement of stability or the functioning of all the interacting partners. For instance, symmetric DNA binding modules are inactive separately and need interactions among them to perform their necessary function (Andreani and Guerois 2014; Ghadie et al. 2018; Makino and Gojobori 2007; Schoenrock et al. 2017).

Some other scenarios may also be observed during PPI evolution. For instance, proteins that accidentally contact each other in the cell may evolve an interaction without having any functional role. The resulting complex might eventually survive if there is no selection pressure to eliminate it from the interactome. Therefore, cases that have failed to rationalize the function of protein oligomerization may be deemed as "happenstance" oligomerization (Nooren and Thornton 2003). The evolution of protein complexes is not very clear and can often provide a very obscure picture. However, it may have some relation with the oligomer folding. In complexes that follow two-state folding, the constituting protomers are folded and oligomerized concurrently (Xu et al. 1998). On the other hand, in interactions that are non-obligate in nature, each of the constituting protomers folds independently. Therefore, evolution has presumably led to the formation of the interaction site on the stable monomer's surface. There are still other oligomers where evolution has occurred via domain swapping. In such PPI evolution, domains have been rearranged to replace interactions between domains by those between monomers (Bennett et al.

1994). Further clues to PPI evolution can be extracted from analyzing the variation in the oligomeric structures or states within a homologous protein family.

Pereira-Leal and colleagues analyzed protein complexes in *Saccharomyces cerevisiae*, pairwise protein interaction clusters in networks of various organisms, and protein complexes from the Protein Data Bank that possess well-known 3D structures. The authors observed that duplication of homomeric interactions constituted a major class of PPIs. This duplication resulted in complex formation between paralogous proteins with high frequency. The authors concluded that the duplication method was a common route for the evolution of PPI clusters and protein complexes. Their results were further validated via the modeling of network evolution (Pereira-Leal et al. 2007).

Makino et al. demonstrated some interesting findings on PPI evolution. Their studies showed that a protein having more interacting partners evolves at a slower rate than one having fewer interactors. Moreover, they observed that ratios corresponding to duplicated pairs that shared interactors were significantly lesser than those for pairs that did not share partners. The authors reasoned that when a duplicated pair did not share any interacting partner, it might be plausible that the interactions take place with protein partners that have different functions. The results of their studies clearly demonstrated that duplicated gene pairs are functionally differentiated via PPI gains and/or losses. The results further elucidated those different functional (DF) proteins that interacted with proteins from different functional classes evolved more slowly than same function (SF) proteins that interacted with partners from the same functional class. This implicated that interacting proteins that had roles in multi-different cellular processes were under strong constraints of functionality. Moreover, it was observed that sparse part (SP) proteins that were located in the sparse parts of the network had a slower evolutionary rate in comparison with their counter dense part (DP) proteins. This observation implied that weaker the relationship was among the interacting partners of proteins; slower the evolution rate was of the interacting partners (Makino and Gojobori 2007).

2.11 Protein Subunit–Subunit to Domain–Domain Interactions

PPIs can be mediated either via subunits or domains. It has been observed that in certain cases subunit-mediated protein interactions in one species can transform to domain-mediated interactions in some other species. This occurs by the formation of the Rosetta stone or a fusion protein via an evolutionary process known as gene fusion (Bornberg-Bauer et al. 2005). In such an event, an operon-like behavior is observed that leads to creation of protein–protein interfaces and properties like alternative splicing across diverse evolutionary distances and multiple functions (Yiting et al. 2004). Due to gene fusion, a functional interface that is usually formed by two subunits in one species is formed by two domains in a fused protein in another species. In other words, a subunit–subunit protein complex transforms to a "fused form" with two domains in a second species. This fusion protein comprises a

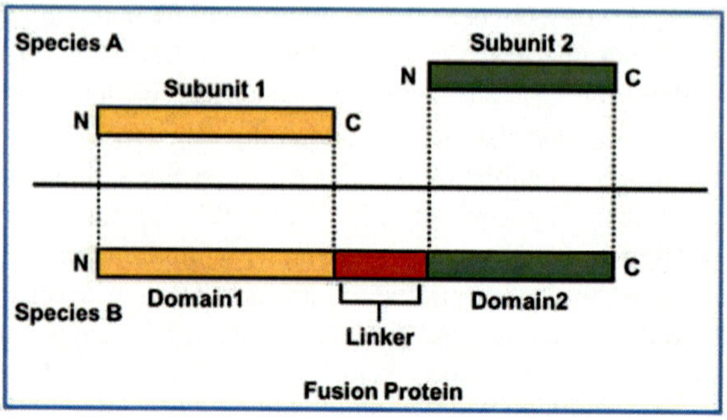

Fig. 2.9 An illustration depicting "Rosetta stone" or "fusion protein." In species A, a subunit–subunit protein complex possessing a protein–protein interaction is found to be in a "fused form" containing a linker stretch forming a fusion protein containing two domains in species B

linker stretch. It should be kept in mind that in this "Rosetta Stone" subunit 1 and domain 1 are homologous, while subunit 2 and domain 2 are homologous (Fig. 2.9).

The phenomenon of protein subunit–subunit to domain–domain interaction is very clearly illustrated in the case of imidazole glycerol phosphate synthetase (IGPS) protein complex that comprises cyclase unit (HisH) and glutaminase unit (HisF) (Yiting et al. 2006). In *S. cerevisiae*, the HisF-HisH domains are linked to form a Rosetta stone and the functional interface is formed between these two domains. However, they are unfused in *T. thermophilus*, where the interface is formed between two subunits. It should however be noted that the IGPS proteins in these two species are homologous. In other words, the molecular function of the protein is analogous despite different states (either domain–domain or subunit–subunit) of functional interface. As it is expected, the difference in the states of the interface is reflected by the difference in parameters of gap volume, interface area, and radius of gyration (Rg). Simulations have proven that the values of these parameters are indeed greater in *S. cerevisiae* than their counterparts in *T. thermophilus*. However, there is a remarkable exception with respect to one parameter—the gap index. It is same for interfaces in both the species. It simply implies that a fused protein has greater gap volume, interface area, and Rg, but conserved gap volume is more favored than its unfused equivalents. This is interesting, as it paves the way toward the understanding of molecular evolution of fusion proteins. It has been hypothesized that the preference for Rosetta stone architecture is evolutionarily driven by biochemical pressure in organisms (Veitia 2002; Yiting et al. 2006).

Another interesting example of the Rosetta stone is the succinyl CoA transferase fusion protein found in humans. It is formed by the fusion of the protein's α- and β-subunits present in *E. coli*. At the N terminal, the α-subunit holds 62% similarity to the fused protein, while at the C-terminal, the β-subunit holds 66% similarity to the fused protein (Heider 2001).

2.12 Motif-Mediated Interactions

It is well-known that several proteins comprise units called domains that have distinct functions and are capable of independently folding into their 3D structures. However, there are also certain segments in the protein's peptide chain that do not take up a globular 3D structure. They are commonly referred to as "disordered protein regions." It is now known that around 20–50% of all eukaryotic proteins have a certain degree of disorder and around 17% of all proteins in a eukaryotic cell are entirely disordered (Dunker et al. 2000). It is however vital to note that the disordered regions are not mere spacer regions connecting protein domains, but are rather the regions of important linear motifs that have several functional implications. Several important PPIs are mediated by the binding of small recognition domains with peptides exhibiting such sequence motifs. In such an interaction, only the domain takes up a globular 3D structure, whereas the motif remains mostly linear. Such domain–motif interactions have been widely found in cell compartment targeting, signaling mechanisms, and post-translational modifications (Diella et al. 2008).

Lymphocyte-specific tyrosine-kinase Lck is an important example of a protein that employs motif-mediated PPIs for its regulation (Fig. 2.10). The kinase domain of this protein exhibits enzymatic activity. On the other hand, the remaining two domains (SH2, SH3) are adaptor-domains that play vital roles in regulation. These two domains recognize specific sequence motifs. For interaction with SH2 domain, the corresponding motifs comprise phosphorylated tyrosine residues, while for SH3 domain the motifs contain sequence stretches that are proline-rich. These interactions are vital for activating the enzymatic (kinase) activity. There are still other motif-mediated interactions in Lck. They include the interaction between the catalytic domain and the substrates as well as binding of Lck's amino-terminal region to CD4 (Palacios and Weiss 2004). Extensive research has shown how indispensable motif-mediated PPIs are to the sustenance of the living cell. It is evident from the fact that mutations in such linear motifs represent the root cause of hereditary disorders such as Usher's syndrome, Noonan syndrome, and Liddle's syndrome (Eudy and Sumegi 1999; Furuhashi et al. 2005; Pandit et al. 2007).

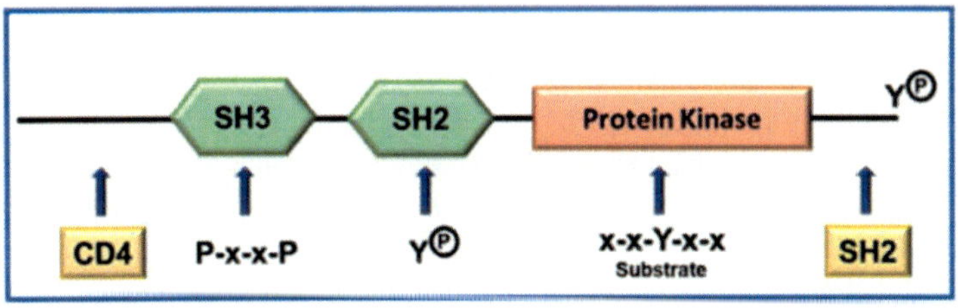

Fig. 2.10 Schematic depicting the principle of motif-mediated interactions using human tyrosine-kinase Lck. The linear sequence motifs and domains binding to distinct regions of Lck are shown in the second line. The SH3 domain interacts and binds to a P-x-x-P consensus motif, while the SH2 domain interacts with phosphorylated tyrosine residue containing motifs

Additionally, functional analyses and genome sequencing have demonstrated that viral effector proteins are frequently dependent on linear sequence motifs to carry out viral replication and persistence. Well-known examples include Tip protein from *Herpesvirus saimiri* and Nef protein from HIV-1 (Dinkel and Sticht 2010).

2.13 Conclusions

This chapter has provided a very detailed and comprehensive account of the fundamental characteristic properties of PPIs and the host of cellular functions that they perform. Since the study of structural features of PPIs gives one better insight into its functional aspects, extensive research has been performed in this field. Although the protein interaction interfaces are variable and depend largely on the type of PPI, the basic interface features have been summarized. Interfaces usually make up a considerable fraction of the monomer surfaces in a protein complex, with around 20 residues participating in the interface from each monomer. The interface residue composition can be approximated as "intermediate" between that of protein exteriors and protein interiors. Despite the variability in residue composition among different interfaces, aromatic and arginine amino acids have been observed to be favored interface residues. No strong preference for a specific type of secondary structure/motif has been found in PPI interfaces.

The driving forces at the interface are usually diverse, starting from hydrogen bonding to electrostatic interactions to salt bridges and hydrophobic interactions. However, hydrophobicity is considered to be key in PPI stabilization. The interface is studded with transient pockets that disappear on binding with a protein partner. The shape complementarity between the interfaces of the partners that is measured with respect to the gap volume parameter is dependent on the type of PPI in question. The water molecules present at the interface not only aid in mediating interactions between the protein partners, but also extend buffering actions. Protein interfaces have been found to be plastic or flexible to accommodate interactions with more than one partner. Conformational changes associated with protein binding have been mainly found to be accountable for this plasticity and several models have been proposed to describe the mechanism of these changes. PPIs have also evolved with time and their evolutionary rates have been found to be greatly dependent on the type of the interaction. Some evolutionary events such as gene fusion have transformed subunit interactions in one species to domain interactions in another species. There are several PPIs that do not belong to the traditional domain–domain or subunit–subunit interaction categories, but instead employ linear motifs to mediate interactions.

References

Aboulwafa M, Zhang Z, Saier MH, Jr. (2019). Protein:Protein interactions in the cytoplasmic membrane apparently influencing sugar transport and phosphorylation activities of the e. coli phosphotransferase system. *PloS one*, *14*, e0219332.

Abrams WR, Cohen AB, Damiano VV, Eliraz A, Kimbel P, Meranze DR, Weinbaum G. (1981). A model of decreased functional alpha-1-proteinase inhibitor. Pulmonary pathology of dogs exposed to chloramine T. *The Journal of clinical investigation*, *68*, 1132-1139.

Acuner Ozbabacan SE, Engin HB, Gursoy A, Keskin O. (2011). Transient protein–protein interactions. *Protein engineering, design, selection*, *24*, 635-648.

Ahmed MH et al. (2011). Bound Water at Protein-Protein Interfaces: Partners, Roles and Hydrophobic Bubbles as a Conserved Motif. *PloS one*, *6*, e24712.

Aloy P, Russell RB. (2006). Structural systems biology: modelling protein interactions. *Nature reviews. Molecular cell biology*, *7*, 188-197.

Andreani J, Guerois R. (2014). Evolution of protein interactions: from interactomes to interfaces. *Archives of biochemistry and biophysics*, *554*, 65-75.

Argos P. (1988). An investigation of protein subunit and domain interfaces. *Protein engineering*, *2*, 101-113.

Arnold H, Pette D. (1970). Binding of Aldolase and Triosephosphate Dehydrogenase to F-Actin and Modification of Catalytic Properties of Aldolase. *Eur J Biochem*, *15*, 360-&.

Arroyo JD, Hahn WC. (2005). Involvement of PP2A in viral and cellular transformation. *Oncogene*, *24*, 7746-7755.

Atwell S, Ultsch M, De Vos AM, Wells JA. (1997). Structural plasticity in a remodeled protein-protein interface. *Science*, *278*, 1125-1128.

Aurenhammer F. (1987). Power Diagrams - Properties, Algorithms and Applications. *Siam J Comput*, *16*, 78-96.

Bahadur RP, Chakrabarti P, Rodier F, Janin J. (2003). Dissecting subunit interfaces in homodimeric proteins. *Proteins*, *53*, 708-719.

Bahar I, Jernigan RL. (1997). Inter-residue potentials in globular proteins and the dominance of highly specific hydrophilic interactions at close separation. *Journal of molecular biology*, *266*, 195-214.

Banga I, Szent-Györgyi A. (1942). Preparation and properties of myosin A and B. *Stud. Inst. Med. Chem. Univ. Szeged*, *1*, 5-15.

Barile E, Pellecchia M. (2014). NMR-based approaches for the identification and optimization of inhibitors of protein–protein interactions. *Chem Rev*, *114*, 4749-4763.

Behr J-P (2008) The lock-and-key principle: the state of the art--100 years on vol 17. John Wiley & Sons,

Bennett MJ, Choe S, Eisenberg D. (1994). Domain Swapping - Entangling Alliances between Proteins. *Proceedings of the National Academy of Sciences of the United States of America*, *91*, 3127-3131.

Berggård T, Linse S, James P. (2007). Methods for the detection and analysis of protein–protein interactions. *Proteomics*, *7*, 2833-2842.

Berlow RB, Dyson HJ, Wright PE. (2015). Functional advantages of dynamic protein disorder. *FEBS letters*, *589*, 2433-2440.

Betts MJ, Sternberg MJE. (1999). An analysis of conformational changes on protein-protein association: implications for predictive docking. *Protein engineering*, *12*, 271-283.

Bogan AA, Thorn KS. (1998). Anatomy of hot spots in protein interfaces. *Journal of molecular biology*, *280*, 1-9.

Bornberg-Bauer E, Beaussart F, Kummerfeld SK, Teichmann SA, Weiner J. (2005). The evolution of domain arrangements in proteins and interaction networks. *Cellular and Molecular Life Sciences CMLS*, *62*, 435-445.

Bosshard HR. (2001). Molecular recognition by induced fit: how fit is the concept? *Physiology*, *16*, 171-173.

Braun P, Gingras AC. (2012). History of protein-protein interactions: From egg-white to complex networks. *Proteomics*, *12*, 1478-1498.

Bregman DB, Bhattacharyya N, Rubin CS. (1989). High affinity binding protein for the regulatory subunit of cAMP-dependent protein kinase II-B. Cloning, characterization, and expression of cDNAs for rat brain P150. *The Journal of biological chemistry*, *264*, 4648-4656.

Brown AM, Crothers DM. (1989). Modulation of the stability of a gene-regulatory protein dimer by DNA and cAMP. *Proceedings of the National Academy of Sciences of the United States of America*, *86*, 7387-7391.

Burgert HG, Kvist S. (1985). An adenovirus type 2 glycoprotein blocks cell surface expression of human histocompatibility class I antigens. *Cell*, *41*, 987-997.

Burz DS, Dutta K, Cowburn D, Shekhtman A. (2006). Mapping structural interactions using in-cell NMR spectroscopy (STINT-NMR). *Nature methods*, *3*, 91-93.

Byrum S, Smart SK, Larson S, Tackett AJ (2012) Analysis of stable and transient protein–protein interactions. In: Chromatin Remodeling. Springer, pp 143-152

Cavanagh J, Akke M. (2000). May the driving force be with you - whatever it is. *Nature structural biology*, *7*, 11-13.

Cazals F, Proust F, Bahadur RP, Janin J. (2006). Revisiting the Voronoi description of protein-protein interfaces. *Protein Sci*, *15*, 2082-2092.

Chakrabarti P, Janin J. (2002). Dissecting protein-protein recognition sites. *Proteins*, *47*, 334-343.

Chatr-Aryamontri A, Ceol A, Licata L, Cesareni G. (2008). Protein interactions: integration leads to belief. *Trends in biochemical sciences*, *33*, 241-242; author reply 242-243.

Chaurasia G, Futschik ME (2010) Interactomics and Cancer. In: An Omics Perspective on Cancer Research. Springer, pp 167-182

Chen J, Sawyer N, Regan L. (2013). Protein–protein interactions: General trends in the relationship between binding affinity and interfacial buried surface area. *Protein Science*, *22*, 510-515.

Chen P, Tian J, Kovesdi I, Bruder JT. (1998). Interaction of the adenovirus 14.7-kDa protein with FLICE inhibits Fas ligand-induced apoptosis. *The Journal of biological chemistry*, *273*, 5815-5820.

Chen R, Mintseris J, Janin J, Weng ZP. (2003). A protein-protein docking benchmark. *Proteins-Structure Function and Genetics*, *52*, 88-91.

Chothia C, Janin J. (1975). Principles of Protein-Protein Recognition. *Nature*, *256*, 705-708.

Clackson T, Ultsch MH, Wells JA, de Vos AM. (1998). Structural and functional analysis of the 1: 1 growth hormone: receptor complex reveals the molecular basis for receptor affinity. *Journal of molecular biology*, *277*, 1111-1128.

Clackson T, Wells JA. (1995). A hot spot of binding energy in a hormone-receptor interface. *Science*, *267*, 383-386.

Clodfelter KH, Waxman DJ, Vajda S. (2006). Computational solvent mapping reveals the importance of local conformational changes for broad substrate specificity in mammalian cytochromes P450. *Biochemistry*, *45*, 9393-9407.

Cooper GM, Hausman R. (2000). A molecular approach. *The Cell. 2nd ed. Sunderland, MA: Sinauer Associates.*

Couzens AL et al. (2013). Protein interaction network of the mammalian Hippo pathway reveals mechanisms of kinase-phosphatase interactions. *Science signaling*, *6*, rs15.

Cowell IG, Hurst HC. (1996). Protein-protein interaction between the transcriptional repressor E4BP4 and the TBP-binding protein Dr1. *Nucleic acids research*, *24*, 3607-3613.

Creighton TE (1993) Proteins: structures and molecular properties. Macmillan,

Dainty M, Kleinzeller A, Lawrence A, Miall M, Needham J, Needham DM, Shen S-C. (1944). Studies on the anomalous viscosity and flow-birefringence of protein solutions: Iii. Changes in these properties of myosin solutions in relation to adenosinetriphosphate and muscular contraction. *The Journal of general physiology*, *27*, 355.

Davies DR, Padlan EA, Sheriff S. (1990). Antibody-antigen complexes. *Annual review of biochemistry*, *59*, 439-473.

De Las Rivas J, Fontanillo C. (2010). Protein–protein interactions essentials: key concepts to building and analyzing interactome networks. *PLoS computational biology*, *6*, e1000807.
De Las Rivas J, Fontanillo C. (2012). Protein–protein interaction networks: unraveling the wiring of molecular machines within the cell. *Briefings in functional genomics*, *11*, 489-496.
DeLano WL, Ultsch MH, de Vos AM, Wells JA. (2000). Convergent solutions to binding at a protein-protein interface. *Science*, *287*, 1279-1283.
DeLisi C (2013) Antigen antibody interactions vol 8. Springer Science & Business Media,
Dick FA, Rubin SM. (2013). Molecular mechanisms underlying RB protein function. *Nat Rev Mol Cell Bio*, *14*, 297-306.
Diella F et al. (2008). Understanding eukaryotic linear motifs and their role in cell signaling and regulation. *Front Biosci-Landmrk*, *13*, 6580-6603.
Dill KA, Chan HS. (1997). From Levinthal to pathways to funnels. *Nature structural biology*, *4*, 10-19.
Dilworth S Cell alterations induced by the large T-antigens of SV40 and polyoma virus. In: Seminars in cancer biology, 1990. vol 6. pp 407-414
Dinkel H, Sticht H (2010) Motif-mediated protein interactions and their role in disease. In: Protein-Protein Complexes: Analysis, Modeling and Drug Design. World Scientific, pp 272-294
Dundas J, Ouyang Z, Tseng J, Binkowski A, Turpaz Y, Liang J. (2006). CASTp: computed atlas of surface topography of proteins with structural and topographical mapping of functionally annotated residues. *Nucleic acids research*, *34*, W116-118.
Dunker AK, Cortese MS, Romero P, Iakoucheva LM, Uversky VN. (2005). Flexible nets. The roles of intrinsic disorder in protein interaction networks. *The FEBS journal*, *272*, 5129-5148.
Dunker AK et al. Protein disorder and the evolution of molecular recognition: theory, predictions and observations. In: Pac Symp Biocomput, 1998. pp 473-484
Dunker AK et al. (2001). Intrinsically disordered protein. *Journal of molecular graphics & modelling*, *19*, 26-59.
Dunker AK, Obradovic Z, Romero P, Garner EC, Brown CJ. (2000). Intrinsic protein disorder in complete genomes. *Genome informatics. Workshop on Genome Informatics*, *11*, 161-171.
Dyson HJ, Wright PE. (2005). Intrinsically unstructured proteins and their functions. *Nat Rev Mol Cell Bio*, *6*, 197-208.
Echols N, Milburn D, Gerstein M. (2003). MolMovDB: analysis and visualization of conformational change and structural flexibility. *Nucleic acids research*, *31*, 478-482.
Edelsbrunner H, Mücke EP. (1994). Three-dimensional alpha shapes. *ACM Transactions on Graphics*, *13*, 43-72.
England P, Bregegere F, Bedouelle H. (1997). Energetic and kinetic contributions of contact residues of antibody D1.3 in the interaction with lysozyme. *Biochemistry*, *36*, 164-172.
Eudy J, Sumegi J. (1999). Molecular genetics of Usher syndrome. *Cellular and Molecular Life Sciences CMLS*, *56*, 258-267.
Eyrisch S, Helms V. (2007). Transient pockets on protein surfaces involved in protein-protein interaction. *J Med Chem*, *50*, 3457-3464.
Eyrisch S, Helms V. (2009). What induces pocket openings on protein surface patches involved in protein–protein interactions? *Journal of computer-aided molecular design*, *23*, 73.
Fauchere J. (1983). Hydrophobic parameters πof amino-acid side chains from the partitioning of N-acetyl-amino-acid amides. *Eur. J. Med. Chem.*, *18*, 369-375.
Feher VA, Cavanagh J. (1999). Millisecond-timescale motions contribute to the function of the bacterial response regulator protein Spo0F. *Nature*, *400*, 289-293.
Ferreon AC, Ferreon JC, Wright PE, Deniz AA. (2013). Modulation of allostery by protein intrinsic disorder. *Nature*, *498*, 390-394.
Ferreon JC, Martinez-Yamout MA, Dyson HJ, Wright PE. (2009). Structural basis for subversion of cellular control mechanisms by the adenoviral E1A oncoprotein. *Proceedings of the National Academy of Sciences of the United States of America*, *106*, 13260-13265.
Fischer EH. (2010). Phosphorylase and the origin of reversible protein phosphorylation. *Biological chemistry*, *391*, 131-137.

Flicek P et al. (2014). Ensembl 2014. *Nucleic acids research*, *42*, D749-755.

Fredrickson DS, Wyngaarden JB, Stanbury JB (1989) The metabolic basis of inherited disease. Springer,

Freilich R, Arhar T, Abrams JL, Gestwicki JE. (2018). Protein–protein interactions in the molecular chaperone network. *Accounts of chemical research*, *51*, 940-949.

Frieden C. (1971). Protein-protein interaction and enzymatic activity. *Annual review of biochemistry*, *40*, 653-696.

Furuhashi M et al. (2005). Liddle's syndrome caused by a novel mutation in the proline-rich PY motif of the epithelial sodium channel β-subunit. *The Journal of Clinical Endocrinology & Metabolism*, *90*, 340-344.

Gadek JE, Fells GA, Crystal RG. (1979). Cigarette smoking induces functional antiprotease deficiency in the lower respiratory tract of humans. *Science*, *206*, 1315-1316.

Gao M, Skolnick J. (2010). Structural space of protein-protein interfaces is degenerate, close to complete, and highly connected. *Proceedings of the National Academy of Sciences of the United States of America*, *107*, 22517-22522.

Ge H, Liu ZH, Church GM, Vidal M. (2001). Correlation between transcriptome and interactome mapping data from Saccharomyces cerevisiae. *Nat Genet*, *29*, 482-486.

Gell DA, Mackay JP (2008) NMR spectroscopy in the analysis of protein-protein interactions. In: Modern Magnetic Resonance. Springer, pp 1339-1346

Gellatly BJ, Finney JL. (1982). Calculation of protein volumes: an alternative to the Voronoi procedure. *Journal of molecular biology*, *161*, 305-322.

Gellman SH. (1991). On the role of methionine residues in the sequence-independent recognition of nonpolar protein surfaces. *Biochemistry*, *30*, 6633-6636.

Ghadie MA, Coulombe-Huntington J, Xia Y. (2018). Interactome evolution: insights from genome-wide analyses of protein–protein interactions. *Curr Opin Struc Biol*, *50*, 42-48.

Glaser F, Steinberg DM, Vakser IA, Ben-Tal N. (2001). Residue frequencies and pairing preferences at protein-protein interfaces. *Proteins*, *43*, 89-102.

Goh CS, Milburn D, Gerstein M. (2004). Conformational changes associated with protein-protein interactions. *Curr Opin Struc Biol*, *14*, 104-109.

Goh KI, Cusick ME, Valle D, Childs B, Vidal M, Barabasi AL. (2007). The human disease network. *Proceedings of the National Academy of Sciences of the United States of America*, *104*, 8685-8690.

Goodsell DS, Olson AJ. (2000). Structural symmetry and protein function. *Annual review of biophysics and biomolecular structure*, *29*, 105-153.

Granelli-Piperno A, Reich E. (1978). A study of proteases and protease-inhibitor complexes in biological fluids. *The Journal of Experimental Medicine*, *148*, 223-234.

Guba F. (1943). Observations on myosin and actomyosin. *Stud. Inst. Med. Chem. Univ. Szeged*, *3*, 40-45.

Guharoy M, Chakrabarti P. (2007). Secondary structure based analysis and classification of biological interfaces: identification of binding motifs in protein-protein interactions. *Bioinformatics*, *23*, 1909-1918.

Hanukoglu I (1996) Electron transfer proteins of cytochrome P450 systems. In: Advances in molecular and cell biology, vol 14. Elsevier, pp 29-56

Hayakawa T, Hirashima M, Ide S, Hamada M, Okabe K, Koike M. (1966). Mammalian α-Keto Acid Dehydrogenase Complexes I. ISOLATION, PURIFICATION, AND PROPERTIES OF PYRUVATE DEHYDROGENASE COMPLEX OF PIG HEART MUSCLE. *Journal of Biological Chemistry*, *241*, 4694-4699.

Headd JJ, Ban YEA, Brown P, Edelsbrunner H, Vaidya M, Rudolph J. (2007). Protein-protein interfaces: Properties, preferences, and projections. *J Proteome Res*, *6*, 2576-2586.

Heider J. (2001). A new family of CoA-transferases. *FEBS letters*, *509*, 345-349.

Holmes KC, Angert I, Kull FJ, Jahn W, Schroder RR. (2003). Electron cryo-microscopy shows how strong binding of myosin to actin releases nucleotide. *Nature*, *425*, 423-427.

Horton TM et al. (1991). Adenovirus E3 14.7K protein functions in the absence of other adenovirus proteins to protect transfected cells from tumor necrosis factor cytolysis. *Journal of virology, 65*, 2629-2639.

Hsu WL et al. (2013). Exploring the binding diversity of intrinsically disordered proteins involved in one-to-many binding. *Protein Sci, 22*, 258-273.

Ideker T, Sharan R. (2008). Protein networks in disease. *Genome research, 18*, 644-652.

Ingram VM. (1957). Gene mutations in human haemoglobin: the chemical difference between normal and sickle cell haemoglobin. *Nature Genetics, 180*, 326-328.

Jaenicke R. (1991). Protein folding: local structures, domains, subunits, and assemblies. *Biochemistry, 30*, 3147-3161.

Jaiswal N et al. (2019). Molecular interaction between human SUMO-I and histone like DNA binding protein of Helicobacter pylori (Hup) investigated by NMR and other biophysical tools. *International journal of biological macromolecules, 123*, 446-456.

Jakus MA, Hall CE. (1947). Studies of actin and myosin. *The Journal of biological chemistry, 167*, 705-714.

Janin J. (1997). Specific versus non-specific contacts in protein crystals. *Nature structural biology, 4*, 973-974.

Janin J (2010) X-ray Study of Protein–Protein Complexes and Analysis of Interfaces. In: Protein-protein Complexes: Analysis, Modeling and Drug Design. World Scientific, pp 1-24

Janin J, Bahadur RP, Chakrabarti P. (2008). Protein-protein interaction and quaternary structure. *Quarterly reviews of biophysics, 41*, 133-180.

Janin J, Chothia C. (1990). The structure of protein-protein recognition sites. *The Journal of biological chemistry, 265*, 16027-16030.

Janin J, Miller S, Chothia C. (1988). Surface, subunit interfaces and interior of oligomeric proteins. *Journal of molecular biology, 204*, 155-164.

Jansen R et al. (2003). A Bayesian networks approach for predicting protein-protein interactions from genomic data. *Science, 302*, 449-453.

Jensen LJ et al. (2009). STRING 8-a global view on proteins and their functional interactions in 630 organisms. *Nucleic acids research, 37*, D412-D416.

Jeronimo C et al. (2007). Systematic analysis of the protein interaction network for the human transcription machinery reveals the identity of the 7SK capping enzyme. *Molecular cell, 27*, 262-274.

Ji TH, Grossmann M, Ji I. (1998). G protein-coupled receptors I. Diversity of receptor-ligand interactions. *Journal of Biological Chemistry, 273*, 17299-17302.

Jiang Y, Huang T, Chen L, Gao YF, Cai Y, Chou KC. (2013). Signal propagation in protein interaction network during colorectal cancer progression. *BioMed research international, 2013*, 287019.

Johnson D, Travis J. (1979). The oxidative inactivation of human al-proteinase inhibitor. *J. Biol. Chem, 254*, 4022-4026.

Jones S, Thornton JM. (1995). Protein-protein interactions: a review of protein dimer structures. *Progress in biophysics and molecular biology, 63*, 31-65.

Jones S, Thornton JM. (1996). Principles of protein-protein interactions. *Proceedings of the National Academy of Sciences of the United States of America, 93*, 13-20.

Jones S, Thornton JM. (1997). Analysis of protein-protein interaction sites using surface patches. *Journal of molecular biology, 272*, 121-132.

Jubb HC, Pandurangan AP, Turner MA, Ochoa-Montaño B, Blundell TL, Ascher DB. (2017). Mutations at protein-protein interfaces: Small changes over big surfaces have large impacts on human health. *Progress in biophysics and molecular biology, 128*, 3-13.

Kadenbach B, Hüttemann M. (2015). The subunit composition and function of mammalian cytochrome c oxidase. *Mitochondrion, 24*, 64 76.

Kar G, Gursoy A, Keskin O. (2009). Human cancer protein-protein interaction network: a structural perspective. *PLoS computational biology, 5*, e1000601.

Karplus M. (1997). The Levinthal paradox: yesterday and today. *Folding & design*, 2, S69-75.

Kay LE, Muhandiram DR, Farrow NA, Aubin Y, FormanKay JD. (1996). Correlation between dynamics and high affinity binding in an SH2 domain interaction. *Biochemistry, 35*, 361-368.

Keskin O, Gursoy A, Ma B, Nussinov R. (2008). Principles of protein– protein interactions: What are the preferred ways for proteins to interact? *Chem Rev, 108*, 1225-1244.

Keskin O, Ma B, Nussinov R. (2005). Hot regions in protein–protein interactions: the organization and contribution of structurally conserved hot spot residues. *Journal of molecular biology, 345*, 1281-1294.

Keskin O, Nussinov R. (2007). Similar binding sites and different partners: implications to shared proteins in cellular pathways. *Structure, 15*, 341-354.

Keskin O, Tsai CJ, Wolfson H, Nussinov R. (2004). A new, structurally nonredundant, diverse data set of protein-protein interfaces and its implications. *Protein Sci, 13*, 1043-1055.

Keskin O, Tuncbag N, Gursoy A. (2016). Predicting Protein-Protein Interactions from the Molecular to the Proteome Level. *Chem Rev, 116*, 4884-4909.

Kesztyüs L, Nikodemusz S, Szilagyi T. (1949). Antigenic activity of myosin and actin. *Nat Genet, 163*, 136-136.

Kim PM, Lu LJ, Xia Y, Gerstein MB. (2006). Relating three-dimensional structures to protein networks provides evolutionary insights. *Science, 314*, 1938-1941.

Kirkin V, Joos S, Zörnig M. (2004). The role of Bcl-2 family members in tumorigenesis. *Biochimica et Biophysica Acta (BBA)-Molecular Cell Research, 1644*, 229-249.

Koike M, Reed LJ, Carroll WR. (1960). α-Keto acid dehydrogenation complexes I. Purification and properties of pyruvate and α-ketoglutarate dehydrogenation complexes of Escherichia coli. *Journal of Biological Chemistry, 235*, 1924-1930.

Korn AP, Burnett RM. (1991). Distribution and complementarity of hydropathy in mutisunit proteins. *PROTEINS: Structure, Function, Bioinformatics, 9*, 37-55.

Koshland Jr D. (1958). Application of a theory of enzyme specificity to protein synthesis. *Proceedings of the National Academy of Sciences of the United States of America, 44*, 98.

Lage K et al. (2007). A human phenome-interactome network of protein complexes implicated in genetic disorders. *Nature biotechnology, 25*, 309-316.

Lan W, Wang J, Li M, Peng W, Wu F. (2015). Computational approaches for prioritizing candidate disease genes based on PPI networks. *Tsinghua Science and Technology, 20*, 500-512.

Lawrence MC, Colman PM (1993) Shape complementarity at protein/protein interfaces. Elsevier,

Levy ED, Teichmann SA (2013) Structural, evolutionary, and assembly principles of protein oligomerization. In: Progress in molecular biology and translational science, vol 117. Elsevier, pp 25-51

Li X, Keskin O, Ma BY, Nussinov R, Liang J. (2004). Protein-protein interactions: Hot spots and structurally conserved residues often locate in complemented pockets that pre-organized in the unbound states: Implications for docking. *Journal of molecular biology, 344*, 781-795.

Lim J et al. (2006). A protein-protein interaction network for human inherited ataxias and disorders of Purkinje cell degeneration. *Cell, 125*, 801-814.

Lin SL, Tsai C-J, Nussinov R. (1995). A study of Four-helix Bundles: Investigating Protein FoldingviaSimilar Architectural Motifs in Protein Cores and in Subunit Interfaces. *Journal of molecular biology, 248*, 151-161.

Lisova O, Belkadi L, Bedouelle H. (2014). Direct and indirect interactions in the recognition between a cross- neutralizing antibody and the four serotypes of dengue virus. *J Mol Recognit, 27*, 205-214.

Livnah O et al. (1998). An antagonist peptide–EPO receptor complex suggests that receptor dimerization is not sufficient for activation. *Nature structural biology, 5*, 993-1004.

Lo Conte L, Chothia C, Janin J. (1999). The atomic structure of protein-protein recognition sites. *Journal of molecular biology, 285*, 2177-2198.

London N, Movshovitz-Attias D, Schueler-Furman O. (2010). The structural basis of peptide-protein binding strategies. *Structure, 18*, 188-199.

Ma B, Kumar S, Tsai CJ, Nussinov R. (1999). Folding funnels and binding mechanisms. *Protein engineering, 12*, 713-720.

Mackay JP, Sunde M, Lowry JA, Crossley M, Matthews JM. (2007). Protein interactions: is seeing believing? *Trends in biochemical sciences, 32*, 530-531.

Makino T, Gojobori T (2007) Evolution of protein-protein interaction network. In: Gene and Protein Evolution, vol 3. Karger Publishers, pp 13-29

Mani R, St Onge RP, Hartman JLt, Giaever G, Roth FP. (2008). Defining genetic interaction. *Proceedings of the National Academy of Sciences of the United States of America, 105*, 3461-3466.

Mattaj IW, Englmeier L (1998) Nucleocytoplasmic transport: the soluble phase. Annual Reviews 4139 El Camino Way, PO Box 10139, Palo Alto, CA 94303-0139, USA,

Mayer BJ (2006) Protein-protein interactions in signaling cascades. In: Transmembrane Signaling Protocols. Springer, pp 79-99

Menche J, Sharma A, Kitsak M, Ghiassian SD, Vidal M, Loscalzo J, Barabasi AL. (2015). Uncovering disease-disease relationships through the incomplete interactome. *Science, 347*.

Mian IS, Bradwell AR, Olson AJ. (1991). Structure, function and properties of antibody binding sites. *Journal of molecular biology, 217*, 133-151.

Miller S, Lesk AM, Janin J, Chothia C. (1987). The accessible surface area and stability of oligomeric proteins. *Nature, 328*, 834-836.

Miller SI, Kukral AM, Mekalanos JJ. (1989). A two-component regulatory system (phoP phoQ) controls Salmonella typhimurium virulence. *Proceedings of the National Academy of Sciences of the United States of America, 86*, 5054-5058.

Mintseris J, Weng ZP. (2005). Structure, function, and evolution of transient and obligate protein-protein interactions. *Proceedings of the National Academy of Sciences of the United States of America, 102*, 10930-10935.

Monod J, Wyman J, Changeux JP. (1965). On the Nature of Allosteric Transitions: A Plausible Model. *Journal of molecular biology, 12*, 88-118.

Mor A, Philips MR. (2006). Compartmentalized Ras/MAPK signaling. *Annual review of immunology, 24*, 771-800.

Ngounou Wetie AG, Sokolowska I, Woods AG, Roy U, Loo JA, Darie CC. (2013). Investigation of stable and transient protein–protein interactions: past, present, and future. *Proteomics, 13*, 538-557.

Nishi H, Hashimoto K, Panchenko AR. (2011). Phosphorylation in protein-protein binding: effect on stability and function. *Structure, 19*, 1807-1815.

Nooren IMA, Thornton JM. (2003). Diversity of protein-protein interactions. *Embo J, 22*, 3486-3492.

O'Connell MR, Gamsjaeger R, Mackay JP. (2009). The structural analysis of protein–protein interactions by NMR spectroscopy. *Proteomics, 9*, 5224-5232.

Oliver S. (2000). Guilt-by-association goes global. *Nature, 403*, 601-603.

Ou HD, May AP, O'Shea CC. (2011). The critical protein interactions and structures that elicit growth deregulation in cancer and viral replication. *Wires Syst Biol Med, 3*, 48-73.

Padlan EA. (1990). On the Nature of Antibody Combining Sites - Unusual Structural Features That May Confer on These Sites an Enhanced Capacity for Binding Ligands. *Proteins, 7*, 112-124.

Palacios EH, Weiss A. (2004). Function of the Src-family kinases, Lck and Fyn, in T-cell development and activation. *Oncogene, 23*, 7990-8000.

Pandit B et al. (2007). Gain-of-function RAF1 mutations cause Noonan and LEOPARD syndromes with hypertrophic cardiomyopathy. *Nat Genet, 39*, 1007-1012.

Pawłowski K, Bierzyński A, Godzik A. (1996). Structural diversity in a family of homologous proteins. *Journal of molecular biology, 258*, 349-366.

Pelka P, Ablack JN, Fonseca GJ, Yousef AF, Mymryk JS. (2008). Intrinsic structural disorder in adenovirus E1A: a viral molecular hub linking multiple diverse processes. *Journal of virology, 82*, 7252-7263.

Peng XQ, Wang JX, Peng W, Wu FX, Pan Y. (2017). Protein-protein interactions: detection, reliability assessment and applications. *Brief Bioinform, 18*, 798-819.

Pereira-Leal JB, Levy ED, Kamp C, Teichmann SA. (2007). Evolution of protein complexes by duplication of homomeric interactions. *Genome biology*, *8*, R51.

Plowman SJ, Hancock JF. (2005). Ras signaling from plasma membrane and endomembrane microdomains. *Biochimica et biophysica acta*, *1746*, 274-283.

Pogoryelov D, Krah A, Langer JD, Yildiz O, Faraldo-Gomez JD, Meier T. (2010). Microscopic rotary mechanism of ion translocation in the F(o) complex of ATP synthases. *Nature chemical biology*, *6*, 891-899.

Prieto C, Risueno A, Fontanillo C, De las Rivas J. (2008). Human gene coexpression landscape: confident network derived from tissue transcriptomic profiles. *PloS one*, *3*, e3911.

Pujana MA et al. (2007). Network modeling links breast cancer susceptibility and centrosome dysfunction. *Nat Genet*, *39*, 1338-1349.

Rayment I, Holden HM, Whittaker M, Yohn CB, Lorenz M, Holmes KC, Milligan RA. (1993). Structure of the actin-myosin complex and its implications for muscle contraction. *Science*, *261*, 58-65.

Richards FM. (1977). Areas, volumes, packing and protein structure. *Annual review of biophysics and bioengineering*, *6*, 151-176.

Romano D, Nguyen LK, Matallanas D, Halasz M, Doherty C, Kholodenko BN, Kolch W. (2014). Protein interaction switches coordinate Raf-1 and MST2/Hippo signalling. *Nature cell biology*, *16*, 673-684.

Rosenberg RR. (1993). A causal role for amyloid in Alzheimer's disease: The end of the beginning 1993 American Academy of Neurology Presidential Address. *Neurology*, *43*, 851-851.

Rosenmund C, Carr DW, Bergeson SE, Nilaver G, Scott JD, Westbrook GL. (1994). Anchoring of Protein-Kinase-a Is Required for Modulation of Ampa/Kainate Receptors on Hippocampal-Neurons. *Nature*, *368*, 853-856.

Safari-Alighiarloo N, Taghizadeh M, Rezaei-Tavirani M, Goliaei B, Peyvandi AA. (2014). Protein-protein interaction networks (PPI) and complex diseases. *Gastroenterology and Hepatology from bed to bench*, *7*, 17.

Schoenrock A et al. (2017). Evolution of protein-protein interaction networks in yeast. *PloS one*, *12*, e0171920.

Seedorf M, Damelin M, Kahana J, Taura T, Silver PA. (1999). Interactions between a nuclear transporter and a subset of nuclear pore complex proteins depend on Ran GTPase. *Mol Cell Biol*, *19*, 1547-1557.

Sheinerman FB, Norel R, Honig B. (2000). Electrostatic aspects of protein-protein interactions. *Curr Opin Struc Biol*, *10*, 153-159.

Slingsby C, Bateman OA, Simpson A. (1993). Motifs involved in protein-protein interactions. *Molecular biology reports*, *17*, 185-195.

Smith-Gill SJ. (1991). Protein—protein interactions: structural motifs and molecular recognition. *Current Opinion in Biotechnology*, 2, 568-575.

Smith FD, Scott JD. (2006). Anchored cAMP signaling: Onward and upward - A short history of compartmentalized cAMP signal transduction. *Eur J Cell Biol*, *85*, 585-592.

Somlyay M, Ledolter K, Kitzler M, Sandford G, Cobb SL, Konrat R. (2020). 19F NMR Spectroscopy Tagging and Paramagnetic Relaxation Enhancement-Based Conformation Analysis of Intrinsically Disordered Protein Complexes. *ChemBioChem*, *21*, 696-701.

Stein RW, Corrigan M, Yaciuk P, Whelan J, Moran E. (1990). Analysis of E1A-mediated growth regulation functions: binding of the 300-kilodalton cellular product correlates with E1A enhancer repression function and DNA synthesis-inducing activity. *Journal of virology*, *64*, 4421-4427.

Sternberg MJ, Gullick WJ. (1989). Neu receptor dimerization. *Nature*, *339*, 587.

Stites WE. (1997). Protein-protein interactions: Interface structure, binding thermodynamics, and mutational analysis. *Chem Rev*, *97*, 1233-1250.

Stubbs MT, Laber B, Bode W, Huber R, Jerala R, Lenarcic B, Turk V. (1990). The refined 2.4 AX-ray crystal structure of recombinant human stefin B in complex with the cysteine proteinase papain: a novel type of proteinase inhibitor interaction. *The EMBO journal*, *9*, 1939-1947.

Sundberg EJ (2009) Structural basis of antibody–antigen interactions. In: Epitope Mapping Protocols. Springer, pp 23-36

Sundberg EJ, Mariuzza RA. (2000). Luxury accommodations: the expanding role of structural plasticity in protein-protein interactions. *Structure, 8*, R137-R142.

Taylor IW et al. (2009). Dynamic modularity in protein interaction networks predicts breast cancer outcome. *Nature biotechnology, 27*, 199-204.

Taylor SS, Buechler JA, Yonemoto W. (1990). cAMP-dependent protein kinase: framework for a diverse family of regulatory enzymes. *Annual review of biochemistry, 59*, 971-1005.

Terakita A, Yamashita T, Nimbari N, Kojima D, Shichida Y. (2002). Functional interaction between bovine rhodopsin and G protein transducin. *Journal of Biological Chemistry, 277*, 40-46.

Tobi D, Bahar I. (2005). Structural changes involved in protein binding correlate with intrinsic motions of proteins in the unbound state. *Proceedings of the National Academy of Sciences of the United States of America, 102*, 18908-18913.

Tollefson AE et al. (1998). Forced degradation of Fas inhibits apoptosis in adenovirus-infected cells. *Nature, 392*, 726-730.

Tollefson AE et al. (2001). Inhibition of TRAIL-induced apoptosis and forced internalization of TRAIL receptor 1 by adenovirus proteins. *Journal of virology, 75*, 8875-8887.

Tonddast-Navaei S, Skolnick J. (2015). Are protein-protein interfaces special regions on a protein's surface? *The Journal of chemical physics, 143*, 12B631_631.

Tong AH et al. (2004). Global mapping of the yeast genetic interaction network. *Science, 303*, 808-813.

Tourette C, Li B, Bell R, O'Hare S, Kaltenbach LS, Mooney SD, Hughes RE. (2014). A large scale Huntingtin protein interaction network implicates Rho GTPase signaling pathways in Huntington disease. *The Journal of biological chemistry, 289*, 6709-6726.

Tripathi DK, Poluri KM. (2020). Molecular insights into kinase mediated signaling pathways of chemokines and their cognate G protein coupled receptors. *Frontiers in bioscience (Landmark edition), 25*, 1361-1385.

Tsai CJ, Kumar S, Ma B, Nussinov R. (1999). Folding funnels, binding funnels, and protein function. *Protein Sci, 8*, 1181-1190.

Tsai CJ, Lin SL, Wolfson HJ, Nussinov R. (1996). A dataset of protein-protein interfaces generated with a sequence-order-independent comparison technique. *Journal of molecular biology, 260*, 604-620.

Tuncbag N, Gursoy A, Guney E, Nussinov R, Keskin O. (2008). Architectures and functional coverage of protein-protein interfaces. *Journal of molecular biology, 381*, 785-802.

Tuncbag N, Kar G, Gursoy A, Keskin O, Nussinov R. (2009). Towards inferring time dimensionality in protein-protein interaction networks by integrating structures: the p53 example. *Mol Biosyst, 5*, 1770-1778.

Tuncbag N, Keskin O, Gursoy A. (2010). HotPoint: hot spot prediction server for protein interfaces. *Nucleic acids research, 38*, W402-406.

Valdar WS, Thornton JM. (2001). Conservation helps to identify biologically relevant crystal contacts. *Journal of molecular biology, 313*, 399-416.

van der Lee R et al. (2014). Classification of intrinsically disordered regions and proteins. *Chem Rev, 114*, 6589-6631.

Varadi M, Vranken W, Guharoy M, Tompa P. (2015). Computational approaches for inferring the functions of intrinsically disordered proteins. *Frontiers in molecular biosciences, 2*, 45.

Veitia RA. (2002). Rosetta Stone proteins:" chance and necessity"? *Genome biology, 3*, interactions1001. 1001.

Venkatesan K et al. (2009). An empirical framework for binary interactome mapping. *Nature methods, 6*, 83-90.

Vidal M, Cusick ME, Barabasi AL. (2011). Interactome networks and human disease. *Cell, 144*, 986-998.

Volkman BF, Lipson D, Wemmer DE, Kern D. (2001). Two-state allosteric behavior in a single-domain signaling protein. *Science, 291*, 2429-2433.

Walls PH, Sternberg MJ. (1992). New algorithm to model protein-protein recognition based on surface complementarity. Applications to antibody-antigen docking. *Journal of molecular biology, 228*, 277-297.

Wang ET et al. (2008). Alternative isoform regulation in human tissue transcriptomes. *Nature, 456*, 470-476.

Wang JX, Peng XQ, Peng W, Wu FX. (2014). Dynamic protein interaction network construction and applications. *Proteomics, 14*, 338-352.

Wang SS, Esplin ED, Li JL, Huang L, Gazdar A, Minna J, Evans GA. (1998). Alterations of the PPP2R1B gene in human lung and colon cancer. *Science, 282*, 284-287.

Watowich SJ, Gross LJ, Josephs R. (1993). Analysis of the Intermolecular Contacts within Sickle Hemoglobin Fibers - Effect of Site-Specific Substitutions, Fiber Pitch, and Double-Strand Disorder. *J Struct Biol, 111*, 161-179.

Westermarck J, Ivaska J, Corthals GL. (2013). Identification of protein interactions involved in cellular signaling. *Molecular & cellular proteomics: MCP, 12*, 1752-1763.

Williamson MP. (2013). Using chemical shift perturbation to characterise ligand binding. *Progress in nuclear magnetic resonance spectroscopy, 73*, 1-16.

Wright PE, Dyson HJ. (1999). Intrinsically unstructured proteins: re-assessing the protein structure-function paradigm. *Journal of molecular biology, 293*, 321-331.

Xu D, Tsai CJ, Nussinov R. (1998). Mechanism and evolution of protein dimerization. *Protein Sci, 7*, 533-544.

Xu ZH, Horwich AL, Sigler PB. (1997). The crystal structure of the asymmetric GroEL-GroES-(ADP)(7) chaperonin complex. *Nature, 388*, 741-750.

Yan C, Wu F, Jernigan RL, Dobbs D, Honavar V. (2008). Characterization of protein-protein interfaces. *Protein J, 27*, 59-70.

Yengo CM, De La Cruz EM, Chrin LR, Gaffney DP, 2nd, Berger CL. (2002). Actin-induced closure of the actin-binding cleft of smooth muscle myosin. *The Journal of biological chemistry, 277*, 24114-24119.

Yiting Y, Chaturvedi I, Meow LK, Kangueane P, Sakharkar MK. (2004). Can ends justify the means? Digging deep for human fusion genes of prokaryotic origin. *Front Biosci, 9*, 2964-2971.

Yiting Y, Lei L, Sakharkar MK, Kangueane P. (2006). Insight into gene fusion from molecular dynamics simulation of fused and un-fused IGPS (Imidazole Glycerol Phosphate Synthetase). *J Bioinformation, 1*, 99.

Young L, Jernigan RL, Covell DG. (1994). A Role for Surface Hydrophobicity in Protein-Protein Recognition. *Protein Sci, 3*, 717-729.

Yusupov MM, Yusupova GZ, Baucom A, Lieberman K, Earnest TN, Cate J, Noller HF. (2001). Crystal structure of the ribosome at 5.5 Å resolution. *Science, 292*, 883-896.

Zhao Z, Roose BW, Zemerov SD, Stringer MA, Dmochowski IJ. (2020). Detecting protein–protein interactions by Xe-129 NMR. *Chemical Communications, 56*, 11122-11125.

Energetic Aspects of Protein–Protein Interactions (PPIs)

3

Abstract

The increasing demand to build holistic models of protein–protein interactions in the cell requires understanding them at the structural level. Structure in turn is elucidated by furnishing biophysical information, that is, thermodynamic and kinetic parameters of the interaction. The thermodynamic parameters that are usually calculated to understand association between proteins are the variations in free energy, entropy, heat capacity, and enthalpy. Similarly, rates of association are also determined under kinetics in order to gain better insight into these interactions. This chapter also highlights the high selectivity of PPIs and outlines basic principles of energetics and some mutagenesis studies that have been performed to get insights into the high affinity protein interactions.

Keywords

Energetics · Thermodynamics · Kinetics · Gibbs free energy · Enthalpy · Entropy · Affinity · Selectivity

3.1 Introduction

Thorough investigation has been performed regarding aspects related to thermodynamics of PPIs. Detailed thermodynamic study of any protein complex usually encompasses the association constant determination for the binding of the protein partners involved in forming the complex. Modern instrumentation has enabled research groups not only to determine association constants with enhanced accuracy, precision, and speed, but also readily calculate on and off rates. It is vital to calculate association constants as they evaluate the concentrations of proteins that are critical for complex formation. Further, they are of great practical utility when the rates and energetic aspects for the protein complex formation are theoretically correlated to the structural features of the complex (Stites 1997).

K. M. Poluri et al., *Protein-Protein Interactions*,
https://doi.org/10.1007/978-981-16-1594-8_3

The free energy difference (ΔG) between the unassociated and associated forms of the protein partners can be utilized to find out the association constant for the given PPI. The free energy change associated with the complex formation comprises only a certain aspect of PPI thermodynamics. Other aspects of protein–protein association include changes in enthalpy (ΔH), heat capacity (ΔC), and entropy (ΔS). ΔS is computed using the combination of ΔG and ΔH values. Heat capacity at constant pressure (ΔC_p) is the first derivative of ΔH with regard to temperature which can be evaluated by determining the binding enthalpies at various temperatures.

Initially, it was not practical for most of the researchers to directly calculate the ΔH for an association between proteins. Values of ΔH were frequently inferred from Van't Hoff analysis by exploiting the change of association constant with temperature. However, it was later observed that due to the unfitting utilization of the simplified version of the Van't Hoff equation (Weber 1993, 1995, 1996), or the inappropriate assumption that there is no considerable conformational variation in the protein's unfolded state over an extended range of temperature (Jelesarov and Bosshard 1996), disagreement between Van't Hoff enthalpies and calorimetric enthalpies began to crop up even more frequently in the experimental reports (Liu and Sturtevant 1995; Naghibi et al. 1995). Later, the advent of the calorimetric instrumentation aided in determining the binding enthalpies directly, thus resulting in a massive boost to the researchers in this field (Ladbury and Chowdhry 1996; Wiseman et al. 1989).

3.2 Overview of PPI Energetics

In order to get better insights into the energetics of PPIs, it is essential to have an overview of the several important thermodynamic parameters, and also how they are related in the context of interactions either between two proteins or between a protein and a peptide molecule.

3.2.1 Thermodynamics of Associated Proteins

Many studies have been performed on the thermodynamics of association for protein complexes (Banerjee et al. 2006; Basdevant et al. 2006; Deng et al. 2020; Jaiswal et al. 2019; Jemimah and Gromiha 2020; Wang et al. 2009). The enthalpy values for these protein–protein systems have been evaluated at a constant temperature using calorimetric approaches. These approaches include techniques like batch calorimetry, isothermal titration calorimetry, etc. Often, the associated free energies of these studied systems have also been evaluated calorimetrically. However, the free energy of formation for stable protein complexes has been difficult to calculate using calorimetric techniques. The free energy changes for such complexes have thus been determined by non-calorimetric techniques like surface plasmon resonance.

Table 3.1 Average values of thermodynamic parameters calculated to distinguish between protein–protein and protein–peptide interactions

Thermodynamic parameter	Values for protein–protein interactions	Values for protein–peptide interactions
ΔG	−7.0 to −17.2 kcal/mol	−5.3 to −11.7 kcal/mol
ΔS	+78.6 to −188.4 cal/(molK)	+95.7 to −109 cal/(molK)
ΔH	+12.6 to −66.7 kcal/mol	+19.9 to −41.9 kcal/mol
ΔC_p	2 to −767 cal/(mol K)	−100 to −1200 cal/(molK)

Some research groups have asserted that PPIs are driven enthalpically (Makhatadze and Privalov 1995), whereas other researchers have debated that they are entropically driven (Weber 1993). It has now been clarified that for larger ΔC_p values, the values of ΔH and ΔS become dependent on temperature. Therefore, it is vital to note a temperature where the association reaction becomes enthalpy or entropy driven. Several studies on PPI energetics have determined that at 25 °C, the primary driving force behind protein interactions is neither enthalpy nor entropy. Some studies have demonstrated that while association enthalpy in a PPI is favorable, it is opposed entropically. Similarly, studies involving other PPIs have proven the reverse phenomenon.

Only in a few studies, both entropy and enthalpy have been shown to be favorable. No obvious correlation has been found for ΔS or ΔH with ΔG. On a similar note, these three parameters have also not been proven to correlate with values of ΔC_p. Albeit meaningless from a thermodynamic perspective, the values of these parameters have been averaged and put into different ranges to have an overall or comparative idea about the two types of interactions—either protein–protein or protein–peptide Table 3.1 (Stites 1997).

The comprehensive findings from several studies based on energetic aspects of PPIs and protein–peptide interactions have been summarized in a tabulated form by Stites (1997). Apart from the values of ΔG, ΔS, ΔC_p, and ΔH, the different experimental conditions (pH, temperature, buffer, ionic strength, etc.) under which the complexes have been studied have also been listed by Stites (1997). Moreover, the data also comprises studies on the thermodynamics of complexes that are in a very close relation with each other i.e., for example a series of single mutations on a particular site of one protein interacting with another protein.

3.2.2 Restriction of Conformation and Entropy

It might be remarkable to note that protein–protein interactions slightly differ from that of the interactions between protein and peptide in terms of their average thermodynamic parameters. There are two very obvious differences between these two classes of interactions. The first is the size difference of one of the interacting partners in the interaction. However, it is not always very evident how much big interacting domain is involved in a PPI. Therefore, in certain cases, a PPI might not involve residues more than the residues that are indulged in a protein–peptide interaction.

For instance, in the interaction of GroEL with an unfolded, destabilized subtilisin BPN′ mutant, where both the polypeptides that are interacting are large, and the interaction is categorized into a protein–peptide interaction (Krueger et al. 2003). Therefore, it implies that the major factor to consider while classifying an interaction is whether both the interactors possess a well-defined 3D structure prior to the interaction or not. An unfolded protein or peptide should lose several degrees of freedom for associating with an interacting partner. Therefore, it is expected for PPIs, change in entropy upon association should be more favorable than for protein–peptide interactions. However, these two classes of interactions do not show major differences in values of ΔS. This is because other entropy-significant factors such as rotation, vibration, reorganization of solvent, and side chain's conformational freedom loss influence the entropy so as to offset the differences in the entropy associated with the conformational restrictions of the main chain (Sturtevant 1977).

Determining the values of ΔS and ΔH under a single set of experimental conditions is extremely arduous for interpretation with respect to hydration and conformational changes of the interacting system (Fisher and Singh 1995). Nevertheless, Lederet et al. demonstrated the thermodynamic signal associated with the major conformational change involved in a protein–peptide interaction (Leder et al. 1995). The authors raised two antibodies, namely 29AB and 13AD against a 29-amino acid peptide LZ, that had the potential to form coiled coils. These antibodies were found to cross react with the LZ(7P14P) homologous peptide. This peptide, which differs by substitution of two proline residues did not demonstrate any helical signal in solution when subjected to CD spectroscopy. However, it was proved by other spectroscopic methods that the peptide acquires a coiled coil conformation upon binding to either of the two antibodies. It was shown that the binding entropy was even more unfavorable for LZ(7P14P) peptide than its LZ counterpart. The authors hypothesized that the unfavorable change in entropy was due to the coiled coil formation from a random coil, reflecting the loss in both rotational and translational degrees of freedom. The authors also made another interesting observation about ΔG which was not seen to vary greatly for the two different peptides. They reasoned that the change in ΔG was limited by a compensation of the change in ΔH that was favorable for the coiled coil formation.

Petrella et al. also studied the protein–peptide system by analyzing the interaction of profilin with various polyproline peptides (Petrella et al. 1996). The authors determined that prior to binding, the peptide was disordered. They also demonstrated that when proline residues were substituted by either alanine or glycine, the interaction enthalpy was not as much affected as the associated entropy. These studies therefore proved that restriction of conformation upon bonding can be discerned in suitable circumstances. However, it should be noted that one cannot confidently state that unfavorable interaction entropies are always associated with only conformational restrictions, when other pertinent information is not present in detail.

3.2.3 Contribution of Various Parameters to Protein–Protein Interactions

Researchers have long wondered about the extent of hydrophobicity contribution in the protein–protein association. When a hydrophobic entity is transferred to a non-polar solution at RT (room temperature) from water, the associated change in entropy is favorable, but that of enthalpy is not. However, it is not possible to understand the contribution of hydrophobicity to PPIs just via analyzing the values of ΔH and ΔS as several other factors contribute to these parameters.

A large negative ΔC_p is distinct thermodynamic feature that marks the presence of buried hydrophobic surface (Spolar et al. 1992; Tayyari 2013). When polar surface is buried, ΔC_p is positive and has a low magnitude (Murphy and Gill 1991). It has been observed that ΔC_p of PPIs and protein folding is dominated by hydrophobic effects (Privalov and Gill 1988; Sturtevant 1977). ΔC_p is determined by performing an interaction experiment at different temperatures. Research groups have carried out such binding experiments by varying temperatures and measured the ΔC_p of association to be negative with large magnitude. Compared to the ΔC_p accompanying protein folding, the ΔC_p of PPIs is smaller as lesser hydrophobic surface is buried in protein–protein association (Murphy and Freire 1992). These findings thus imply that the hydrophobic effect is quite significant with respect to PPIs.

Nevertheless, some prominent exceptions observed between hydrophobic surface and ΔCp values have led research groups to reassess their existing relationship in case of PPIs. When calmodulin is present in its Ca^{2+} saturated form, it binds to plasmin protein. Performing several preliminary experiments, Miloset et al. determined that ΔC_p for the PPI was approximately zero (Milos et al. 1988). In another experiment, where the interaction between the anti-idiotypic antibody E225 and monoclonal antibody D1.3 was studied, ΔC_p was found to have a very low value (Tello et al. 1994). In an interesting case, where cytochrome c-cytochrome c peroxidase interaction was studied, it was initially thought that the binding is greatly influenced by hydrophobicity (Kresheck et al. 1995). However, an exhaustive analysis revealed that the value of ΔC_p was remarkably close to zero. This led authors to conclude that the PPI is driven by other factors apart from hydrophobicity.

This study opens a new plausible fact as other PPIs related to the electron transport chain are strongly influenced by ionic strength (Fredericks and Hamilton 1996). The authors concluded that the driving force behind this PPI was probably van der Waals or electrostatic interactions. They have also examined the thermodynamics of the binding as a function of ionic strength and revealed that electrostatic forces dominate the PPI. However, it should be noted that this is not a property universally applicable to all other PPIs of the electron transport chain. For instance, interactions of four different antibodies with cytochrome c were studied by two distinct research groups (Murphy et al. 1995; Raman et al. 1995). It was observed that the association enthalpy is not independent of temperature in all of these 4 interaction reactions. This might seemingly implicate the significance of hydrophobic interactions in all these four PPIs. However, there are stark differences between these four binding reactions, as two of them have higher magnitudes of ΔC_p and two have lower magnitude values. Therefore, this brings up an important question as to whether the variation in ΔC_p for other interaction reactions provides any kind of specific or quantitative information about the relative contribution of hydrophobic interactions to PPIs and the stability of the resulting complexes. There seems to exist a correlation between the extent of hydrophobic surface area buried and the value of ΔC_p in case of protein folding reactions (Gómez and Freire 1995; Murphy and Freire 1992).

However, in order to find a correlation for a PPI complex, one has to attain knowledge about both the value of ΔC_p and the complex's crystal structure. From the few studies that have been conducted on this, results have not been conclusive and it has not been clearly ascertained if there is an existing correlation between the two parameters for PPIs. For some PPIs such as angiotensin II-Fab (Murphy et al. 1993), HyHEL5-hen egg white lysozyme (Hibbits et al. 1994), and endothiapepsin-pepstatin (Gómez and Freire 1995), good correlations between buried hydrophobic surface area and ΔC_p values have been found. In contrast, the ΔC_p values in ribonuclease *S* system showed substantial variation and did not have any apparent correlation to the quantity of buried non-polar surface (Varadarajan et al. 1992). Mutation in the interfacial region of the human growth hormone-hGH receptor yielded more negative values (by 0.2 kcal/(mol K)) of ΔC_p. This was a remarkable observation given that large interfacial hydrophobic residues were deleted in some of these mutations. This implied that the real changes in the buried hydrophobic surface that were measured from the crystal complexes were different from those calculated from changes in the values of ΔC_p (Pearce et al. 1996).

Disparities were similarly observed for antibody mutants that were raised against a receptor known as epidermal growth factor receptor. However, mutant crystal structures were not available in this case to affirm the buried surface area variations (Kelley et al. 1992). Bhat et al. observed that the ΔC_p for antibody D1.3-hen egg white lysozyme interaction was in very well concordant with the roughly calculated buried surface area value (Bhat et al. 1994; Ito et al. 1993). However, the authors concluded the agreement to be accidental and argued that the remaining part of their data did not indicate hydrophobic effect to be the propulsive force for the PPI. Indeed, when they carried out a more detailed calculation subsequently, they found that the calculated value of ΔC_p was way lower than the experimental value (Schwarz et al. 1995; Tello et al. 1993).

It is therefore a pertinent question as to why the value of ΔC_p often does not tune well with the quantity of buried nonpolar surface area in a PPI. Another case study might answer this question and interpret the physical meaning of ΔC_p with respect to reactions pertaining interactions. Guinto and Di Cera demonstrated ΔC_p of -1.1 kcal/(mol K) for the association between the single sodium ion to thrombin (Guinto and DiCera 1996). Such a large, negative value of ΔC_p is larger than most of the ΔC_p values calculated in PPI studies. The authors have hypothesized that sodium binding does not lead to major variations at conformational level in the protein's structure that might in turn assist in the burial of a great amount of hydrophobic surface. Instead, they have speculated that the buried waters in the structure have a less heat capacity compared to the bulk solvent, and their consequent burial results in the large change in heat capacity during interaction. Cavities found in PPI interfaces are often water-filled and might be a plausible explanation for the discrepancies in the ΔC_p-buried hydrophobic surface correlation.

In a nutshell, experiments based on calorimetry evidence that hydrophobic interactions do not exert much influence on PPIs as they do on protein folding. However, other case studies have also demonstrated that hydrophobicity is not negligible and is one of the major energetic patrons to PPIs and assistant in the creation of stable protein complexes. However, these concluding remarks are weakened by the complications in the interpretation of ΔC_p values for these association reactions.

Recently, researchers have studied the hydrophobic, ionic, hydrogen bonding interactions in the PPI formation and multimerization of the proteins using the molecular dynamics approaches. These computational approaches have also provided deep insights into the thermodynamic parameters that regulate the formation/attenuation of these protein–protein interactions.

Chong and Ham deciphered the contribution of interfacial water during the protein complex formation and their implications on thermodynamics of binding of proteins (Chong and Ham 2017). Authors conducted MD simulations and also performed statistical analyses in order to study the thermodynamic characteristics of the hydration water molecules that surround the barnase-barstar complex and unraveled the dynamic features of interfacial water molecules which appear upon the complex formation and compared them with that of dynamics related to hydration water that surrounds the free proteins (Fig. 3.1). They focused on rearrangement of hydrogen bonding patterns between the water and protein surfaces. It has been observed that water molecules forming single hydrogen bonds with protein showed slower relaxation dynamics in comparison with bulk water molecules (those water molecules forming 2–3 hydrogen bonds with protein). The interfacial water molecules that are bridged between two proteins through concurrent hydrogen bonding showed 1000 times slower relaxation dynamics than bulk water. This implies slower hydrogen bond rearrangements of special bridging water molecules. The major reason behind such slower bridging water is the number of hydrogen bond involved and also due to the additional stability conferred through electrostatic field between the protein binding surfaces which are of electrostatic complementarity in nature. This demonstrates that for the meaningful estimation of binding affinity between the proteins, it is essential to take into account both the interacting protein partners and interfacial water molecules uniformly (Chong and Ham 2017).

Tse et al. explored free energy landscape and thermodynamics for the multimerization activity of HIV-1 integrase (IN) that is induced by allosteric HIV-1 integrase inhibitors (ALLNIs) (Tse et al. 2020). Authors employed umbrella sampling and unbiased molecular dynamic simulations to unravel the detailed thermodynamics and kinetic view for the multimerization of IN. It has been observed that association of two dimers of IN into an aberrant tetramer is a spontaneous process. This process of association is highly short ranged which implies that two

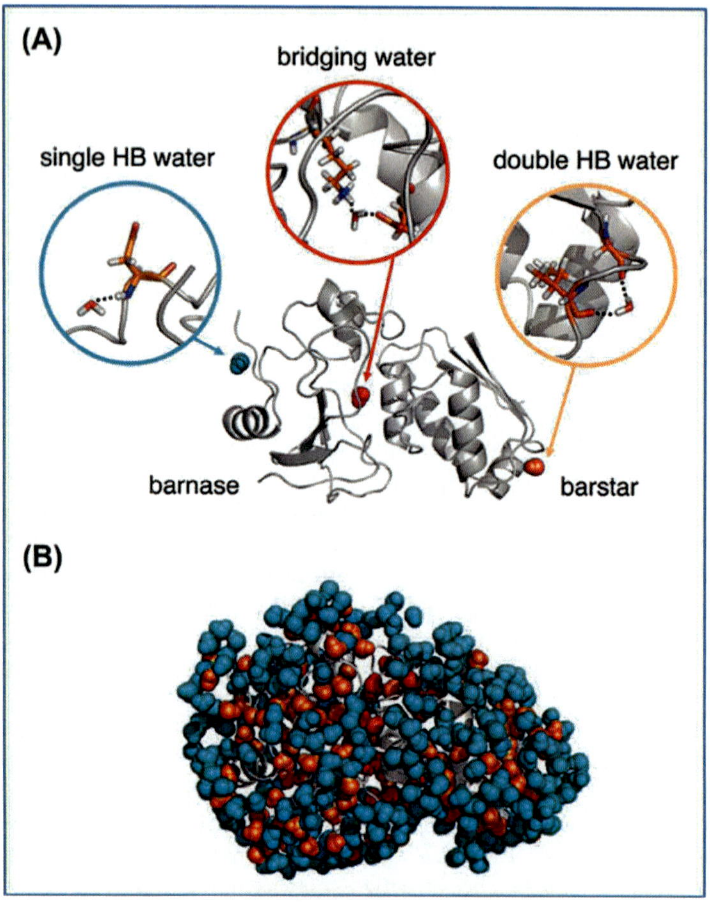

Fig. 3.1 Figure depicting hydration pattern of barnase-barstar complex. (**a**) shows the three different classes of hydration water, namely, (1) Single HB water: bonding of a water molecule with a residue of barstar/barnase by single hydrogen bond; (2) Bridging water: the water molecule is bound to both barnase and barstar creating a "bridge" between them; and (3) Double HB water: bonding of a water molecule with two residues of barstar/barnase. (**b**) illustrates the hydrated barstar-barnase complex. The single HB water is shown in cyan, bridging water is depicted in red while double HB water is depicted by orange (Adapted from Chong and Ham 2017)

dimeric proteins associate fast with each other in tens of nanoseconds when they are less than 4.3 Å apart. When this distance increases, the kinetics of association of proteins become diffusion controlled and take longer time to associate. Authors also found the significant effect of presence of ALLNIs on the free energy profile of association of proteins and also mark that several mutations at (C-terminal domain) CTD make the protein resistant to the inhibitor-induced multimerization process.

It has been hypothesized that when proteins are present at small distance apart, their association process involves two phases that involve differential thermodynamic profiles. Firstly, there is a small increment in the solvent's entropy by the release of inter-protein water molecules, which is compensated by the drop in entropy of side chains when the two proteins come close to each other for association. When two proteins are present nearby, then the dry binding surfaces of proteins adjust themselves for the optimization of their interaction energy at the cost of additional loss in the entropy associated with their configurations. Furthermore, interestingly, the binding interfaces of IN involve clustering of hydrophobic residues. The overall binding process is augmented by the enthalpy and does not involve entropy. Exploration of such thermodynamic parameters is essential to gain understanding of basic biophysical processes associated with PPIs. Such kinetic and thermodynamics studies also provide a wealth of information for designing novel inhibitors for the aberrant multimerization of proteins (Tse et al. 2020).

3.3 Estimation of Protein–Protein Interaction Affinity and Stability

PPIs are highly variable with respect to not only structure, but also energetics. The magnitudes of affinities of functionally relevant PPIs vary by more than ten orders. Therefore, the resulting dissociation equilibrium constants (Kd) vary between 10^{-5} M to 10^{-14} M and the Gibb's free energy ($\Delta G = \mathrm{RTlnKd} = \mathrm{RTlnKa}$) ranges from -7 to -20 kcal/mol (Zhou 2003). It is essential to know the chemical aspects of PPIs in order to predict and analyze them better. These aspects not only include parameters like shape complementarity, but also the contribution of physicochemical and energetic factors to the stability of PPIs (Keskin et al. 2008).

Most of the structure-based methodologies that evaluate PPI stability require information about the complex's 3D structure. In contrast, the assessment of PPI stability from a systems-oriented perspective, such as via the use of in-silico mutagenesis, is quite promising. This is because such a method aids in a quick screening of effects of mutation on the stability measures and dynamic features of protein complexes, when it is applied with sufficient accuracy. Mutagenesis screenings are pathophysiologically relevant as minute alterations in the binding affinities of interacting protein partners can have drastic, harmful outcomes. For example, when fibroblast growth factor receptor 2 (FGFR2) is mutated such that it has enhanced affinity to bind to FGF2 and that resulted in the manifestation of Apert syndrome (Anderson et al. 1998). In this diseased condition, the skull is malformed and the affected individual has mental deficiency and syndactyly. Different research groups have utilized various approaches to predict the affinity of PPI complexes not only in their natural state, but also in the presence of point mutations that affect the thermodynamics of association (Krueger et al. 2003; Li et al. 2014; Weinkam et al. 2013).

3.3.1 Solvent Accessible Surface Area Approaches

Various structure-based empirical approaches have cropped up to study PPI energetics. One of the most commonly used approaches that have been employed for several years was developed by Murphy and coworkers (Baker and Murphy 1998; Luque et al. 1998). This approach assumes that in a PPI both entropy and enthalpy change, and thus ΔG can be related in a linear fashion to the alterations in the solvent accessible surface area (ΔASA) when polar and non-polar atoms belonging to proteins interact. The linear dependence can in turn be established via empirical relations. In this method, one calculates the enthalpic term first. This is done by dividing the non-polar and polar components of the change in heat capacity upon interaction. This is followed by the calculation of the entropic term.

One needs to consider all the contributions that add up to the entropy in a ΔASA-based manner. These contributing factors include: (a) term accounting for conformational changes in the backbone and side chains, (b) term accounting for restructuring of solvent around polar groups, and (c) "mixing" term accounting for variations in vibrational, translational, and rotational degrees of freedom upon interaction (Murphy 1999). This method has proven to be effective in studies where the stability of enzyme–inhibitor or antibody–antigen complexes were studied (Baker and Murphy 1998; Lavigne et al. 2000; Luque et al. 1998; Murphy 1999). However, this method did not perform well for some wild-type and mutated PPI interfaces and showed little agreement with experimentally obtained data (Dell'Orco et al. 2005, 2006; Henriques et al. 2000).

A second ASA-based methodology based on a linear free energy relationship was formulated by Horton and Lewis (Horton and Lewis 1992). This approach relies on the various additive terms that lead to ΔG. These terms include (a) isomerization term (b) interaction term (calculated from solvation parameters), and (c) term corresponding to the loss in rotational and translational degrees of freedom upon interaction. This methodology has proven to be successful for quite a few PPI studies; however, the very simplified assumptions have often proved to be unrealistic (Brooijmans et al. 2002).

In a third approach, developed by Audie and Scarlata, a regression equation is utilized to estimate binding affinities of PPIs (Audie and Scarlata 2007). The binding affinity is measured in terms of contributions of (a) hydrogen bonds, (b) water-mediated interactions, (c) salt bridges between receptor-ligand, (d) desolvation of charged and hydrophobic moieties, and (e) immobilization of interfacial torsion. This analysis helped to derive the immediate physical meaning of six parameters and the regression equation proved to be highly capable of predicting ΔG (correlation coefficient $R = 0.98$). The authors also employed 35 protein–protein complexes as a test set for the prediction of affinity of binding and revealed an excellent correlation ($R = 0.89$) between the predicted and experimental ΔG values for all the protein complexes, barring four of them.

3.3.2 Energy Function-based Models

Principles of energy function have also been exploited to probe the stability of PPI complexes. An energy function model known as FOLD-X was developed by Guerois et al. to study both protein folding and association (Guerois et al. 2002). This energy function included terms that were significant contributors to PPI stability such as (a) hydrogen bond term accounting for ΔG of formation of intramolecular in comparison with intermolecular hydrogen bonds, (b) van der Waals interaction term, summed up from the all atom's van der Waals contributions, (c) solvation term, accounting for the differential solvation energy for both polar and non-polar groups while transitioning to a folded state, (d) electrostatic effect term, accounting for the contributions of charged groups, (e) water bridge term that accounts for an additional stabilizing ΔG, catered by the formation of more than one hydrogen bond between a water molecule and the protein, and (f) entropic effects for the side chains and backbone.

To calculate the backbone entropy, one needs to statistically analyze the protein structure database to derive the secondary structure amino acid preferences. On the other hand, to calculate side chain entropy, one needs to estimate values after accounting for the side chain's mobility loss upon burying (Guerois et al. 2002). The Fanelli group, on the other hand, used a completely distinct methodology for the effective assessment of the relative and absolute values of ΔG in a PPI (Dell'Orco et al. 2005). Their approach utilized a docking algorithm that was originally developed for the prediction of the 3D structures of PPI complexes and initially starts from individual protein structures. The authors observed that when the roto-translational space of proteins with regard to one another was exhaustively sampled, ZDOCK 2.3's scoring function (Chen et al. 2003) had the potential to empirically elucidate the free energy function for protein–protein interactions (Dell'Orco et al. 2007, 2005, 2006).

The conditions that are required to apply the docking-based approach to study PPI energetics are: (a) absence of major conformational changes upon interactions and (b) structural model of the complex formed between wild-type versions of the interactors (Dell'Orco et al. 2005). ZDOCK looks up for complementarity feature between the proteins, providing a value/score for every docked protein complex. The score depends not only on complementarity, but also on other factors such as atom desolvation and electrostatics. This pairwise shape complementarity function does not rely on protein surface curvature and is thus unique in comparison with other docking algorithms. While computing the total number of target-probe pairs of atoms within a threshold distance, it eliminates the potential bad contacts arising from neglecting side-chain rearrangements during interaction (Chen et al. 2003).

In a PPI energetics study, Dell'Orco et al. chose a set of structurally different PPI complexes from the PDB (Protein Data Bank) that were all known to interact with minimal conformational changes (Dell'Orco et al. 2005).These complexes, which formed the training set, varied broadly in their physiologically relevant affinity values (10–17 kcal/mol). Starting from separate structures of the interacting proteins, the authors ran docking simulations on the entire training set to build the native 3D protein complexes. To their surprise, the authors observed that the docking scores from these runs furnished an index (ZD-s) that had an excellent linear correlation ($R = 0.94$; probability for R (p) = 0.00015) with the ΔG° of the respective PPI complexes of the training set. This correlation was thus utilized successfully for the prediction of energetics of calbindin D9k reconstitution from complementary fragments of wild-type and mutated forms (Dell'Orco et al. 2005). The authors concluded that under conditions of minimal conformational changes in the interacting partners, this method was widely applicable. Till date, this approach has been successfully employed for predicting the effects of point mutations on ΔG in several macromolecular association reactions such as homo-dimerization of transmembrane proteins (Dell'Orco et al. 2007, 2008).

3.3.3 Mutational Approaches

An overwhelming number of studies that have exploited mutagenesis to probe PPIs have been performed. It is obvious that mutation is not the only experimental way for probing PPI energetics. For instance, nicotinic acetylcholine receptor-κ-bungarotoxin interface (McLane et al. 1991), HIV tat dimer interface (Frankel et al. 1988), and HIV-1 protease dimer interface (Schramm et al. 1993) have been analyzed using peptides that duplicated a part of their interface. However, mutagenesis is more reliable, efficient, and applicable in comparison with other methods.

During the initial few efforts to study PPI via mutagenesis, the interface was often disrupted either by removal or introduction of charged groups. When isolated charges (Jones et al. 1985) or ion pairs (Ward et al. 1987) were introduced into the interface of tryrosyl-tRNA synthetase, the former enhanced dissociation tendencies, while the latter did not. When a lysine residue was mutated to aspartate in ribulose-1,5-bisphosphate carboxylase, its heterodimeric association was precluded (Lee et al. 1987). Similarly, an elimination of naturally occurring ion pair from aspartate transcarbamoylase interfacial region resulted in its lower stability (Stebbins and Kantrowitz 1989). However sometimes, unexpected results can be yielded by disturbing interfaces. For example, when electrostatically important residues lysine and glutamate were mutated to neutral ones in lysozyme, its interaction with an antibody was strengthened unexpectedly (Roberts et al. 1987).

3.3.3.1 Interaction Disruption

Mutational studies are widespread with respect to the extent of incorporated mutations and the analysis of their outcomes. One approach with the central theme of interface disruption has aimed to create stable proteins but with lesser tendency for association. Disrupting an interface can be done relatively easily; however, looking for the exact group of mutations that can lead to the formation of a stably folded monomer is more difficult. It is easier to remove an interaction interface in case it is segregated from the majority of the rest of the protein. In an alternate approach, the interface reactions can be sustained via the addition of another copy of the interacting domain to the monomer. This alternative approach was utilized by Mossing and Sauer where a partial copy of the β ribbon interface contained in λ cro was linked to the end of an integral copy via a β turn (Mossing and Sauer 1990). This enabled the added copy for an interaction with the remaining part of the protein by looping back. The interaction in turn assisted in the formation of a stable monomer.

The first strategy, whereby one aims to remove much of the binding domain, was utilized to generate a stable monomer of triosephosphate isomerase. A 15 amino acid residues long that comprising of majority of the interfacial amino acid residues was replaced with a loop containing eight residues (Borchert et al. 1993). However, since removing a domain involved in binding is not always preferred or desirable, researchers proposed another approach to manipulate side chains to destabilize the association. This is done by either removal of stabilizing side chains or introduction of destabilizing side chains. For instance, the conversion of tetrameric form of fructose-1,6-bis-phosphate aldolase into a dimeric form by substituting an aspartate residue in the interface with other residues (Beernink and Tolan 1994).

Triosephosphate isomerase was also utilized as a model system by Mainfroid et al. who replaced the interface arginine and methionine with glutamine (Mainfroid et al. 1996). Each mutation, respectively, resulted in the protein's monomeric and dimeric versions. However, the double mutant yielded only the monomer. Moreover, the mutations resulted in proteins with marginal stability of folding. Subsequently, this research group also demonstrated that the folding stability of the monomer can be enhanced by still other mutations (Mainfroid et al. 1996). A stable monomer can be generated from an oligomer if the single domain is segregated from the multi-domain protein and stabilized. Leistler and Perham excised the INT domain of glutathione reductase and replaced seven hydrophobic residues in the domain. The interface amino acid residues were indulged in interaction with rest of the other two domains of the protein. Substitutions with hydrophilic residues generated a stable and soluble protein (Leistler and Perham 1994).

Research groups have explored the contribution of interfacial residues of associating proteins in conferring the specificity and stability to the protein complex. It is very well established that interfacial residues interact with side chain as well as main chain atoms of the interacting protein partner. However, the relative extent of involvement of interfacial residues in inter-protein and intra-protein interactions is not yet much clarified. This bifurcated role of interface residues in both inter- and intra- protein interactions has been probed by Jayashree et al. (2019) (Fig. 3.2). Authors analyzed a set of 45 transient protein–protein complex structures and found the involvement of around 75% of the interfacial amino acids in bifurcated interactions. They were the first to notice that interfacial residues apart from inter-protein interactions are also extensively indulged in intra-protein interactions. Hence, it has been proposed that the environment for such interfacial residues is pre-formed in the structure of the protein and that will not be disturbed upon the formation of complex with its cognate interacting protein partner during the transient interactions.

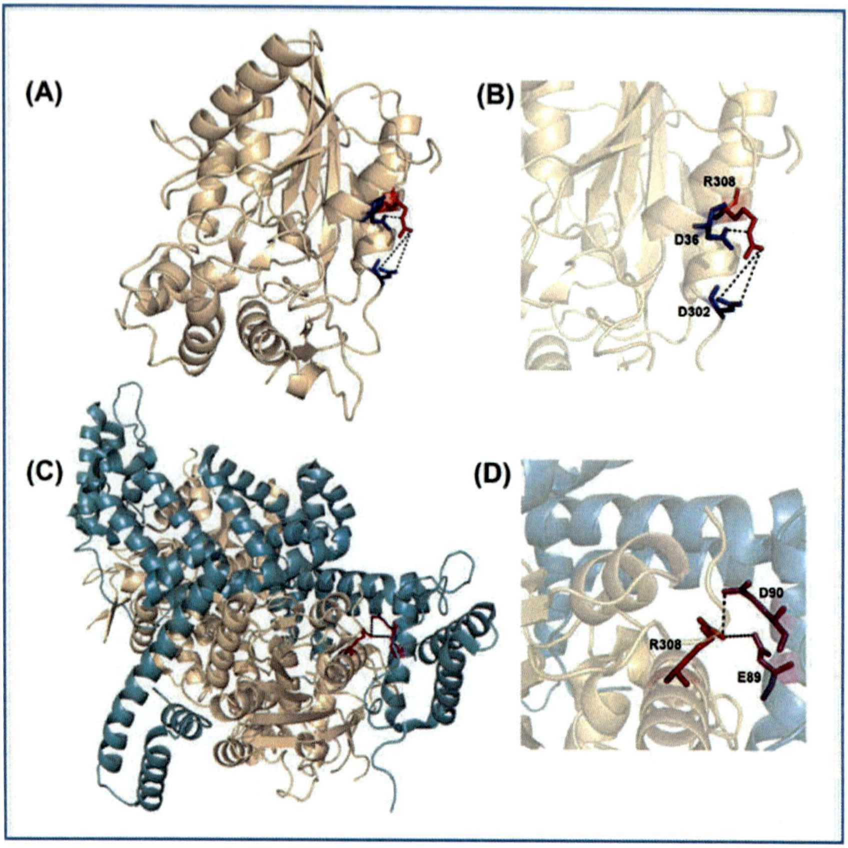

Fig. 3.2 Figure depicting the role of interface residues in intra-protein and inter-protein interactions for unbound lipase and lipase-foldase complex. (**a**) and (**b**) depict Arg308 (red) residue of lipase involved in intra-protein interactions with Asp36 and Asp302 (blue). While (**c**) and (**d**) depict the same Arginine residue (Arg308) (red) involved in inter-protein interactions with Asp90 and Glu89 (Purple) residues of foldase (Adapted and redrawn from Jayshree et al. 2019)

Propensities of differential residues for their simultaneous involvement in both intra- and inter-protein interactions have also been calculated. It is observed that except lysine and glutamine, rest of the 18 amino acids have propensity greater than 1 for their involvement in bifurcated interactions. Hydrophobic residues such as Leu, Trp, Met, and Phe, and also the polar amino acids including Glu, Asp, Arg and His, which tend to be present at protein–protein interface show high propensity for bifurcated interactions. It has also been noticed that the residues at interface contribute substantially to the energy of stabilization to the protein complex through both intra- and inter-protein interaction energies (Jayashree et al. 2019). This implicates that one should consider the involvement of an interface residues in both intra- and inter-protein interactions while designing mutants or inhibitors to tinker the specificity or stability of the protein complex and in rational/*in silico*/de novo designing of protein complexes.

3.3.3.2 Side-Chain Contributions to Interaction Stability

A second important strategy with respect to mutagenesis in PPI studies is to mutate the interface systematically. This is done to find out the relative contribution of specific residues and the associated interactions that are accountable for the stability of the formed complex. One of the major strategies has the central theme of replacing amino acid residues one after the other and then observing their outcomes. Alanine is mostly the residue that is chosen to neutrally replace the residues. This is because alanine is likely to remove majority or all of the side-chain interactions without interfering with the main chain conformations. This technique, also known as "alanine scanning" has been utilized extensively to map protein–protein interactions and to dissect the energetically important residues behind the interaction (Wells 1991).

Alanine scanning has been employed in the classical interaction studies involving human growth hormone (hGH) and 21 different monoclonal antibodies (Jin et al. 1992). The sites where mutations to alanine caused considerable change in the binding affinity were defined as the functional epitopes. The epitopes that endured a twofold effect on binding affinity were considered as functional epitopes. Despite such a lenient threshold, only about 8 residues, on an average, comprised the functional epitopes for each of the 21 interfaces. This number is lower than the number apparently to be in contact across the interfacial region. It was found that only three amino acid residues were extremely critical to the energetics of the interaction, as their mutation to alanine caused a drastic 20-fold reduction in the binding affinity. This classical study validated the fact that mutagenesis can reveal detailed data on PPI energetics.

hGH–receptor interactions have also been the great area of research at Genentech (Wells and de Vos 1996). This hormone structure encompasses two binding sites for its receptor and their study focused exclusively on the first site. When residues were mutated to alanine in hGH (Cunningham and Wells 1989, 1993) and hGH receptor (Bass et al. 1991; Clackson and Wells 1995), the relative importance of the residues with respect to binding affinity was identified. When the first binding site in the complex's crystal structure was studied, it was observed that there are 30 contact residues in the receptor and 31 contact residues in hGH. However, only few of them were significant from an energetic viewpoint. Three residues of hGH lowered the binding affinity by 1.0–1.5 kcal/mol, while five other residues lowered it by 1.6–2.4 kcal/mol. Together, these 8 residues contributed for over 85% of the binding energy. The situation was similar for residues in the receptor counterpart. Two residues reduced binding affinity by more than 4.5 kcal/mol, 1 residue by 3.5 kcal/mol while 7 others reduced by 1.0–2.0 kcal/mol. These hotspot residues in both the hormone and receptor were found to be highly hydrophobic. Even if they were surrounded by hydrophilic residues, they were not very important energetically. Therefore, electrostatic forces were concluded to perform only a nominal role in these interactions.

Thermodynamics of the hormone-receptor association was studied in even more detail with the aid of 3 alanine-substitution mutants of hGH (Pearce et al. 1996). The three residues were phenylalanine (F25), tyrosine (Y42), and glutamine (Q46). However, on mutation, these three residues did not make great differences in the binding affinity. Although ΔG of the interaction varied slightly for the F25A mutant, the entropy and enthalpy of the association demonstrated huge changes that were compensatory in nature. In spite of this, the Y42 and Q46 mutants caused considerable changes in ΔC_p, similar to F25. Although the data obtained was detailed and sufficient, the researchers failed to understand the effects completely.

On a similar note, Castro and Anderson have performed an analysis of the thermodynamics of the interaction of bovine pancreatic trypsin inhibitor (BPTI) with β-trypsin and α-chymotrypsin (Castro and Anderson 1996). They replaced 15 residues in BPTI's reactive surface with alanine. In line with previous studies, they found that majority of the residues had no significant contribution toward the binding energy. For BPTI/trypsin complex, it was found that only 3 amino acid residues, namely, Glycine12, Isoleucine18, and Glycine 36, reduced the binding affinity by more than 4 kcal/mol. One lysine residue (K15) reduced the binding affinity by 10 kcal/mol since its side chain was indulged in interaction with the primary site of the protease specifically. The interaction of BPTI with chymotrypsin was significantly less favorable, but showed similar results in binding affinity changes for its K15A mutant. Therefore, this and other studies portrayed a general feature that loss of binding energy was greatly associated with the increased rates of dissociation (Hawkins et al. 1993; Kelley and O'Connell 1993; Wells and de Vos 1996).

Nevertheless, exceptions to this feature have been discovered. For instance, studies on the interaction of turkey lysozyme with antibody D1.3 have revealed that there is a considerable conformational rearrangement of the antibody during interaction, as compared to its binding to chicken lysozyme. Furthermore, the rate of association with turkey lysozyme was slower as compared to that for chicken lysozyme (Braden et al. 1996). Since complexes having "subtractive" mutations have been demonstrated to rarely show conformational rearrangement and/or influence rates of association, it has been proposed that drop in association rates implies that the interactions of the mutants require conformational rearrangements and crossing of considerable energy barriers. However, one must remember that association rates may be decreased by other factors as well. For example, in the trypsin–BPTI K15A mutant interaction, the drop in association rate was in all probability due to a loss in electrostatic interactions in the primary site rather than a change in the conformation (Castro and Anderson 1996).

Another strategy, that is orthogonal to alanine scanning mutagenesis, may also be employed. In this approach, a small number of interfacial sites are selected and substituted by various residues. This approach has also generated remarkably interesting data. For example, 1 partially buried and 3 contact residues from the binding interface of hen egg white lysozyme have been chosen and mutated to a large extent (Kam-Morgan et al. 1993). The effects of the mutations have then been studied with respect to binding affinity. One of the mutations involved a partially buried residue, and the ΔG of association was not affected much. The remaining two mutations demonstrated large variations, but in distinct ways. For the 9 substitutions that were done for Asp 101, a good correlation was observed between drop in binding affinity and volume of side chain. When this site was mutated, the affinity was practically the same as the native lysozyme. On the other hand, when it was mutated to alanine, the binding affinity decreased by 1.2 kcal/mol; this would have seemingly marked this residue as partially imperative, in case alanine scanning had been performed. When ΔΔG was plotted against side chain volume, the 9 substitutions generated a nearly straight line with only wild-type aspartate as the significant outlier. It was not clear whether asparagine and glutamate would form stable complexes. Considerable destabilization was brought in by arginine which was even more than what its volume suggested. The effects of serine were also in line with arginine. In contrast, all the 7 substitutions at Arg21 generated a uniform effect, resulting in the drop of affinity by 2.2 kcal/mol. The authors proposed that the drop was equivalent to the energetic cost of converting the two H-bonds of the guanidinium group to two (OH) hydroxyl groups of tyrosine in the antibody.

3.3.3.3 Computational Approaches to Probe Mutational Effects

Computational predictors have also been developed to probe the point mutations effects on the PPIs affinity in solution and in membranes. For example, FOLDEF algorithm has been utilized to predict the mutation-induced changes in free energies of binding in 82 mutants (Guerois et al. 2002). The structural models of the mutants were generated after virtual mutation of the X-ray structures of the native complex. A test set was created which is comprised of: (a) 20 mutations in p53's tetrameric domain, (b) 21 mutations in IL-4-receptor complex, (c) 37 mutations (either single or double or triple or quadruple) in BLIP-TEM-1 β-lactamase complex, and (d) 10 single to quadruple mutations in an SH3-peptide complex. When all the 82 mutants were considered, the coefficient of correlation between the free energy changes of binding determined experimentally and those predicted by FOLDEF was 0.64. However, when 4 outliers were eliminated, the coefficient increased to 0.80. The structures of the mutants being essentially same as the wild-type complexes, except for the side chains of the mutated residues, imposed a stringent criterion for the good performance of the algorithm.

According to this criterion, the mutations should not induce a large, significant structural change in the wild type. Nevertheless, this condition is more lenient than the criterion of the docking-based method since the mutant complexes are predicted by the docking algorithm (Dell'Orco et al. 2007, 2005). The mutant structures however differ slightly from the wild type accounting for differential modes of binding. The performances of both the methods in the prediction of ΔΔG on a large scale of PPIs were compared quantitatively. It was seen that the dataset from the docking-based method extended considerably on a wider range of affinities than for FOLDEF algorithm (1–12 kcal/mol versus 1–3 kcal/mol). Further, the docking algorithm included many cases of several (maximum 4), simultaneous point mutations at the interaction interfaces of soluble proteins. Additionally, the docking approach was used more extensively in a general context of physicochemical environments including challenging PPIs of membrane proteins (Dell'Orco et al. 2007, 2008).

Another effective method to elucidate the mutational effects on PPIs of water soluble proteins has been floated by Kortemme and Baker (2002). It is based on the principles of energy function and has been successfully employed for the prediction of results of alanine scanning experiments. It has been utilized to study the effects of 233 mutations at 19 protein–protein interfaces. When the performance of this method was compared with the docking-based approach for a set of mutants in the Bn-Bs PPI, the workers obtained similar trends.

A very notable case was the structure-based energy prediction for the interaction between BPTI and b-trypsin (b-Tryp) (Dell'Orco et al. 2006). In this study, the important K15 residue of BPTI was subjected to replacement by 8 distinct residues that remarkably changed the physicochemical properties of the interfacial region. The X-ray structures of each mutant as well as the effects of each mutation on the binding thermodynamics were available. Comparative study using ΔASA-based approach and the docking-based methodology was carried out to determine their ability to predict the effects of single amino acid replacements in this high affinity PPI. Although the prediction of affinity changes by the ΔASA-based methodology demonstrated a fairly accurate trend for some of the substitutions, the method failed for the total set of mutants. Furthermore, it predicted the overall change in affinity to be about 1 pK unit, which was way off from the experimentally determined value of 4 pK units. On the other hand, the docking-based approach furnished a rational estimation of binding affinity changes for each mutant, irrespective of the different physicochemical natures of the mutant interfaces.

It is challenging to predict the binding thermodynamics of PPIs involving membrane proteins because of the very different physicochemical atmosphere. The docking-based approach has been proved to be quite effective among the computational techniques that have been floated to analyze PPI energetics. Using structures from solid-state NMR, 32 in-silico alanine replacements were done on a transmembrane helical region of the glycophorin A (GpA) protein. GpA protein natively forms homodimers in the membranes of erythrocytes. The consequences of the mutations pertaining to the GpA system were analyzed computationally by other research groups too. Performing a regression analysis among the empirical parameters and their propensities of dimerization of the mutants, it has been inferred that the role of hydrophobicity is minimal in the interactions of transmembrane helices (Finger et al. 2006; Hénin et al. 2005). Docking-based approach has also been utilized for other membrane protein interactions. This study measured the ΔG of binding for both wild type and mutants of OMPLA, a bacterial dimeric phospholipase membrane protein (Dell'Orco et al. 2008).

3.4 Fast Evaluation of Protein–Protein Binding Kinetic Parameters

It is essential to estimate kinetic parameters of PPIs either experimentally or computationally in order to mathematically model the biochemical pathways. Parameters such as catalytic turnover, enzyme specificity, and ligand-binding affinity are hidden in the 3D protein structure. Kinetic parameters influence the free energy profile of the reactions that are enzyme-catalyzed. Both rate-limiting step of the reaction and their free energy profiles vary with factors such as environmental conditions (pH, temperature) and nature of substrate (Stein et al. 2007). If amino acid residues are substituted, the kinetic effects can be drastically altered so as to disturb the complete network to which the given interaction belongs. Keeping in mind the computational point of view, dynamic simulations (such as Brownian dynamics simulations) have been greatly utilized for analyzing protein association reactions between enzyme–inhibitor complex.

These simulations have also been proven effective to discern the outcomes of point mutations and variations in viscosity, pH, and ionic strength on the diffusional association rate constants (Stein et al. 2007). In an interesting study, Wade with his colleagues employed similarity indices for the comparison of molecular interaction fields (MIFs) of structurally related proteins (De Rienzo et al. 2000). If one determines the kinetic parameters of some of the related proteins, then MIFs of such proteins can be subjected to comparison with that of protein whose kinetic parameter is unavailable. Then, on the basis of extent of similarity or dissimilarity, one can assign a value to the unavailable kinetic parameter. Another approach based on this principle which is known as quantitative Protein Interaction Property Similarity Analysis (qPIPSA) approach, takes into assumption that MIF is a very significant factor for determining the kinetic parameters (Gabdoulline et al. 2007). The webPIPSA proves to be promising technique in the prediction of mutations in PPI kinetics (Richter et al. 2008).

Alsallaq and Zhou introduced a completely different approach for kinetic studies of PPI (Alsallaq and Zhou 2007a). Their theory for association rates was built on transition state concept for the interaction. In a nutshell, when the two interacting partners are in a bound state, it is characterized by a deep well in their interaction energy profile. A sharp jump in interaction energy profile points toward the state of transition and the start point of enhanced freedom in relative rotation and translation between the interacting protein partners lies at the outer boundary of this bound state (Alsallaq and Zhou 2007b). If the involved conformational changes are not significant, then the association rate approaches the diffusion-limited rate association constant (k_a). On the other hand, when electrostatic forces are involved in the interactions between associating proteins, a dominant long-range force is generated that can considerably enhance the association rates (Schreiber 2002; Schreiber and Fersht 1996). According to this approach, the enhancement in rate can be predicted using the transition state formula with high accuracy which is as follows:

$$k_a = \overset{\circ}{k}_{\mathrm{i}} \, \exp\left(\Delta G_{\mathrm{el}}{}^{++}/k_{\mathrm{B}}T\right)$$

where $\overset{\circ}{k}_a$ denotes the basal rate (when electrostatic interactions are absent), $\Delta G_{\mathrm{el}}{}^{++}$ indicates the free energy of electrostatic interaction in the transition state, and the term k_{B}T corresponds to the thermal energy (Alsallaq and Zhou 2007b).

This model furnishes a theoretical basis to explain significant effects including ionic strength effects on association rate and has been successfully employed in the study of mutants of four protein complexes analyzed over a huge range of ionic strengths. The PPIs studied using this approach include DNase E9-Im9, Bn–Bs, IL4-fasciculin 2 (Fas), and acetylcholinesterase (AChW) (Dell'Orco 2009). The predictions provided by this model were accurate for association rate constants that range between 10^6 to 10^{10} M^{-1} s^{-1}.

On a similar note, Schreiber et al. developed a method utilizing the concept of electrostatic-enhanced rate of association and Coulombic energy of interaction (Selzer et al. 2000; Selzer and Schreiber 1999). This method was very effective in the prediction of mutational consequences on PPIs. It is essential to understand both the effects of mutations and the short-range interactions within the associating site in order to tailor proteins with slower rate constants of dissociation (k_d). On an alternate note, the affinity of interaction can be increased by enhancing the association rate (k_a). This in turn can be carried out via the optimization of electrostatic attraction among the interactors, without affecting the dissociation constant. The rate of increment in formation of protein complexes has been demonstrated to have a good correlation with the electrostatic energy of association ΔU (Selzer and Schreiber 1999).

It has been observed that when ΔU is increased by 1 kcal/mol of Coulombic interaction energy, the association rate increases by 2.8 times. The linear relationship between association constant and ΔU was observed to be independent of the protein complex. Observing this simple quantitative relationship, Selzer et al. built a program called Protein Association Rate Enhancement (PARE) to calculate association rate constants of mutant protein complexes. The program exploited the Debye-Hückel energy of interaction among the proteins and the value of k_a determined experimentally. This approach utilized the 3D structure of the protein complex and k_a as input and generated the k_a of the target protein complex as output. PARE has been tested for several protein complexes such as hirudin-thrombin, Bn–Bs, and AChW–Fas. The calculated values of association rate constants have shown excellent agreement with the experimentally obtained values for all the three systems (Selzer et al. 2000). Therefore, this method has also been utilized for other systems, such as designing mutants of the Ras effector protein Ral guanine nucleotide dissociation stimulator (RalGDS) (Kiel et al. 2004).

3.5 Statistical Methods and Web-based Tools for Assessment of Protein–Protein Interaction Energetic and Kinetic Parameters

To unravel details on thermodynamics of PPIs, and quickly screen protein interaction interfaces, several methods have been developed. However, they are not exactly based on straightforward computational algorithms which rely on 3D structure of protein complexes for determining their physicochemical properties. Relying on the concept of hotspots, Darnell et al. developed a machine learning tool called the KFC server to predict residues in the interfacial region which are accountable for most of the binding free energy of interaction (Darnell et al. 2008). The server automatically analyzes the protein complex queried by the user and furnishes the visualizations of its predicted hotspot residues. The server has found utility in PPI systems such as bone morphogenetic protein-2 (BMP-2)-BMP receptor-type I (BMPR-IA) and calmodulin-smooth muscle myosin light chain kinase. The generated data showed that the hotspots predicted by this server and the experimental mutations that are associated with significant decrease in binding affinity are in good agreement (Darnell et al. 2007).

On a similar note, Li et al. analyzed the side-chain interactions between atoms in numerous alanine mutated interfaces (Li et al. 2006). They found that from an energetic perspective, the hotspots contribute to the favorable interactions between side chains and are crucial to distinguish non-specific interactions from specific ones. Using another web-based approach, Zhou et al. attempted to predict the energetic landscape in mutated PPIs (Zhou et al. 2005). Utilizing statistical energy function tools option given by the server, decent predictions were attained by Zhang et al. about the binding affinities for PPI complexes (Zhang et al. 2005).

Single point mutations made systematically at the exactly same position in an enzyme-inhibitor complex have resultant variations in the binding affinity which suggests that the substituent effects on the thermodynamic/kinetic constants represent a classical situation and might be applied to a congeneric molecular series. This, in turn, might help to define the theme of a congeneric protein series. Dell'Orco and De Benedetti took two protein inhibitors in consideration, namely turkey ovomucoid third domain (OMTKY3) and BPTI (Dell'Orco and De Benedetti 2008). Both these proteins were mutated in a systematic manner at the binding site (P1). The site was solvent-exposed and the mutation in this region was aimed to severely influence the thermodynamic constants pertaining to their interactions with different serine proteases such as: human neutrophil elastase (HNE); anionic salmon trypsin (AST); bovine a-chymotrypsin (BCHYM) for the BPTI inhibitor, bovine b-trypsin (BT), human leukocyte elastase (HLE), porcine pancreatic elastase (PPE) for OMTKY3 inhibitor, and *Streptomyces griseus* proteinase A (SGPA). Apart from the 20 encoded amino acids, 4 non-coded residues were also used for the systematic mutations at the P1 position of OMTKY3 and they were potent enough to influence the k_a values. Models were built between the chosen descriptors of physicochemical properties of residues introduced at the P1 region and the experimentally determined equilibrium constants. These models were based on linear regressions and demonstrated predictive and explicative ability for the cases.

Wang and Wade floated a similar approach known as COMparative BINding Energy (COMBINE) analysis (Wang and Wade 2002). Using this analysis tool, Tomic et al. studied the specificity and selectivity of interactions of Ras proteins with other interacting protein partners (Tomić et al. 2007). They performed the COMBINE analysis of various native and mutant protein complexes between Ras proteins, Ras-Raf, Ras-Rap, etc. From the analysis, the non-specific hotspot residues of Ras proteins were found that had significant participation in stabilizing Raf–Ras and RalGDS–Ras complexes. Authors found that although the Ras-Raf and RalGDS–Ras complexes differed in size, the built COMBINE model for one complex sufficed for the other. Jemimah et al. have also contributed to the field of prediction of binding free energies of protein complexes by developing a sequence-based approach known as ProAffiMuSeq. This methodology was tested on various sets of test data and it was observed that the experimental and predicted values of binding free energies attained were concordant (Jemimah et al. 2020).

Freyhult et al. employed a different approach that was relevant for studying the kinetics of PPIs (Freyhult et al. 2003). By modeling the interactions of wild-type and mutant antibodies against lysozyme, they describe association/dissociation rates, binding affinity, etc. These models thus helped to identify the residues that had significant contribution to these rates. Similarly, De Genst et al. utilized the same test system and analyzed effects from not only mutations, but also environmental perturbations (De Genst et al. 2002). Using a biosensor, they measured the changes in association and dissociation rates. Then utilizing the experimental dataset, the authors quantitatively constructed models for predicting both affinity and kinetic constants. These models, that were designed based on multivariate experimental data, were very potent for the prediction of parameters in 4 variants of the studied system. In a nutshell, it generated extremely useful data on the physicochemical aspects of PPIs.

3.6 Energetics of Protein–Protein Interaction Networks

Researchers have long aimed to perform a multiscale modeling in systems biology because predicted affinity and/or kinetic constants are not provided in the broader context of a dynamic network. Recently, research groups have integrated the structure-based approaches with the cellular network simulations. Utilizing protein design tools and structural data, Kiel et al. have constructed a network of protein interactions among several subfamilies of Ras (Kiel et al. 2007). The constructed network was validated through experiments by generating 6 Ras binding domains and 2 Ras proteins through cloning and studying them by pull-down experiments to differentiate between interactors and non-interactors. Energy calculations were performed to mark the PPIs that had higher probabilities of being relevant in the in vivo context. Following the analysis, several novel interactions between Ras proteins and effector proteins were found that were relevant from a physiological perspective.

Using a different approach, Gabdoulline et al. employed Brownian Dynamics to analyze a kinetic parameter and utilize it in the simulation of a biochemical network involving the peroxidase–oxidase reaction (Gabdoulline et al. 2003). The network was composed of several reaction steps and the resulting dynamics was complex and oscillatory in nature. The second-order rate constants that were computed were in line with experimental data and enabled one to predict the rate constants at physiological conditions. Their studies provided a very strong basis for the integration of atomic level structural data on proteins and cellular phenomenon such as metabolite oscillation. In another study, Shih et al. analyzed how mutations in the ErbB1 tyrosine kinase receptor protein could lead to oncogenic transformation via the induction of cascading fragility mechanisms and disturbance of the signaling pathway (Shih et al. 2008). They employed MD simulations to study the dimer-mediated receptor activation of ErbB1 and used docking simulations to analyze the interactions of the protein with peptide containing substrate tyrosine. To understand the ErbB1-mediated signaling, they built a network-based, deterministic, kinetic modeling scheme. On the other hand, they utilized hybrid discrete/continuum stochastic dynamics to analyze the initial process of internalization of ErbB1 receptor. The multiscale approach has immense potential to extend over both space (nm–mm) and time (ns–ms) scales and is quite effective to manage the information flux of a complicated signaling pathway.

3.7 Energetic Basis for High Selectivity of Protein–Protein Interactions: A Case Study

High-selectivity and high-affinity PPIs are imperative for cell sustenance and are essential in an evolutionary paradox because it poses a very basic question: When lethal mutations lead to a drop in binding affinity, how does the selectivity of the interaction evolve? To understand PPI selectivity, it is essential to obtain structural data on non-cognate, weak complexes. However, such complexes may be difficult to obtain since they are dynamic and transient in nature. Utilizing an NMR-based docking, Meenan et al. arranged a disulfide-trapping approach on E9DNase-Im2, a non-cognate PPI complex (Meenan et al. 2010). This complex is seven times weaker than the femto-molar E9 DNase-Im9 interaction of cognate nature. The study of the 1.77 Å crystal structure of the non-cognate E9 DNase-Im2 complex revealed a complete non-covalent interface where the lattice was merely supported by intersubunit disulfide bonds.

Combining the structural information with computational alanine scanning of interfacial residues, the authors demonstrated that the driving force behind the interaction was so strong that an immensely unfavorable specificity contact was accepted by the interaction interface. Consequently, the protein complex was subjected to a process of "frustration" or accumulation of non-native contacts and was ultimately weakened. The comparison of the non-cognate structure with previously available cognate complex structures not only rationalized past thermodynamic data, but also highlighted the significance of loops in the development of selectivity and emphasized the various roles played by buried water molecules in the stabilization or amelioration of interfacial contacts (Meenan et al. 2010). In short, the study essentially validated the dual-recognition in colicin DNase-Im protein complex and demonstrated that non-cognate, weakened complexes are primed for high-affinity interactions, by economical, compensating mutations of a limited number of interfacial residues.

3.8 Effects of Viscosity, pH, and Salt on Association of Proteins

It is important to know how different parameters (co-solvents, viscosity, pH) can affect the interactions/kinetics of PPIs to get a holistic picture of the energetics of association of proteins.

3.8.1 Contribution of Co-solvents

When proteins associate in dilute solutions, there is a requirement of a long diffusion step to allow the two proteins for collision within the rigid geometrical constraints of binding. Co-solvents can be employed to perturb this diffusion step. Such co-solvents usually include agents that enhance viscosity of solution, such as sucrose (which allows slower diffusion) or glycerol or salts (that block electrostatic attraction) or PEG, BSA, which act as crowding agents (Ellis 2001).

3.8.2 Effect of Viscosity on k_{on}

It has been demonstrated that the relative association rate constant, in the absence and presence of a viscous agent (k^{o}_{on}/k^{v}_{on}), for a bimolecular association depends in a linear fashion on the medium's relative viscosity (η/η_{o}), with a slope of 1 or less for a reaction that is either completely or partially diffusion-controlled (Berg and von Hippel 1985; Kramers 1940). Thus, the reliance of k_{on} on viscosity was utilized to find out if a PPI is diffusion-controlled. Indeed, for some PPIs, k_{on} and solvent viscosity were simply related with a slope of approximately 1 (Raman et al. 1992; Stone and Hermans 1995). However, for certain PPIs such as HyHEL-5/bob-white quail lysozyme, barnase/barstar, and HyHEL-10/hen-egg lysozyme (HEL), the slopes were observed to be greater than 1 (Xavier et al. 1999; Xavier and Willson 1998). Moreover, the slope of k^{o}_{on}/k^{v}_{on} versus η/η_{o} for the barnase-barstar interaction was found to be dependent on how strong the electrostatic attraction was between the two interacting proteins (Schreiber 2002). Since this phenomenon was not explained from a theoretical point of view, many researchers questioned the usage of the viscosity dependence of k_{on} as a standard for diffusion-controlled PPIs.

3.8.3 Ionic Strength Dependence of k_{on}

Research groups have long recognized favorable electrostatic attraction as one of the major forces driving fast association (Sheinerman et al. 2000). If one alters the ionic strength of solution, the magnitude of electrostatic attraction can be changed. It has been observed that the relation between k_{on} and ionic strength obeys the Debye–Hückel's equation for interaction energy of a PPI which is as follows (Selzer and Schreiber 1999; Vijayakumar et al. 1998):

$$\ln k_{on} = \ln k^{o}{}_{on} - (U/RT)(1/(1 + \kappa_{\alpha}))$$

where k_{on} and $k^{o}{}_{on}$ are the association rates when electrostatic forces are present and absent, respectively, U denotes the interaction's electrostatic energy, κ implies the inverse Debye length, and α marks the minimal distance of approach. Hence, k_{on} is the sum of the electrostatic forces between the interacting proteins and the basal rate of association in the absence of electrostatic forces ($k^{o}{}_{on}$)

The former can be attended via changing the value U (by mutation) or via changing the conditions of the solution. Debye–Hückel's equation suggests that if k_{on} is plotted against ($1/(1+\kappa_{\alpha})$), the plot would be linear and have a slope equal to −(U/RT) (the electrostatic contribution of the two proteins when salt is absent). The intercept of the line, when ($1/(1+\kappa_{\alpha})$) equals 0, corresponds to the basal rate ($\ln k^{o}{}_{on}$), i.e. when salt shields electrostatic forces. The intercept, when ($1/(1+\kappa_{\alpha})$) equals 1, corresponds to $\ln k_{on}$, i.e. when the salt is absent and the electrostatic forces are maximized. The linear relationship was demonstrated for PPIs such as TEM1–BLIP, hirudin–thrombin, interferon–receptor, barnase–barstar, etc. (Piehler and Schreiber 1999; Schreiber and Fersht 1996; Selzer et al. 2000; Selzer and Schreiber 1999). Interestingly, the basal rate calculated for barnase-barstar via the extrapolation of ionic strength to infinity had a value different from the one obtained from mutational data (Selzer and Schreiber 1999).

3.8.4 pH Dependence

Histidine is the one and only amino acid residue that has pKa almost equal to the physiological pH. When the pKa values of histidine residues are altered, they provide valuable data about their environment. R67 dihydrofolate reductase (DHFR)'s rate of association has a pKa of 6.6, due to the presence of H62. In contrast, the pKa of the dissociation reaction is less than 5.5 (Méjean et al. 2001). Owing to short-range interactions during complex formation, the pKa of H62 is shifted and it occurs after the rate-determining step or the slowest step of the reaction. H102 of barnase similarly faces a major pKa shift upon interaction with barstar (Schreiber and Fersht 1995).

3.9 Energetic Pathway of Protein Complex Formation

The formation of an AB protein complex from its interacting partners A and B can be explained using a four-state model (Fig. 3.3):

$$A + B \leftrightarrow AB* \leftrightarrow AB ** \leftrightarrow AB$$

The forward and backward reaction rates for the 3 reversible steps in the reaction are k_1/k_{-1}, k_2/k_{-2}, and k_3/k_{-3} respectively. Two proteins A and B form an unstable encounter complex via the process of diffusion (AB*). This complex has a tendency to re-dissociate, i.e. $k_{-1}>>k_2$. Further, the transition of the diffusion complex AB* into an intermediate AB** complex, bears the commitment of forming the final complex, i.e., $k_3 >> k_{-2}$. If a PPI reaction is described in terms of a four-state model, it enables one to validate the presence of an encounter complex, prior to both the transition state and the intermediate state that follows this transition state. Experimentally, both intermediate and encounter complexes have been observed for some PPIs (Schreiber 2002). However, it should be borne in mind that some other interactions might even follow a two-state or three-state model.

In a diffusion-limited association reaction, when AB* is absent, k_{obs} is expected to depend linearly on protein concentrations. However, k_{obs} is recognized to depend hyperbolically on the protein concentration (Fig. 3.4). This would imply that $k_{-1} >> k_2$ and the reaction is not controlled by diffusion. For example, when the interaction between bovine β-trypsin and soybean trypsin inhibitor was analyzed, it was observed that the diffusion encounter complex formation step was rate limiting (Luthy et al. 1973). Sydor et al. have revealed that interaction between Ras and c-Raf1 also involves the formation of diffusion encounter complex (Sydor et al. 1998). They suggested a two-step association process where an isomerization step followed an initial rapid equilibrium step. For both of these PPIs, the intermediate

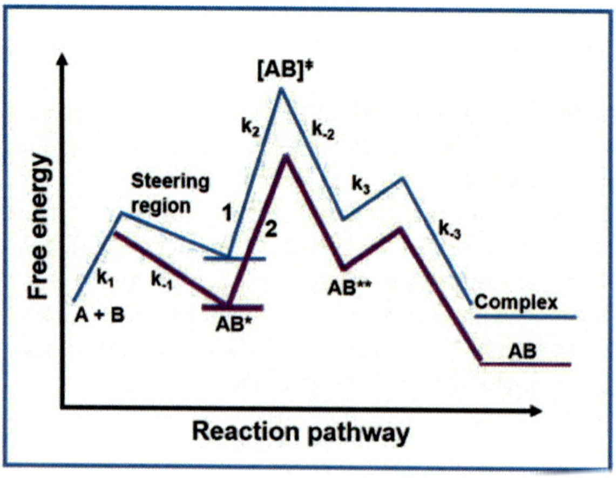

Fig. 3.3 The figure illustrates the four-state model for formation of protein complex model. Initially protein A and B form an encounter complex (AB*) which can dissociate easily. The AB* complex then undergoes a transition [AB]‡ which forms an intermediate complex (AB**). This intermediate complex then further forms the protein complex (AB) (Adapted and redrawn form Schreiber 2002)

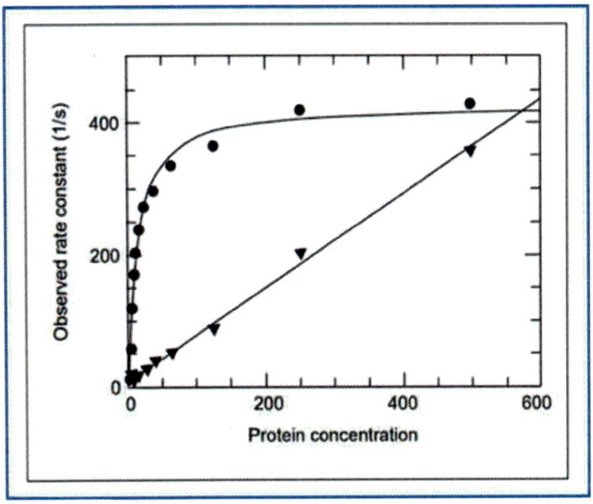

Fig. 3.4 Figure illustrated linear and hyperbolic dependencies of enzymatic inhibition on protein concentration (Adapted from Schreiber 2002)

(AB*) was implied to dissociate sooner as compared to complex formation ($k_{-1} >> k_2$). For majority of the PPIs, research groups have not been able to detect a diffusion encounter complex. However, this does not necessarily refute the existence of AB* as this complex can be observed only at higher concentrations of proteins (that can range from 10's to 100's μM).

The intermediate (AB**) does not influence the association's overall rate since it is formed after the rate-limiting step ($k_3 >> k_{-2}$). The intermediate is essentially a complex that is partially formed and needs reorganization for the formation of the final complex. This stage of reorganization can be faster, as depicted for cystatin A–papain interaction (230 s^{-1}) (Estrada et al. 2000, 1999), or slower, as attained for HEL-HyHEL-10 or HEL-HyHEL-26 interaction (~10–3 s^{-1}) (Li et al. 2001).

The main obstacle that one faces to investigate this intermediate is to search a probe which has the ability to independently monitor the intermediate formation versus the final complexation. For an interaction between cystatin A and papain, the inhibition rate of the catalytic activity of papain was utilized to monitor the intermediate complex formation. On the other hand, spectroscopy was utilized to observe the final complex formation. The enzymatic inhibition demonstrated a linear dependency of k_{obs} on the concentration of a protein. However, according to the spectroscopic probe k_{obs} depends hyperbolically on protein concentrations (Fig. 3.4). It is not clear as to whether the hyperbolic dependence corresponds to the build-up of the encounter complex or the intermediate reorganization. This type of mechanism has also been suggested for the thrombin–hirudin association (Myles et al. 2001).

In a reaction pathway, transition state is referred to as the most unstable species. In a reaction coordinate diagram, it occurs at the peak. This state represents the formation and breaking of chemical bonds. The theory on transition state was originally proposed for unimolecular reactions. However, if one considers the relative change of this state due to alteration of the association rate (by mutation/changing solvent conditions), then it can be utilized for bimolecular reactions such as

PPIs. Several studies have been carried out to study the nature of transition state for the association of proteins (Baerga-Ortiz et al. 2000; Frisch et al. 2001; Hopkins et al. 2000; Li et al. 2001; Piehler and Schreiber 1999; Radić et al. 1997; Rajpal et al. 1998; Selzer et al. 2000; Selzer and Schreiber 1999; Taylor et al. 1998; Xavier et al. 1999; Xavier and Willson 1998).

Direct data on the transition state's structure was revealed when activation interaction energies were measured utilizing double mutant cycles. In case the change in k_{on} induced by a pair of mutations is additive, the two residues are not found to be indulged in interaction during the transition state. In contrast, if the change is lower than the additive, the residues might be assumed to interact. When the structures of the activated complexes of thrombin–hirudin and barnase–barstar were probed using this method, only the charged residues in close vicinity in the final complex already interacted during the activated state. The increase in salt concentration masked only some of these specific long-range electrostatic interactions. The retained pairwise charge–charge implied the existence of the activated complex even at higher salt concentrations for these complexes (Frisch et al. 2001; Myles et al. 2001; Schreiber and Fersht 1996).

Φ value analysis has also been employed in transition state studies, where $\Phi = \Delta\Delta G\ddagger_{on}/\Delta\Delta G_D$ ($\Delta\Delta G\ddagger_{on}$ equals to the free energy difference between the unbound proteins and the transition state for a mutant versus the wild type, and $\Delta\Delta G_D$ is the free energy difference between the unbound and bound proteins for a mutant versus the wild type) (Selzer and Schreiber 2001). A Φ value near 1 represents the existence of specific interaction at the transition state. In contrast, a Φ value near 0 implies the formation of the interaction after the transition state. Making multiple replacements at two residue positions in the HyHEL-10 FAB complex, Φ values close to 0 were observed for most of the replacements. This implied an early occurrence of transition state that is before the formation of short-range interactions. It was speculated that short-range interactions influenced k_{off}, and long-range electrostatic interactions influenced k_{on}.

To test this, charged mutations were introduced near the binding site between TEM1 and BLIP. Although the mutations increased k_{on} by 250-folds, the k_{off} was not affected (Selzer et al. 2000). These data implied that owing to the participation of the long-range electrostatic interactions in the rates of association, the free energy of the transition state was lowered by a magnitude equal to the equilibrium constant. Moreover, it was observed that k_{off} was unaltered with the increment in long-range electrostatic steering. This suggested that k_{-1} (which decreases with increasing electrostatic forces) is very crucial for an early step along the reaction, prior to the transition (Selzer and Schreiber 2001). All the generated data from the studies indicate that the transition during association utilizes electrostatic interactions for stabilization and the structure of the transition state appears like a more solvated version of the final complex (Fig. 3.5). The formation of specific short-range interactions, desolvation, and structural rearrangement usually generate the high-energy barriers. Further, the data indicates that the encounter complex (AB*), occurring before the transition state, simulates the latter with respect to structure and that electrostatic forces contribute immensely to stabilize the encounter complex

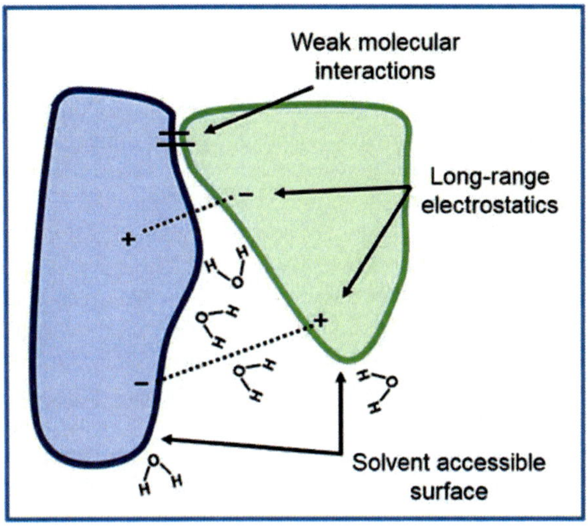

Fig. 3.5 Schematic illustrating the involvement of electrostatic interactions in the transition state of protein complex formation. For a stable association of protein complex both specific long-range electrostatics and weak molecular interaction such as van der Waals and hydrophobic interactions are involved, while majorly the protein surface is solvated (Adapted and redrawn from Schreiber 2002)

(Selzer and Schreiber 2001). It has to be borne in mind that both the data on encounter complex and transition state are generated by analyzing a specific set of proteins. Therefore, the kinetic pathway may vary for other proteins.

The association rates of a number of PPIs have been simulated using both transition state theory and Brownian dynamic simulations (Elcock et al. 2001). While the simulations succeeded in prediction of the relative variation in k_{on} upon mutation or alteration in ionic strength, they were not beneficial for the calculation of absolute values (Altobelli and Subramaniam 2000; Selzer et al. 2000; Vijayakumar et al. 1998). The association reaction was considered to be diffusion controlled in majority of the Brownian dynamic simulations and the discrepancies in absolute rates of association between experiment and simulation were accounted for structural rearrangement or slow desolvation steps (Elcock et al. 1999; Gabdoulline and Wade 2001). This would help to validate experimental data that even for extremely fast associating PPI complexes, the reaction may not be controlled by diffusion alone.

3.10 Biological Significance of Binding Kinetics

It is not yet known whether binding affinity or association or dissociation rates govern biological activity. Recent research has demonstrated that kinetic constants hold biological significance. For instance, researchers have carried out a kinetic study of the interaction between a cytokine and receptor and have revealed that the signal transduction cascade is initiated by binding. It has been seen that for certain interactions such as those between human growth hormone, interferon, and interleukin-4 along with their corresponding receptors, the dissociation rate depends linearly on the biological activities magnitude under the stringent criteria of k_{off} >

0.02 s^{-1} (Pearce Jr. et al. 1999; Piehler et al. 2000; Wang et al. 1997). However, it was also observed that the bioactivity was not enhanced by slower dissociation (Pearce Jr. et al. 1999; Wang et al. 1997).

When researchers studied how the B cell response varied according to concentration of an antigen, they found a similar threshold and a maximal value for the off-rates (Batista and Neuberger 1998; Foote and Eisen 1995). Majority of the protein complexes have an association rate that is comparable with a diffusion-limited reaction as they are bound by geometrical constraints. However, it must be noted that certain other PPIs have much higher association rates. Such fast interaction rates are usually observed when the electrostatic attraction forces between the proteins are optimized (Botti et al. 1998; Schreiber et al. 1994). Some classical examples include the enzyme protein inhibitor interactions and here the rate of inhibition essentially yields an evolutionary pressure on the system (Lee et al. 1989; Radić et al. 1997; Schreiber and Fersht 1996; Stone et al. 1989; Wallis et al. 1995). On the other hand, studies on Ras signal transduction have shown that it is mostly controlled by association (Sydor et al. 1998).

3.11 Challenges in Studying PPI Energetics

Ever since high quality instrumentation has been made available commercially it has attracted several research groups to perform thermodynamic studies on PPIs. This field is extensively studied and has promising future for attracting even more research groups. However, it should be noted that not all researchers who are involved in these studies are specifically trained in calorimetry. This does not essentially denote a bad situation since it is not always necessary to include a specialist for thermodynamics, secondly in majority of the papers the energetic and thermodynamic aspects of PPIs have served secondary to other aspects. On the one hand, it is very easy to mix two protein solutions and determine values of ΔG, ΔH, and ΔS in a calorimeter. However, it is very arduous to perform the experiments with a large number of experimental variations and controls. Both experimental variations and number of controls are necessary for the validation of results and in certain cases it might be very difficult for research groups to exactly pinpoint the experimental variations and controls that should be chosen to generate appropriate results.

For instance, if one wants to understand the importance of electrostatic interactions in PPIs, one must perform several studies on the effects of ionic strengths. On a similar note, non-ideal conditions might crop up in calorimetric experiments when protein concentrations are relatively high. In such a case, a calorimetrist should always determine the results using different protein concentrations. Thirdly, when a protein–protein complex is formed, protons are either uptaken or released, such protons may lead to deprotonation or protonation events in the buffer. Care should be taken to recognize such ionization events in the buffer as the enthalpy of ionization can contribute to the total measured enthalpy (Christensen et al. 1976). Furthermore, ionization in the buffer can also contribute

ΔC_p values to the total ΔC_p. Therefore, a researcher should always carry out the thermodynamic studies using various buffers at different pH in order to ensure if there is any connected protonation or deprotonation event (Baker and Murphy 1996; Doyle et al. 1995).

Moreover, it is not always possible to exactly determine factors like solvent conditions, buffer counter ions, etc. that are used in the experiments. One case study on PPIs has highlighted some of the difficulties that researchers face while performing energetic studies on PPIs. When Yakovlev et al. (1995) analyzed an interaction of barnase with barstar, they reported different results from the work that had been done by Martinez et al. (1995). Yakovlev et al. reported that the ΔH value was −26.4 kcal/mol whereas Martinez et al. reported ΔH value of −13.9 kcal/mol. Furthermore, the former group reported a ΔS value of −0.034 kcal/(mol K) whereas the value reported by the latter was +0.012 kcal/(mol K). It is not yet known why the studies conducted by these two groups had such profound disagreement. It is assumed that there is an error in one of the two research studies. This error might either originate from difference in pH, protein concentration, or ionic strength. It is essential to perform a detailed study to investigate the real reason behind this disagreement.

3.12 Conclusions

In PPI, critical residues of one protein interact with their binding partner with the help of critical residues present across the interface. For this interaction to occur, it is very essential to have a limited critical residues present in each of the binding interfacial regions. Whether it is a residue of hydrophobic nature or it is the hydrophobic part of a residue with hydrophilic nature, the critical interaction that occurs between the critical residues turns out to be hydrophobic. Detailed examination of thermodynamics showed that, upon mutation, residues that have lesser contribution to binding affinity start showing variations in entropy and enthalpy of association. However, the implication of such variations remains unclear. If extensive mutations are carried out at a single point, no conclusions can be drawn except one that is similar to experiments that scan alanine. The conclusion that could be derived is that, upon mutation, changes in ΔS, ΔH, and C_P seem to be the rule, though association binding energy remains unchanged and any rule that controls their changes are yet to become obvious.

The association pathway leads to the generation of a diffusion encounter complex that is unstable in nature, but can undergo evolution for the final complex formation or an intermediate complex which is predetermined for final complex formation. Experimentally, it is very difficult to track these pre complexes and hence it is reasonable to think that they are more copious than has been brought to notice so far. Efforts have been put in to characterize the transition state which forms during the association reaction. Studies indicate that long-range electrostatic forces perform an imperative role toward stabilization of the transition state. The estimation of kinetic and affinity constants has been made easier with the knowledge of 3D structures. In regard to systems biology, the effectiveness of docking methods to predict the interactions among proteins has been brought into the lime light for better understanding of PPI energetics.

References

Alsallaq R, Zhou HX. (2007a). Energy landscape and transition state of protein-protein association. *Biophysical journal, 92*, 1486-1502.

Alsallaq R, Zhou HX. (2007b). Prediction of protein-protein association rates from a transition-state theory. *Structure, 15*, 215-224.

Altobelli G, Subramaniam S. (2000). Kinetics of association of anti-lysozyme monoclonal antibody D44.1 and hen-egg lysozyme. *Biophysical journal, 79*, 2954-2965.

Anderson J., Burns HD, Enriquez-Harris P, Wilkie AO, Heath JK. (1998). Apert syndrome mutations in fibroblast growth factor receptor 2 exhibit increased affinity for FGF ligand. *J Hum Mol Genet, 7*, 1475-1483.

Audie J, Scarlata S. (2007). A novel empirical free energy function that explains and predicts protein–protein binding affinities. *J Biophysical chemistry, 129*, 198-211.

Baerga-Ortiz A, Rezaie AR, Komives EA. (2000). Electrostatic dependence of the thrombin-thrombomodulin interaction. *Journal of molecular biology, 296*, 651-658.

Baker BM, Murphy KP. (1996). Evaluation of linked protonation effects in protein binding reactions using isothermal titration calorimetry. *Biophysical journal, 71*, 2049-2055.

Baker BM, Murphy KP. (1998). Prediction of binding energetics from structure using empirical parameterization. *Method Enzymol, 295*, 294-315.

Banerjee A, Hu J, Goss DJ. (2006). Thermodynamics of Protein– Protein Interactions of cMyc, Max, and Mad: Effect of Polyions on Protein Dimerization. *Biochemistry, 45*, 2333-2338.

Basdevant N, Weinstein H, Ceruso M. (2006). Thermodynamic basis for promiscuity and selectivity in protein– protein interactions: PDZ domains, a case study. *Journal of the American Chemical Society, 128*, 12766-12777.

Bass SH, Mulkerrin MG, Wells JA. (1991). A systematic mutational analysis of hormone-binding determinants in the human growth hormone receptor. *Proceedings of the National Academy of Sciences of the United States of America, 88*, 4498-4502.

Batista FD, Neuberger MS. (1998). Affinity dependence of the B cell response to antigen: a threshold, a ceiling, and the importance of off-rate. *Immunity, 8*, 751-759.

Beernink PT, Tolan DR. (1994). Subunit interface mutants of rabbit muscle aldolase form active dimers. *Protein science : a publication of the Protein Society, 3*, 1383-1391.

Berg OG, von Hippel PH. (1985). Diffusion-controlled macromolecular interactions. *Annual review of biophysics and biophysical chemistry, 14*, 131-160.

Bhat TN et al. (1994). Bound water molecules and conformational stabilization help mediate an antigen-antibody association. *Proceedings of the National Academy of Sciences of the United States of America, 91*, 1089-1093.

Borchert TV, Abagyan R, Kishan KV, Zeelen JP, Wierenga RK. (1993). The crystal structure of an engineered monomeric triosephosphate isomerase, monoTIM: the correct modelling of an eight-residue loop. *Structure, 1*, 205-213.

Botti SA, Felder CE, Sussman JL, Silman I. (1998). Electrotactins: a class of adhesion proteins with conserved electrostatic and structural motifs. *Protein engineering, 11*, 415-420.

Braden BC et al. (1996). Crystal structure of the complex of the variable domain of antibody D1. 3 and turkey egg white lysozyme: a novel conformational change in antibody CDR-L3 selects for antigen. *Journal of molecular biology*.

Brooijmans N, Sharp KA, Kuntz ID. (2002). Stability of macromolecular complexes. *Proteins, 48*, 645-653.

Castro MJM, Anderson S. (1996). Alanine point-mutations in the reactive region of bovine pancreatic trypsin inhibitor: effects on the kinetics and thermodynamics of binding to β-trypsin and α-chymotrypsin. *Biochemistry*, *35*, 11435-11446.

Chen R, Li L, Weng ZP. (2003). ZDOCK: An initial-stage protein-docking algorithm. *Proteins-Structure Function and Bioinformatics*, *52*, 80-87.

Chong S-H, Ham S. (2017). Dynamics of hydration water plays a key role in determining the binding thermodynamics of protein complexes. *Scientific reports*, *7*, 1-10.

Christensen JJ, Hansen LD, Izatt RM (1976) Handbook of proton ionization heats and related thermodynamic quantities. Wiley,

Clackson T, Wells JA. (1995). A hot spot of binding energy in a hormone-receptor interface. *Science*, *267*, 383-386.

Cunningham BC, Wells JA. (1989). High-resolution epitope mapping of hGH-receptor interactions by alanine-scanning mutagenesis. *Science*, *244*, 1081-1085.

Cunningham BC, Wells JA. (1993). Comparison of a structural and a functional epitope. *Journal of molecular biology*, *234*, 554-563.

Darnell SJ, Page D, Mitchell JC. (2007). An automated decision-tree approach to predicting protein interaction hot spots. *Proteins-Structure Function and Bioinformatics*, *68*, 813-823.

Darnell SJ, LeGault L, Mitchell JC. (2008). KFC Server: interactive forecasting of protein interaction hot spots. *Nucleic acids research*, *36*, W265-269.

De Genst E, Areskoug D, Decanniere K, Muyldermans S, Andersson K. (2002). Kinetic and affinity predictions of a protein-protein interaction using multivariate experimental design. *J Biol Chem*, *277*, 29897-29907.

De Rienzo F, Gabdoulline R, Menziani MC, Wade R. (2000). Blue copper proteins: a comparative analysis of their molecular interaction properties. *Protein Science*, *9*, 1439-1454.

Dell'Orco D. (2009). Fast predictions of thermodynamics and kinetics of protein–protein recognition from structures: from molecular design to systems biology. *Mol Biosyst*, *5*, 323-334.

Dell'Orco D, De Benedetti PG. (2008). Quantitative structure–activity relationship analysis of canonical inhibitors of serine proteases. *Journal of Computer-Aided Molecular Design*, *22*, 469-478.

Dell'Orco D, Xue W-F, Thulin E, Linse S. (2005). Electrostatic contributions to the kinetics and thermodynamics of protein assembly. *Biophysical journal*, *88*, 1991-2002.

Dell'Orco D, De Benedetti P, Fanelli F. (2006). From computational biophysics to systems biology workshop. *NIC series*, 67.

Dell'Orco D, Casciari D, Fanelli F. (2008). Quaternary structure predictions and estimation of mutational effects on the free energy of dimerization of the OMPLA protein. *Journal of structural biology*, *163*, 155-162.

Dell'Orco D, Seeber M, De Benedetti PG, Fanelli F. (2005). Probing fragment complementation by rigid-body docking: in silico reconstitution of calbindin D9k. *Journal of chemical information and modeling*, *45*, 1429-1438.

Dell'Orco D, De Benedetti PG, Fanelli F. (2007). In silico screening of mutational effects on enzyme-proteic inhibitor affinity: a docking-based approach. *BMC structural biology*, *7*, 37.

Deng N, Tse C, Wickstrom L, Kvaratskhelia M, Gallicchio E, Levy R. (2020). Exploring the Free Energy Landscape and Thermodynamics of Protein-Protein AssociationAssociation: HIV-1 Integrase Multimerization Induced by an Allosteric Inhibitor. *BioRxiv*.

Doyle ML, Louie G, Dal Monte PR, Sokoloski TD (1995) [8] Tight binding affinities determined from thermodynamic linkage to protons by titration calorimetry. In: Methods in enzymology, vol 259. Elsevier, pp 183-194

Elcock AH, Gabdoulline RR, Wade RC, McCammon JA. (1999). Computer simulation of protein-protein association kinetics: acetylcholinesterase-fasciculin. *Journal of molecular biology*, *291*, 149-162.

Elcock AH, Sept D, McCammon JA (2001) Computer simulation of protein– protein interactions. ACS Publications,

Ellis RJ. (2001). Macromolecular crowding: an important but neglected aspect of the intracellular environment. *Current opinion in structural biology, 11*, 114-119.

Estrada S, Pavlova A, Bjork I. (1999). The contribution of N-terminal region residues of cystatin A (stefin A) to the affinity and kinetics of inhibition of papain, cathepsin B, and cathepsin L. *Biochemistry, 38*, 7339-7345.

Estrada S, Olson ST, Raub-Segall E, Bjork I. (2000). The N-terminal region of cystatin A (stefin A) binds to papain subsequent to the two hairpin loops of the inhibitor. Demonstration of two-step binding by rapid-kinetic studies of cystatin A labeled at the N-terminus with a fluorescent reporter group. *Protein Science, 9*, 2218-2224.

Finger C, Volkmer T, Prodöhl A, Otzen DE, Engelman DM, Schneider D. (2006). The stability of transmembrane helix interactions measured in a biological membrane. *Journal of molecular biology, 358*, 1221-1228.

Fisher HF, Singh N (1995) [9] Calorimetric methods for interpreting protein—Ligand interactions. In: Methods in enzymology, vol 259. Elsevier, pp 194-221

Foote J, Eisen HN. (1995). Kinetic and affinity limits on antibodies produced during immune responses. *Proceedings of the National Academy of Sciences of the United States of America, 92*, 1254-1256.

Frankel AD, Chen L, Cotter RJ, Pabo CO. (1988). Dimerization of the tat protein from human immunodeficiency virus: a cysteine-rich peptide mimics the normal metal-linked dimer interface. *Proceedings of the National Academy of Sciences of the United States of America, 85*, 6297-6300.

Fredericks J, Hamilton AJ. (1996). Comprehensive Supramolecular Chemistry. *9*, 565.

Freyhult EK, Andersson K, Gustafsson MG. (2003). Structural modeling extends QSAR analysis of antibody-lysozyme interactions to 3D-QSAR. *Biophysical journal, 84*, 2264-2272.

Frisch C, Fersht AR, Schreiber G. (2001). Experimental assignment of the structure of the transition state for the association of barnase and barstar. *Journal of molecular biology, 308*, 69-77.

Gabdoulline RR, Wade RC. (2001). Protein-protein association: investigation of factors influencing association rates by brownian dynamics simulations. *Journal of molecular biology, 306*, 1139-1155.

Gabdoulline RR, Kummer U, Olsen LF, Wade RC. (2003). Concerted simulations reveal how peroxidase compound III formation results in cellular oscillations. *Biophysical journal, 85*, 1421-1428.

Gabdoulline RR, Stein M, Wade RC. (2007). qPIPSA: relating enzymatic kinetic parameters and interaction fields. *BMC bioinformatics, 8*, 373.

Gómez J, Freire EJ. (1995). Thermodynamic mapping of the inhibitor site of the aspartic protease endothiapepsin. *Journal of molecular biology, 252*, 337-350.

Guerois R, Nielsen JE, Serrano L. (2002). Predicting changes in the stability of proteins and protein complexes: A study of more than 1000 mutations. *Journal of molecular biology, 320*, 369-387.

Guinto ER, DiCera E. (1996). Large heat capacity change in a protein-monovalent cation interaction. *Biochemistry, 35*, 8800-8804.

Hawkins RE, Russell SJ, Baier M, Winter G. (1993). The contribution of contact and non-contact residues of antibody in the affinity of binding to antigen: the interaction of mutant D1. 3 antibodies with lysozyme. *Journal of molecular biology.*

Hénin J, Pohorille A, Chipot C. (2005). Insights into the recognition and association of transmembrane α-helices. The free energy of α-helix dimerization in glycophorin A. *Journal of the American Chemical Society, 127*, 8478-8484.

Henriques DA, Ladbury JE, Jackson RM. (2000). Comparison of binding energies of SrcSH2-phosphotyrosyl peptides with structure-based prediction using surface area based empirical parameterization. *Protein Science, 9*, 1975-1985.

Hibbits KA, Gill DS, Willson RC. (1994). Isothermal titration calorimetric study of the association of hen egg lysozyme and the anti-lysozyme antibody HyHEL-5. *Biochemistry, 33*, 3584-3590.

Hopkins PC, Pike RN, Stone SR. (2000). Evolution of serpin specificity: cooperative interactions in the reactive-site loop sequence of antithrombin specifically restrict the inhibition of activated protein C. *Journal of molecular evolution*, *51*, 507-515.

Horton N, Lewis M. (1992). Calculation of the free energy of association for protein complexes. *Protein science : a publication of the Protein Society*, *1*, 169-181.

Ito W, Iba Y, Kurosawa Y. (1993). Effects of Substitutions of Closely-Related Amino-Acids at the Contact Surface in an Antigen-Antibody Complex on Thermodynamic Parameters. *J Biol Chem*, *268*, 16639-16647.

Jaiswal N et al. (2019). Molecular interaction between human SUMO-I and histone like DNA binding protein of Helicobacter pylori (Hup) investigated by NMR and other biophysical tools. *International journal of biological macromolecules*, *123*, 446-456.

Jayashree S, Murugavel P, Sowdhamini R, Srinivasan N. (2019). Interface residues of transient protein-protein complexes have extensive intra-protein interactions apart from inter-protein interactions. *Biology direct*, *14*, 1-14.

Jelesarov I, Bosshard HR. (1996). Thermodynamic characterization of the coupled folding and association of heterodimeric coiled coils (leucine zippers). *Journal of molecular biology*, *263*, 344-358.

Jemimah S, Gromiha MM. (2020). Insights into changes in binding affinity caused by disease mutations in protein-protein complexes. *Computers in Biology and Medicine*, *123*, 103829.

Jemimah S, Sekijima M, Gromiha MM. (2020). ProAffiMuSeq: sequence-based method to predict the binding free energy change of protein–protein complexes upon mutation using functional classification. *Bioinformatics*, *36*, 1725-1730.

Jin L, Fendly BM, Wells JA. (1992). High resolution functional analysis of antibody-antigen interactions. *Journal of molecular biology*, *226*, 851-865.

Jones DH, McMillan AJ, Fersht AR, Winter G. (1985). Reversible dissociation of dimeric tyrosyl-tRNA synthetase by mutagenesis at the subunit interface. *Biochemistry*, *24*, 5852-5857.

Kam-Morgan LN, Smith-Gill SJ, Taylor MG, Zhang L, Wilson AC, Kirsch JF. (1993). High-resolution mapping of the HyHEL-10 epitope of chicken lysozyme by site-directed mutagenesis. *Proceedings of the National Academy of Sciences of the United States of America*, *90*, 3958-3962.

Kelley RF, O'Connell MP. (1993). Thermodynamic analysis of an antibody functional epitope. *Biochemistry*, *32*, 6828-6835.

Kelley RF et al. (1992). Antigen binding thermodynamics and antiproliferative effects of chimeric and humanized anti-p185HER2 antibody Fab fragments. *Biochemistry*, *31*, 5434-5441.

Keskin O, Gursoy A, Ma B, Nussinov RJ. (2008). Principles of protein– protein interactions: What are the preferred ways for proteins to interact? *Chem Rev*, *108*, 1225-1244.

Kiel C, Selzer T, Shaul Y, Schreiber G, Herrmann C. (2004). Electrostatically optimized Ras-binding Ral guanine dissociation stimulator mutants increase the rate of association by stabilizing the encounter complex. *Proceedings of the National Academy of Sciences of the United States of America*, *101*, 9223-9228.

Kiel C, Foglierini M, Kuemmerer N, Beltrao P, Serrano L. (2007). A genome-wide Ras-effector interaction network. *Journal of molecular biology*, *370*, 1020-1032.

Kortemme T, Baker D. (2002). A simple physical model for binding energy hot spots in protein-protein complexes. *Proceedings of the National Academy of Sciences of the United States of America*, *99*, 14116-14121.

Kramers HA. (1940). Brownian motion in a field of force and the diffusion model of chemical reactions. *Physica*, *7*, 284-304.

Kresheck GC, Vitello LB, Erman JE. (1995). Calorimetric studies on the interaction of horse ferricytochrome c and yeast cytochrome c peroxidase. *Biochemistry*, *34*, 8398-8405.

Krueger S, Gregurick SK, Zondlo J, Eisenstein E. (2003). Interaction of GroEL and GroEL/GroES complexes with a nonnative subtilisin variant: a small-angle neutron scattering study. *Journal of structural biology*, *141*, 240-258.

Ladbury JE, Chowdhry BZ. (1996). Sensing the heat: the application of isothermal titration calorimetry to thermodynamic studies of biomolecular interactions. *Chemistry & biology, 3*, 791-801.

Lavigne P, Bagu JR, Boyko R, Willard L, Holmes CFB, Sykes BD. (2000). Structure-based thermodynamic analysis of the dissociation of protein phosphatase-1 catalytic subunit and microcystin-LR docked complexes. *Protein Science, 9*, 252-264.

Leder L, Berger C, Bornhauser S, Wendt H, Ackermann F, Jelesarov I, Bosshard HR. (1995). Spectroscopic, calorimetric, and kinetic demonstration of conformational adaptation in peptide-antibody recognition. *Biochemistry, 34*, 16509-16518.

Lee EH, Soper TS, Mural RJ, Stringer CD, Hartman FC. (1987). An Intersubunit Interaction at the Active-Site of D-Ribulose-1,5-Bisphosphate Carboxylase Oxygenase as Revealed by Cross-Linking and Site-Directed Mutagenesis. *Biochemistry, 26*, 4599-4604.

Lee FS, Auld DS, Vallee BL. (1989). Tryptophan Fluorescence as a Probe of Placental Ribonuclease Inhibitor Binding to Angiogenin. *Biochemistry, 28*, 219-224.

Leistler B, Perham RN. (1994). Solubilizing buried domains of proteins: a self-assembling interface domain from glutathione reductase. *Biochemistry, 33*, 2773-2781.

Li YL, Lipschultz CA, Mohan S, Smith-Gill SJ. (2001). Mutations of an epitope hot-spot residue alter rate limiting steps of antigen-antibody protein-protein associations. *Biochemistry, 40*, 2011-2022.

Li L, Zhao B, Cui Z, Gan J, Sakharkar MK, Kangueane P. (2006). Identification of hot spot residues at protein-protein interface. *J Bioinformation, 1*, 121.

Li M, Petukh M, Alexov E, Panchenko AR. (2014). Predicting the impact of missense mutations on protein–protein binding affinity. *Journal of chemical theory and computation, 10*, 1770-1780.

Liu Y, Sturtevant JM. (1995). Significant discrepancies between van't Hoff and calorimetric enthalpies. II. *Protein Science, 4*, 2559-2561.

Luque I, Gomez J, Semo N, Freire E. (1998). Structure-based thermodynamic design of peptide ligands: Application to peptide inhibitors of the aspartic protease endothiapepsin. *Proteins-Structure Function and Genetics, 30*, 74-85.

Luthy JA, Praissman M, Finkenstadt WR, Laskowski M. (1973). Detailed mechanism of interaction of bovine β-Trypsin with soybean trypsin inhibitor (Kunitz) I. Stopped flow measurements. *J Biol Chem, 248*, 1760-1771.

Mainfroid V et al. (1996). Three hTIM mutants that provide new insights on why TIM is a dimer. *Journal of molecular biology, 257*, 441-456.

Makhatadze GI, Privalov PL (1995) Energetics of protein structure. In: Advances in protein chemistry, vol 47. Elsevier, pp 307-425

Martinez JC, Filimonov VV, Mateo PL, Schreiber G, Fersht AR. (1995). A calorimetric study of the thermal stability of barstar and its interaction with barnase. *Biochemistry, 34*, 5224-5233.

McLane KE, Wu X, Conti-Tronconi BM. (1991). Amino acid residues forming the interface of a neuronal nicotinic acetylcholine receptor with κ-bungarotoxin: A study using single residue substituted peptide analogs. *Biochemical and biophysical research communications, 176*, 11-17.

Meenan NA et al. (2010). The structural and energetic basis for high selectivity in a high-affinity protein-protein interaction. *Proceedings of the National Academy of Sciences of the United States of America, 107*, 10080-10085.

Méjean A, Bodenreider C, Schuerer K, Goldberg ME. (2001). Kinetic characterization of the pH-dependent oligomerization of R67 dihydrofolate reductase. *Biochemistry, 40*, 8169-8179.

Milos M, Schaer JJ, Comte M, Cox JA. (1988). Microcalorimetric investigation of the interaction of calmodulin with seminalplasmin and myosin light chain kinase. *J Biol Chem, 263*, 9218-9222.

Mossing MC, Sauer RT. (1990). Stable, Monomeric Variants of Lambda-Cro Obtained by Insertion of a Designed Beta-Hairpin Sequence. *Science, 250*, 1712-1715.

Murphy KP. (1999). Predicting binding energetics from structure: looking beyond ΔG. *Medicinal research reviews, 19*, 333-339.

Murphy KP, Freire E (1992) Thermodynamics of structural stability and cooperative folding behavior in proteins. In: Advances in protein chemistry, vol 43. Elsevier, pp 313-361

Murphy KP, Gill SJ. (1991). Solid model compounds and the thermodynamics of protein unfolding. *Journal of molecular biology*, *222*, 699-709.

Murphy KP, Xie D, Garcia KC, Amzel LM, Freire E. (1993). Structural energetics of peptide recognition: angiotensin II/antibody binding. *Proteins*, *15*, 113-120.

Murphy KP, Freire E, Paterson Y. (1995). Configurational Effects in Antibody-Antigen Interactions Studied by Microcalorimetry. *Proteins-Structure Function and Bioinformatics*, *21*, 83-90.

Myles T, Le Bonniec BF, Betz A, Stone SR. (2001). Electrostatic steering and ionic tethering in the formation of thrombin-hirudin complexes: The role of the thrombin anion-binding exosite-I. *Biochemistry*, *40*, 4972-4979.

Naghibi H, Tamura A, Sturtevant JM. (1995). Significant discrepancies between van't Hoff and calorimetric enthalpies. *Proceedings of the National Academy of Sciences*, *92*, 5597-5599.

Pearce KH, Jr., Cunningham BC, Fuh G, Teeri T, Wells JA. (1999). Growth hormone binding affinity for its receptor surpasses the requirements for cellular activity. *Biochemistry*, *38*, 81-89.

Pearce KH, Ultsch MH, Kelley RF, deVos AM, Wells JA. (1996). Structural and mutational analysis of affinity-inert contact residues at the growth hormone-receptor interface. *Biochemistry*, *35*, 10300-10307.

Petrella EC, Machesky LM, Kaiser DA, Pollard TD. (1996). Structural requirements and thermodynamics of the interaction of proline peptides with profilin. *Biochemistry*, *35*, 16535-16543.

Piehler J, Schreiber G. (1999). Biophysical analysis of the interaction of human ifnar2 expressed in E. coli with IFNα2. *Journal of molecular biology*, *289*, 57-67.

Piehler J, Roisman LC, Schreiber G. (2000). New structural and functional aspects of the type I interferon-receptor interaction revealed by comprehensive mutational analysis of the binding interface. *J Biol Chem*, *275*, 40425-40433.

Privalov PL, Gill SJ (1988) Stability of protein structure and hydrophobic interaction. In: Advances in protein chemistry, vol 39. Elsevier, pp 191-234

Radić Z, Kirchhoff PD, Quinn DM, McCammon JA, Taylor P. (1997). Electrostatic influence on the kinetics of ligand binding to acetylcholinesterase distinctions between active center ligands and fasciculin. *J Biol Chem*, *272*, 23265-23277.

Rajpal A, Taylor MG, Kirsch JF. (1998). Quantitative evaluation of the chicken lysozyme epitope in the HyHEL-10 Fab complex: Free energies and kinetics. *Protein Science*, *7*, 1868-1874.

Raman CS, Jemmerson R, Nall BT, Allen MJ. (1992). Diffusion-limited rates for monoclonal antibody binding to cytochrome c. *Biochemistry*, *31*, 10370-10379.

Raman CS, Allen MJ, Nall BT. (1995). Enthalpy of antibody--cytochrome c binding. *Biochemistry*, *34*, 5831-5838.

Richter S, Wenzel A, Stein M, Gabdoulline RR, Wade RC. (2008). webPIPSA: a web server for the comparison of protein interaction properties. *Nucleic acids research*, *36*, W276-W280.

Roberts S, Cheetham JC, Rees AR. (1987). Generation of an antibody with enhanced affinity and specificity for its antigen by protein engineering. *Nature*, *328*, 731-734.

Schramm HJ, Billich A, Jaeger E, Rucknagel KP, Arnold G, Schramm W. (1993). The inhibition of HIV-1 protease by interface peptides. *Biochemical and biophysical research communications*, *194*, 595-600.

Schreiber G. (2002). Kinetic studies of protein-protein interactions. *Current opinion in structural biology*, *12*, 41-47.

Schreiber G, Fersht AR. (1995). Energetics of protein-protein interactions: Analysis ofthe Barnase-Barstar interface by single mutations and double mutant cycles. *Journal of molecular biology*, *248*, 478-486.

Schreiber G, Fersht AR. (1996). Rapid, electrostatically assisted association of proteins. *Nature structural biology*, *3*, 427-431.

Schreiber G, Buckle A, Fersht A. (1994). Stability versus function: two competing forces in the evolution of barstar. *J Structure*, *2*, 945-951.

Schwarz FP, Tello D, Goldbaum FA, Mariuzza RA, Poljak RJ. (1995). Thermodynamics of antigen-antibody binding using specific anti-lysozyme antibodies. *European Journal of Biochemistry, 228*, 388-394.

Selzer T, Schreiber G. (1999). Predicting the rate enhancement of protein complex formation from the electrostatic energy of interaction. *Journal of molecular biology, 287*, 409-419.

Selzer T, Schreiber G. (2001). New insights into the mechanism of protein-protein association. *Proteins-Structure Function and Bioinformatics, 45*, 190-198.

Selzer T, Albeck S, Schreiber G. (2000). Rational design of faster associating and tighter binding protein complexes. *Nature structural biology, 7*, 537-541.

Sheinerman FB, Norel R, Honig B. (2000). Electrostatic aspects of protein-protein interactions. *Current opinion in structural biology, 10*, 153-159.

Shih AJ, Purvis J, Radhakrishnan R. (2008). Molecular systems biology of ErbB1 signaling: bridging the gap through multiscale modeling and high-performance computing. *Mol Biosyst, 4*, 1151-1159.

Spolar RS, Livingstone JR, Record MT. (1992). Use of Liquid-Hydrocarbon and Amide Transfer Data to Estimate Contributions to Thermodynamic Functions of Protein Folding from the Removal of Nonpolar and Polar Surface from Water. *Biochemistry, 31*, 3947-3955.

Stebbins JW, Kantrowitz E. (1989). The importance of the link between Glu204 of the catalytic chain and Arg130 of the regulatory chain for the homotropic and heterotropic properties of Escherichia coli aspartate transcarbamoylase. *J Biol Chem, 264*, 14860-14864.

Stein M, Gabdoulline RR, Wade RC. (2007). Bridging from molecular simulation to biochemical networks. *Current opinion in structural biology, 17*, 166-172.

Stites WE. (1997). Protein-protein interactions: Interface structure, binding thermodynamics, and mutational analysis. *Chem Rev, 97*, 1233-1250.

Stone SR, Hermans JM. (1995). Inhibitory mechanism of serpins. Interaction of thrombin with antithrombin and protease nexin 1. *Biochemistry, 34*, 5164-5172.

Stone SR, Dennis S, Hofsteenge J. (1989). Quantitative evaluation of the contribution of ionic interactions to the formation of the thrombin-hirudin complex. *Biochemistry, 28*, 6857-6863.

Sturtevant JM. (1977). Heat capacity and entropy changes in processes involving proteins. *Proceedings of the National Academy of Sciences of the United States of America, 74*, 2236-2240.

Sydor JR, Engelhard M, Wittinghofer A, Goody RS, Herrmann C. (1998). Transient kinetic studies on the interaction of Ras and the Ras-binding domain of c-Raf-1 reveal rapid equilibration of the complex. *Biochemistry, 37*, 14292-14299.

Taylor MG, Rajpal A, Kirsch JF. (1998). Kinetic epitope mapping of the chicken lysozyme center dot HyHEL-10 Fab complex: Delineation of docking trajectories. *Protein Science, 7*, 1857-1867.

Tayyari F. (2013). Development Of Isotags For Nmr Based Metabolite Profiling And Applications.

Tello D, Goldbaum FA, Mariuzza RA, Ysern X, Schwarz FP, Poljak RJ (1993) Three-dimensional structure and thermodynamics of antigen binding by anti-lysozyme antibodies. Portland Press Ltd.,

Tello D, Eisenstein E, Schwarz FP, Goldbaum FA, Fields BA, Mariuzza RA, Poljak RJ. (1994). Structural and physicochemical analysis of the reaction between the anti-lysozyme antibody D1. 3 and the anti-idiotopic antibodies E225 and E5. 2. *Journal of Molecular Recognition, 7*, 57-62.

Tomić S, Bertoša B, Wang T, Wade RC. (2007). COMBINE analysis of the specificity of binding of Ras proteins to their effectors. *Proteins: Structure, Function, and Bioinformatics, 67*, 435-447.

Tse C, Wickstrom L, Kvaratskhelia M, Gallicchio E, Levy R, Deng N. (2020). Exploring the free-energy landscape and thermodynamics of protein-protein association. *Biophysical journal, 119*, 1226-1238.

Varadarajan R, Connelly PR, Sturtevant JM, Richards FM. (1992). Heat capacity changes for protein-peptide interactions in the ribonuclease S system. *Biochemistry, 31*, 1421-1426.

Vijayakumar M, Wong KY, Schreiber G, Fersht AR, Szabo A, Zhou HX. (1998). Electrostatic enhancement of diffusion-controlled protein-protein association: comparison of theory and experiment on barnase and barstar. *Journal of molecular biology, 278*, 1015-1024.

Wallis R, Moore GR, James R, Kleanthous C. (1995). Protein-protein interactions in colicin E9 DNase-immunity protein complexes. 1. Diffusion-controlled association and femtomolar binding for the cognate complex. *Biochemistry, 34*, 13743-13750.

Wang T, Wade RC. (2002). Comparative binding energy (COMBINE) analysis of OppA-peptide complexes to relate structure to binding thermodynamics. *J Med Chem, 45*, 4828-4837.

Wang Y, Shen B-J, Sebald W. (1997). A mixed-charge pair in human interleukin 4 dominates high-affinity interaction with the receptor α chain. *Proceedings of the National Academy of Sciences, 94*, 1657-1662.

Wang W, Xu W-X, Levy Y, Trizac E, Wolynes P. (2009). Confinement effects on the kinetics and thermodynamics of protein dimerization. *Proceedings of the National Academy of Sciences, 106*, 5517-5522.

Ward WH, Jones DH, Fersht AR. (1987). Effects of engineering complementary charged residues into the hydrophobic subunit interface of tyrosyl-tRNA synthetase. *Biochemistry, 26*, 4131-4138.

Weber G. (1993). Thermodynamics of the Association and the Pressure Dissociation of Oligomeric Proteins. *J Phys Chem-Us, 97*, 7108-7115.

Weber G. (1995). van't Hoff revisited: enthalpy of association of protein subunits. *The Journal of Physical Chemistry B, 99*, 1052-1059.

Weber G. (1996). Persistent confusion of total entropy and chemical system entropy in chemical thermodynamics. *Proceedings of the National Academy of Sciences of the United States of America, 93*, 7452-7453.

Weinkam P, Chen YC, Pons J, Sali A. (2013). Impact of mutations on the allosteric conformational equilibrium. *Journal of molecular biology, 425*, 647-661.

Wells JA (1991) [18] Systematic mutational analyses of protein-protein interfaces. In: Methods in enzymology, vol 202. Elsevier, pp 390-411

Wells JA, de Vos AM. (1996). Hematopoietic receptor complexes. *Annual review of biochemistry, 65*, 609-634.

Wiseman T, Williston S, Brandts JF, Lin LN. (1989). Rapid Measurement of Binding Constants and Heats of Binding Using a New Titration Calorimeter. *Anal Biochem, 179*, 131-137.

Xavier KA, Willson RC. (1998). Association and dissociation kinetics of anti-hen egg lysozyme monoclonal antibodies HyHEL-5 and HyHEL-10. *Biophysical journal, 74*, 2036-2045.

Xavier KA, McDonald SM, McCammon JA, Willson RC. (1999). Association and dissociation kinetics of bobwhite quail lysozyme with monoclonal antibody HyHEL-5. *Protein engineering, 12*, 79-83.

Yakovlev GI et al. (1995). Dissociation constants and thermal stability of complexes of Bacillus intermedius RNase and the protein inhibitor of Bacillus amyloliquefaciens RNase. *FEBS letters, 366*, 156-158.

Zhang C, Liu S, Zhu QQ, Zhou YQ. (2005). A knowledge-based energy function for protein-ligand, protein-protein, and protein-DNA complexes. *J Med Chem, 48*, 2325-2335.

Zhou HX. (2003). Association and dissociation kinetics of colicin E3 and immunity protein 3: convergence of theory and experiment. *Protein science : a publication of the Protein Society, 12*, 2379-2382.

Zhou H, Zhang C, Liu S, Zhou Y. (2005). Web-based toolkits for topology prediction of transmembrane helical proteins, fold recognition, structure and binding scoring, folding-kinetics analysis and comparative analysis of domain combinations. *Nucleic acids research, 33*, W193-197.

Evolution-Structure Paradigm of Protein Complexes

4

Abstract

The cell carries out many sophisticated functions by employing protein complexes that comprised of several interacting partners. Therefore, it is essential to identify these complexes and also to study their structure and mechanism of formation in order to understand the cellular machinery more lucidly. This chapter discusses all such aspects of protein complexes along with some of their applications. Further, it classifies the complexes according to several criteria and provides an account of how they have evolved over time to suit the requirements of the living cell. The chapter also throws light into various patented protein complexes.

Keywords

Homomeric protein complexes · Heteromeric protein complexes · Evolution of protein complexes · Inflammasome · Chaperonin

4.1 Introduction

The environment within a living cell is so crowded that proteins often constantly associate with other protein partners and macromolecules (Ellis 2001; McGuffee and Elcock 2010). The interactions between the proteins and the resulting complexes are highly diverse with respect to factors such as specificity, frequency, and duration. Several proteins are indulged in the formation of heteromeric or homomeric complexes with well-specified molecular functions. These complexes usually being long-lived have been subjected to experimental analysis and characterization. However, numerous proteins in cell are indulged in promiscuous, short-lived interactions owing to crowding in the intracellular environment (Landry et al. 2013; Levy et al. 2012). Such interactions are usually not that relevant biologically and have not been subjected to evolutionary selection. Despite this, there are several

K. M. Poluri et al., *Protein-Protein Interactions*,
https://doi.org/10.1007/978-981-16-1594-8_4

transient protein interactions which play imperative cellular roles, especially in the cell signalling related processes (Perkins et al. 2010; Tompa et al. 2014).

The dynamic nature of the living cell along with various types of molecular interactions makes scientists ponder about the true definition of a protein complex. It is vital to address questions such as: What interactions genuinely denote the formation of protein complexes? How do we view non-specific and/or transient complexes? What should be the threshold while distinguishing stable protein complexes from that of ultra-transient ones? Since it is highly improbable to acquire information about the cell's total interactome, the concept of protein complexes in the present day has been largely dependent on the data furnished by several characterization methods, such as experimental or computational techniques. For instance, researchers from the proteomics field have succeeded in co-purifying proteins and defined the entire set of such co-purified proteins to comprise a protein complex (Gavin et al. 2006; Havugimana et al. 2012; Krogan et al. 2006).

In the current chapter, protein complexes that are capable of being purified and characterized utilizing available structural data have been discussed comprehensively. The chapter is focused on several multi-protein complexes that possess significant cellular functions. Further, the structural organization of protein complexes and the associated dynamics such as the mechanism of assembly of individual subunits to form the final complex have been addressed. Lastly, the chapter also throws light on how the protein complexes have evolved, and the factors that influence the evolution process.

4.2 Molecular Functions of Protein Complexes

Protein complexes play a multitude of roles in the living cells starting from prokaryotic organisms such as bacteria to the most evolved species on the planet. Few well-known protein complexes, along with their fundamental structures and functions are outlined in the following paragraphs.

Stressosome is a protein complex, comprising of multiple protein partners. It acts as a hub to transduce and integrate signals in several bacterial species (Pané-Farré et al. 2017). It regulates gene transcription in the stress response mechanism in bacteria such as *Bacillus subtilis*. After receiving the cognate signal, this protein complex relays the signal for initiating a phosphorylation-reliant partner switching cascade. This ultimately results in the activation of the alternative sigma factor σB, which influences the transcription of more than a hundred stress-response genes. Computational analyses have demonstrated that genetic elements that code for the major stressosome proteins are found in several species of bacteria, thus signifying the advantage of these complexes from an evolutionary aspect (Pané-Farré et al. 2017).

Inflammasome is a multi-protein complex which is indulged in imperative function of inducing an inflammatory response to stimuli that may possess danger to the cell. Ligands derived exogenously and/or endogenously are recognized in the cell cytoplasm that leads to the initiation of conformational changes in the proteins like

NLRP3 (Monie 2017). These proteins are sensor proteins that enable adaptor proteins such as ASC to be recruited. As a result of this recruitment, an inflammatory hub is gradually assembled. The complex assembly process is governed by not only interactions between sensor proteins, but also adaptor–adaptor and sensor–adaptor interactions. The assembly ultimately results in a protein complex that measures around 1 micron in diameter. This complex varies with respect to its stoichiometry and composition. The inflammasome is responsible for triggering effector caspases (caspase-8, caspase-1) in a proximity-induced manner. This induction consequently activates inflammatory cytokines such as pro-IL1β and pro-IL18 and cleaves Gasdermin D, which acts as a critical player of cell death (Monie 2017).

The *ferritin* protein complex is known for its special role in storing iron and retaining its homeostasis in the cell. When the gene that codes for the H subunit of ferritin is deleted, early embryonic death is observed in mice models (Yévenes 2017). On the other hand, when the gene encoding the L subunit in humans is mutated, certain neurodegenerative diseases have been observed to set in. Therefore, it has become imperative for scientists to understand the exact working mechanism of this protein complex. Several research groups have conducted studies on bacterioferritins and their human counterparts. From the studies, the general conclusion is that all ferritins have some common requisites: (a) center for catalyzing oxidation of iron, (b) routes of iron entry to the center, and (c) the capacity of nucleating an iron core. Despite these fundamental similarities, the ferritins differ in their structural features and mechanistic aspects of iron mineralization and oxidation (Yévenes 2017).

Peroxiredoxins (Prxs) are complexes belonging to the peroxidase family. They are regarded as the most vital players in protecting cells from oxidative stress in almost every living organism. Prxs comprise a thioredoxin fold along with an active site containing peroxidatic cysteine that is capable of reducing substrates (peroxide) (Cao and Lindsay 2017). To define the functional unit of Prxs minimally, they exist as either dimers or monomers. However, often several dimers associate to form decameric rings. The structures of these rings can further assist in the formation of different high molecular weight complexes. Several Prxs in eukaryotes comprise a conserved C-terminal YF motif and GGLG. These two components aid in the sensing mechanism of the protein to elevated peroxide levels, resulting in hyperoxidation and inactivation. Enzyme sulfiredoxin can re-reduce inactivated Prxs in an ATP-dependent reaction. Hyperoxidation, followed by reactivation is important in H_2O_2-mediated cell signalling not only in stress, but also in non-stress conditions. Prx protein complexes are also known to display chaperone-like functions under conditions of thermal stress or oxidative stress (Cao and Lindsay 2017).

α2-macroglobulins are multi-domains proteins that act as endopeptidase inhibitors in wide range of organisms, starting from metazoans (invertebrates and vertebrates) to gram-negative bacteria. These proteins are large macromolecules and comprised of 1474 amino acid residues in the human homologs. They are made of 11 domains, out of which the first 7 comprise antiparallel β-sandwiches (Garcia-Ferrer et al. 2017). These 7 domains are referred to as macroglobulin-like

(MG) domains, named from MG1 to MG7. α2-macroglobulins utilize two related mechanisms to carry out their function. The α2-macroglobulin homolog in humans is involved in the regulation of proteolysis in intricate biological processes, including tissue remodelling, cell signalling, and nutrition. Further, it is also involved in the defence against external virulence factors and toxins during circumstances of envenomation and infection. Additionally, it carries out other biological functions via the modification of activities of hormones, cytokines, lipid factors, growth factors, other proteins, etc. (Garcia-Ferrer et al. 2017).

Genetic information in each of the living organism is transmitted via a two-step process. First step involves RNA transcription from the DNA. Second step involves protein translation in which RNA is utilized as template by the cell to synthesize proteins. The first process, i.e., transcription is performed by *RNA polymerases (RNAPs)* not only in bacteria but also in eukaryotes and archae. RNAPs are multi-subunit protein complexes that have a conserved core. These protein complexes are involved in the catalysis of RNA polymerization from nucleotide triphosphate (NTP) building blocks. This is a highly accurate process and employs the DNA as template. Further, all the three steps of transcription (initiation, elongation, and termination) are assisted by various other proteins, commonly known as transcription factors. The fact that this process is highly dynamic and intricate is reflected in the complex network of PPIs in this transcription complex. Furthermore, this complex experiences severe conformational changes as it progresses through various phases of the transcription cycle (Griesenbeck et al. 2017)

Dihydrodipicolinate synthase is a protein complex that carries out the first step in lysine biosynthesis. This enzyme, although found in bacteria, plants, fungi, archaea, is not produced in mammals. The structural characterization of these protein complexes from different plant and bacterial species has demonstrated that they are considerably variable with regard to their quaternary structure, implying their evolution. This protein in its oligomeric state is very vital, both in allosteric regulation and catalysis. In majority of the plants and bacteria, this protein complex exists in the form of tetrameric enzymes, with conserved dimeric structures. However, the arrangement of the dimers is often found to differ (Pearce et al. 2017)

Pyruvate carboxylase is a protein complex involved in cell metabolism. It drives the tricarboxylic acid cycle (TCA), utilizing one of the intermediates of the cycle. Further, it is involved in the first step of gluconeogenesis (Valle 2017). The protein complex is large and multifunctional. It occurs in the form of a tetramer containing 4 identical subunits. Each of its subunits possesses two active sites which are responsible for the catalysis of two successive reactions. This results in pyruvate carboxylation to oxaloacetate, and the generation of a binding site for acetyl-CoA, that functions as an allosteric regulator of the enzyme. The oligomers of this protein complex covalently interact with biotins to mediate carboxyl group transfer between active sites that are distantly located. Research on this complex has revealed that domains undergo large movements that influence even the overall quaternary organization of the tetramers during catalysis (Valle 2017).

Ubiquitination is a type of post-translational modification (PTM) that is highly conserved and occurs in eukaryotes. In this process, there is a degradation of target proteins by the 26S proteasome. Proteins that are intended for proteasomal degradation are chosen by the *E3 ubiquitin ligases*, which are essentially modular complexes. These protein complexes are subject to tight regulation in the cell. The largest and most common superfamily of these ligases are Cullin-RING E3 ubiquitin ligases (CRLs) that have more than 400 members occurring in mammals (Nguyen et al. 2017)

Glyceraldehyde-3-phosphate dehydrogenase (GAPDH) is a multi-subunit protein complex that performs a significant role in glycolysis. Apart from this, it has been observed to play other key cellular functions as well. This protein complex is homotetrameric and is usually regarded as a dimer of dimers (White 2017). It comprised of 4 subunits, that are identical, namely O, P, Q, and R. The functions performed by this complex are driven by bio-molecular interactions, protein oligomerization, variations in subcellular localization, and post-translational modifications. Various regulatory mechanisms and functions of GAPDH are found to overlap with each other and converge while performing its role in intermediary metabolism. The regulation and function of this complex are dictated by the different determinants of its structure. GAPDH is ubiquitously found in almost every domain of life. Furthermore, this protein complex has been associated with several diseases, such as cardiovascular, pathogenic, degenerative, tumorigenic, and diabetic disorders (White 2017).

Spindle assembly checkpoint (SAC) is a system that acts as a vital cellular guardian. When there are aberrations in the cell division process, chromosomes can be gained, lost or even rearranged. If these chromosomal mistakes are severe, it can lead to spontaneous miscarriages and cell lethality. On the other hand, abnormalities in the SAC may result in various disorders such as birth defects, cancer, and even developmental anomalies such as Down's syndrome (Bolanos-Garcia 2017). The SAC is a signalling platform that is self-regulated and evolutionarily conserved. It governs the segregation of chromosomes during mitosis in an accurate and time dependent manner. It is driven by the dynamic and coordinated interactions between a multi-protein complex known as kinetochore and the microtubules of the spindle fiber. The kinetochore is responsible for attaching the chromosomes to the microtubules such that the sister chromatids can be pulled apart during the cell division (Bolanos-Garcia 2017).

Chaperonin is a protein complex which acts as a molecular chaperone and enables proteins to acquire their native conformation. First, it aids in precluding the folding of proteins during membrane translocation or synthesis. Second, it mediates their proper folding post-synthesis in an ATP-dependent step-wise manner (Ishii 2017). In the GroEL–GroES complex, one GroES ring is bound by a single heptameric ring of GroEL when ATP/ADP is present. The double ring tetradecamer of GroEL presents itself two different types of complexes: (a) a "bullet"-shaped, asymmetric GroEL:GroES complex in the 1:1 ratio and (b) a "football" shaped, symmetric GroEL:GroES2 complex in the 1:2 ratio. It is not exactly clear as to which of these two complexes are essential for effective protein folding. Furthermore, the mechanism of co-ordination of GroES with GroEL in the reaction cycle of chaperonin with regard to adenine-nucleotide regulation and substrate protein interplay is not yet fully elucidated (Ishii 2017).

Aminoacyl-tRNA synthetases (AARSs) are protein systems responsible for specifically aminoacylating one tRNA molecule with its related amino acid residue. For every residue, there is one such protein system, i.e. there are 20 types of such complexes in total. These systems couple a residue to a cognate t-RNA, governed by the anticodon or the RNA triplet that interprets the genetic code. Apart from this conventional role in translation, these protein platforms have non-canonical roles. They are moonlighting proteins that have multiple physiological functions. It has been observed that nine synthetases in mammals, specific for residues Arg, Asp, Gln, Glu, Leu, Lys, Ile, Pro, and Met assembled into a complex comprising of multiple aminoacyl-tRNA synthetases. This assembly is believed to be of crucial importance not only in the process of translation, but also in complex regulatory functions of the enzymes. The disassembly and the assembly of this supramolecular complex dictate how the alternative functions are balanced against each other. Therefore, studies on this complex's structural organization have gained importance in the recent times. The native complex with molecular weight of around 1.5 MDa has not yet been elucidated structurally at high resolution. Only low-resolution structures of the multi-aminoacyl-tRNA synthetase complex have been reported. Only for sub-complexes or individual proteins of the complex, high-resolution structural data has been generated (Mirande 2017)

2-oxoacid dehydrogenase complexes (2-OADC) are a family of very important protein complexes. Pyruvate dehydrogenase multi-enzyme complex (PDC) is the most well-known member of this family. These complexes are essentially enormous molecular machines that serve key roles in citric acid cycle flux (OGDC, 2-oxoglutarate dehydrogenase), glucose homeostasis (PDC), and branched-chain amino acid (leucine, valine, and isoleucine) metabolism (BCOADC, branched-chain 2-OADC). These complexes have been studied extensively to serve as model complexes to analyze and understand substrate channelling, enzyme cooperativity, activity regulation by phosphorylation, and PPIs (Byron and Lindsay 2017)

The mechanism behind how proteins have mechanical contribution to a living cell can be easily understood by studying the *clathrin* system, that not only has an unusual structure, but is also capable of assembling and dissembling quickly inside the cells (Halebian et al. 2017). Units called triskelions assemble to assist in the formation of clathrin cages. The function that clathrin performs is dependent on the triskelions' structures as well as the intra- and inter-interactions with adaptor proteins during the process of assembly. Extensive studies have revealed that the three-legged structure serves as the fundamental building unit for assembling clathrin. 3 heavy chain subunits, each having a molecular weight of about 190 kDa are present in clathrin. Each of these subunits is associated with a light chain subunit of 25 kDa molecular weight. As the heavy chains are thin, lengthy and have tight interactions at their C-terminals, the resulting assembly appears like a triskelion or "Isle of Man" three-legged structure. For clathrin-coated vesicles to perform their functions in the cell, it is essential for clathrin triskelions to assemble into clathrin cages. Clathrin triskelions that have been successfully purified from sources such as liver or brain tissue have been found to be capable of spontaneously assembling into cages in presence of calcium/magnesium ions and high proton concentrations. The formation of cages is balanced energetically and necessitates precise interactions between legs of the triskelions. Clathrin-coated vesicles have an average size of 50 to 80 nm in cells, while in vitro, they have been observed to generate cages having variable structures, including some symmetrical ones (Halebian et al. 2017).

4.3 Classification of Protein Complexes

The quaternary structure of a protein complex is usually defined by the way in which the constituting, individual subunits/protein chains are arranged with respect to each other. In simple words, a protein complex is defined as homomeric if it is made of several copies of a single subunit. On the other hand, a protein complex is defined as heteromeric if it consists of more than one type of subunit. The multiple, different subunits in heteromeric complexes are usually coded by disparate genes. However, there are certain heteromers, where different subunits can also arise via the cleavage of a single polypeptide chain.

4.3.1 Protein Self-Assembly Into Homomeric Complexes

Structural data on homomeric protein complexes are more widely available than that of heteromeric complexes. This is not due to the fact that researchers have focused more on the former due to its high inherent propensity of individual protein molecules to self-assemble into homomeric complexes. Most of the homomeric protein complexes that have been characterized have been found to be symmetric (Klotz et al. 1970; Venkatakrishnan et al. 2010). Homomeric complexes studied till date have been categorized into the following few groups such as (1) Twofold dimeric complexes; (2) Cyclic complexes; (3) Dihedral complexes; (4) Cubic complexes; (5) Helical complexes; and (6) Asymmetric homomers as described by Marsh and Teichmann (2015).

4.3.1.1 Two-Fold Dimeric Complexes

These complexes are denoted by *C2* symmetry group. In other words, two subunits in such complexes are related via a twofold rotational symmetry axis. The presence of this twofold axis necessitates that the two involved subunits interact with each other in a symmetric fashion utilizing identical patches on their surfaces. Such interfaces are also known as head-to-head or isologous interfaces (Fig. 4.1).

4.3.1.2 Cyclic Complexes

These complexes are represented by *Cn* ($n>2$) symmetry groups. In other words, they are denoted by rotational symmetry axes of higher orders (for example,

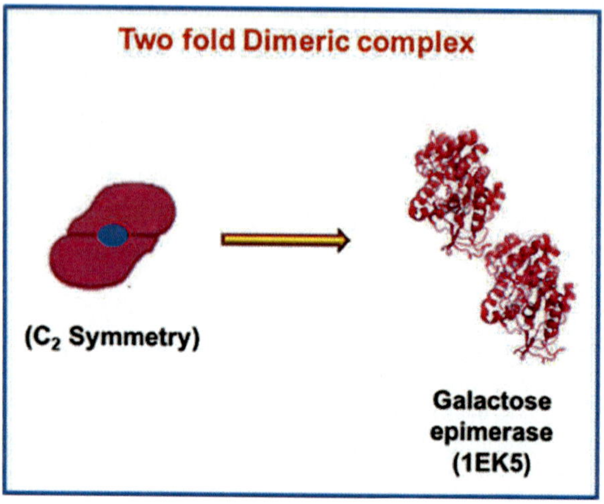

Fig. 4.1 Depicting the quaternary structure topology of twofold dimeric complex in which the interface between the subunits is marked as blue circle and an illustrative protein with C2 symmetry is displayed on the other side of the arrow. PDB ID is mentioned in parenthesis

threefold in *C*3 and fourfold in *C*4 complexes). The interfaces involved in these complexes include interactions that are asymmetric in nature and are mediated by unlike surfaces on the interacting interfaces. These interfaces are also known as head-to-tail or heterologous interfaces. Technically, the twofold dimeric complexes also fall under this group. However, they are put into a separate category as they do not form closed rings, unlike these cyclic complexes (Fig. 4.2).

4.3.1.3 Dihedral Complexes

These complexes belong to *Dn* ($n>1$) symmetry groups and are defined by two symmetry axes that are orthogonal. One of the axes is a twofold rotational symmetry axis, while the other is either a twofold or even higher orders. Thus, *D*2 complexes are in essence, best defined as "dimer of dimers," while the *D*3 complexes either denote a "trimer of dimers" or a "dimer of trimers" (Fig. 4.3).

4.3.1.4 Cubic Complexes

These complexes are represented by octahedral (*O*), tetrahedral (*T*), or icosahedral (*I*) symmetry group. T homomers have twelve subunits having twofold and threefold symmetry axes. O homomers have twenty-four subunits with twofold, threefold, and fourfold axes of symmetry. I homomers have sixty subunits with 2-fold, 3-fold, and 5-fold axes of symmetry (Fig. 4.4).

4.3.1.5 Helical Complexes

These complexes bear helical (*H*) symmetry. Helical symmetry is defined as translational motion combined with rotation, as in a screw. Unlike the other types of complexes, helical complexes are open and do not represent closed symmetry. Theoretically, these complexes have the capacity to indefinitely assemble. Therefore, in real-life conditions, the number of subunits existing in helical complexes is

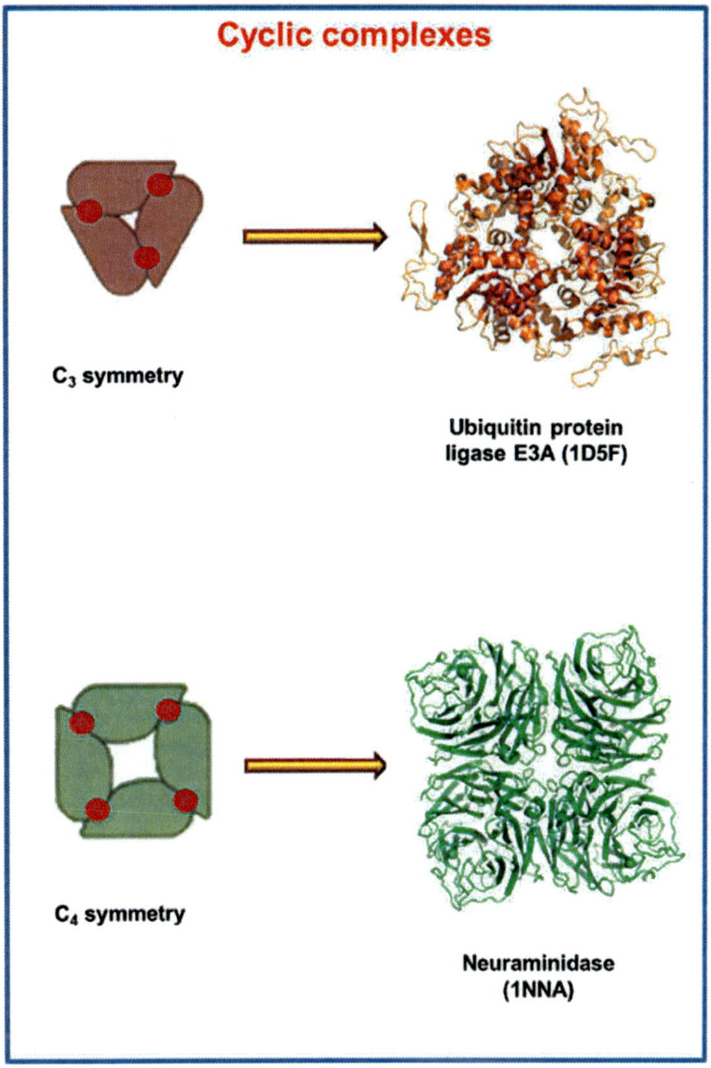

Fig. 4.2 Schematic depicting the quaternary structure topology of cyclic complexes in which the interfaces between the subunits are marked as maroon circles and illustrative proteins with C_3 and C_4 symmetry are displayed on the other side of the arrows. PDB IDs are mentioned in parenthesis

limited by factors such as concentrations of subunits, steric restrictions, and association-dissociation kinetics (Fig. 4.5).

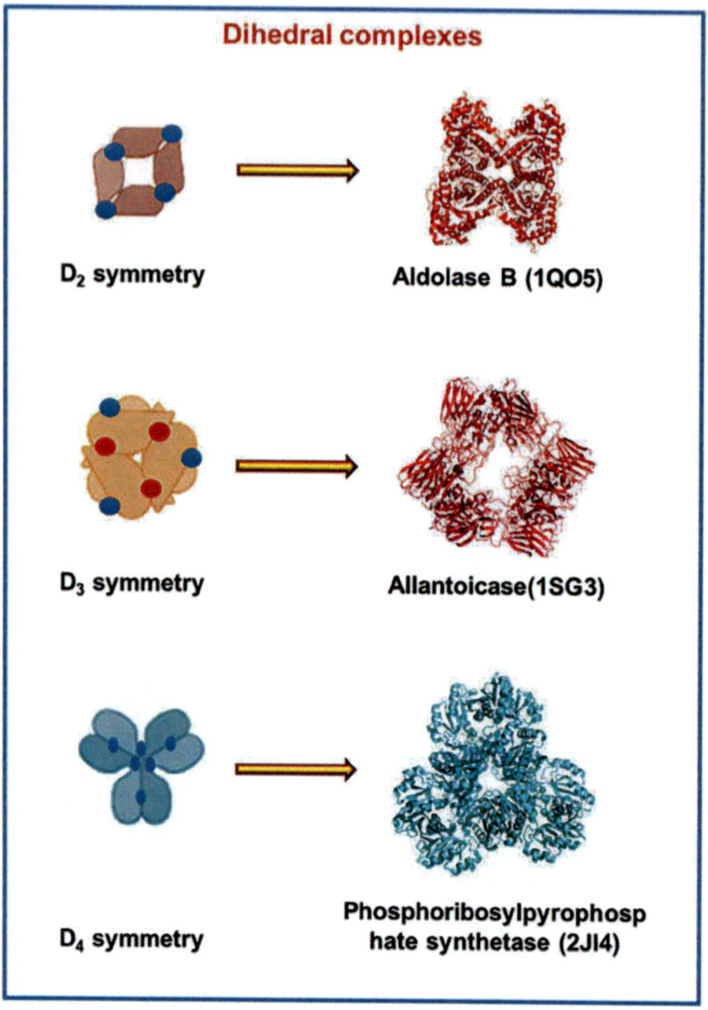

Fig. 4.3 Illustrating the quaternary structure topology of dihedral complexes in which the interfaces between the subunits are marked as blue and maroon circles and the illustrative proteins with D_2, D_3, and D_4 symmetry are displayed on the other side of the arrows. PDB IDs are mentioned in parenthesis

4.3.1.6 Asymmetric Homomers

These protein complexes are denoted by trivial symmetry group $C1$. Such complexes can be formed in many ways, but according to definition, the different types of subunits must be present in non-equivalent positions in such complexes (Fig. 4.6).

It is alluring to note that in homomeric structures the distributions of the different types of symmetry groups vary widely. Symmetric dimers constitute a very small fraction of structures determined by electron microscopy (EM), whereas they

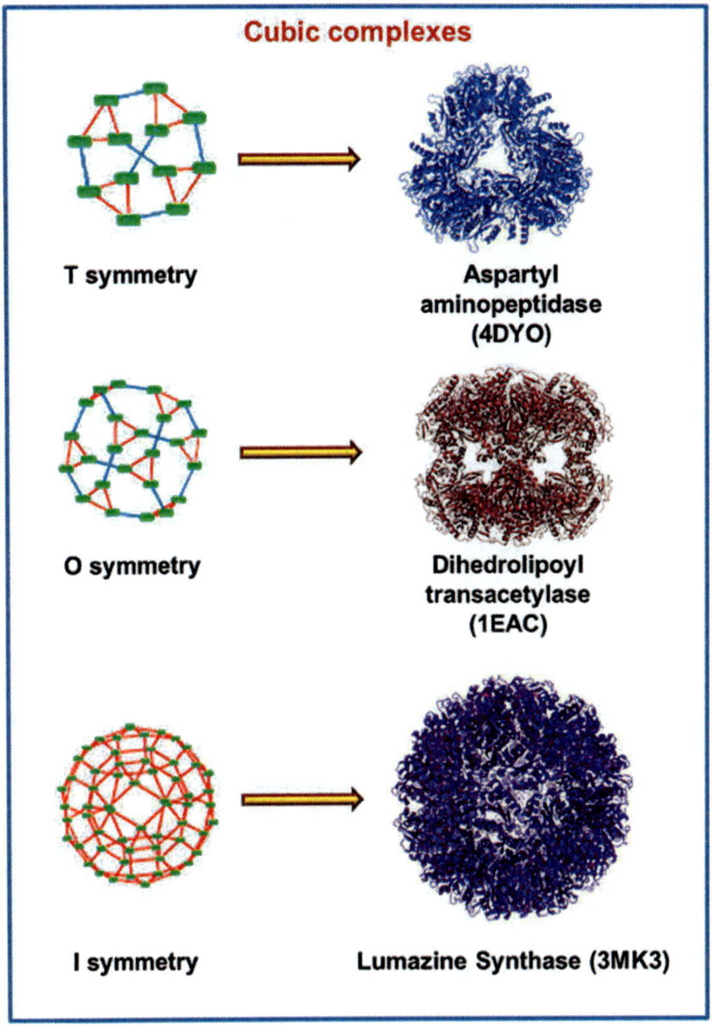

Fig. 4.4 Depicting the quaternary structure topology of cubic complexes in which the interfaces between the subunits are marked as red and blue lines, and illustrative proteins with T, O, and I symmetry are displayed on the other side of the arrows. PDB IDs are mentioned in parenthesis

comprise the majority of NMR and X-ray crystallography-determined structures. Structures of homomers that have cubic and cyclic symmetries are mostly determined using EM. Researchers have reasoned this observation stating that very large structures, usually having repeated subunits, are subjected to EM techniques. In a nutshell, the studies and growing information on these complexes have emphasized the obligation of employing several techniques to furnish data on quaternary structure in even more detail.

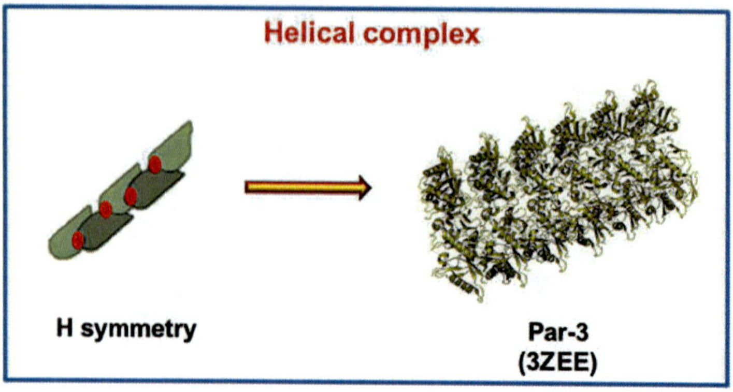

Fig. 4.5 Schematic depicting the quaternary structure topology of helical complex in which the interface between the subunits is marked as red circle and an illustrative protein with H symmetry is displayed on the other side of the arrow. PDB ID is mentioned in parenthesis

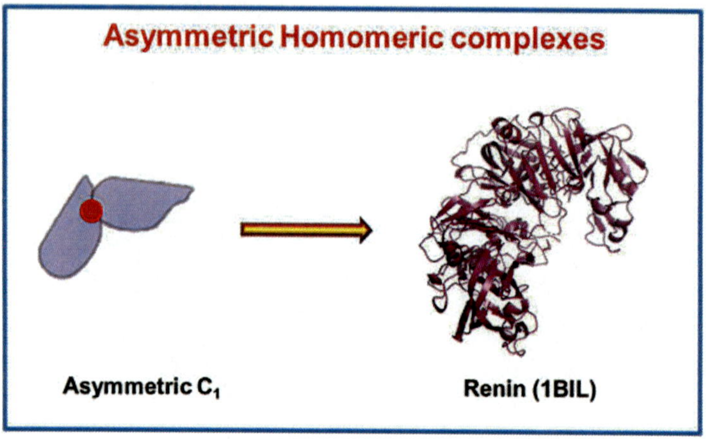

Fig. 4.6 Depicting the quaternary structure topology of asymmetric homomeric complexes in which the interface between the subunits is marked as maroon circle and an illustrative protein with C_1 asymmetry is displayed on the other side of the arrow. PDB ID is mentioned in parenthesis

Furthermore, complexes that have been defined as symmetric globally often comprised of certain deviations from perfect symmetry. For instance, in a protein complex, the various copies of a single subunit can demonstrate either small or large-scale variations structurally although it is very complex to analyze these variations utilizing constraints of non-crystallographic symmetry (Kleywegt 1996). NMR studies on these complexes, however, have not evidenced the existence of any such asymmetry (Godoy-Ruiz et al. 2011). These results imply that fluctuations regarding asymmetry in structure are not uncommon and may occur as a result of energetically preferred states in crystal, that are otherwise averaged out on the NMR timescale in solution. Therefore, it is not unusual for one to observe mixing of local symmetry elements as well. For example, researchers have observed *C*4 symmetry in the intramembrane channel in the tetrameric AMPA subtype glutamate receptor, but *C*2 symmetry in its extracellular domains (Sobolevsky et al. 2009).

4.3.2 Quaternary Structure Diversity of Heteromeric Complexes

Researchers have not yet explored quaternary structures of heteromers in detail and there is much less data available for heteromers in comparison to homomers. However, this necessarily does not reflect the in vivo scenario, since there are several heteromeric protein complexes present in the cell. For instance, it has been observed that protein complexes that have been isolated and purified from native sources (such as tissue) comprise more heteromeric subunits than that generated employing recombinantly synthesized proteins (Marsh and Teichmann 2014b; Perica et al. 2012b). This observation is further strengthened by the fact that several proteins have been erected to interact with other proteins in vivo (Tarassov et al. 2008) or have been co-purified as components of heteromeric complexes (Gavin et al. 2006; Havugimana et al. 2012; Krogan et al. 2006). Even in one of the organisms with the smallest genome, *Mycoplasma pneumoniae*, studies have revealed that only 35% of the soluble protein complexes in the cell were homomeric in nature (Kühner et al. 2009). The authors have speculated that having a relatively small number of different proteins in its cell, the upper limit of homomeric complexes might be somewhere close to 47%, making heteromeric protein complexes still the majority in this organism. Despite such a bias against heteromeric protein complexes, several research groups have conducted their studies on their structural arrangements at the quaternary level. Heteromeric protein complexes have been categorized into different classes, namely: paralogous and non-paralogous heteromers.

The simplest kinds of heteromeric protein complexes are usually formed by the association of paralogous subunits (Fig. 4.7). These *paralogous heteromers* are very common and constitute about 9% of bacterial, 15% of eukaryotic and 19% of archaeal heteromeric protein complex structures (Marsh and Teichmann 2014b). Owing to adoption of symmetric and simple topologies, these paralogous heteromers often look like symmetric homomers. For instance, the Rad9–Hus1-Rad1 heterotrimer in humans consists of 3 paralogous subunits. These subunits resemble the arrangement of a cyclic trimer and have a pseudo-*C*3 symmetry, despite the lack of flawless global symmetry and being classified as *C*1. As a matter of fact, the separate polypeptide chains of this heterotrimer possess pseudo-*C*2 symmetry internally because of the presence of repeated domains. Therefore, at the domain level, this heterotrimeric complex is best described as pseudo-*D*3 (Xu et al. 2009).

On a similar note, in the archaeal group II chaperonin (thermosome) heteromer, the overall symmetry is *D*4, but it converts to pseudo-*D*8 when considered from the point of view of the similarity in structures of the two subunits (Ditzel et al. 1998). However, all paralogous heteromeric complex do not resemble simple homomer-like structural organization. For example, in the Skp1 complex, 3 paralogous subunits, having 10 copies are organized in a *D*5 symmetric manner (Jeyaprakash et al. 2012).

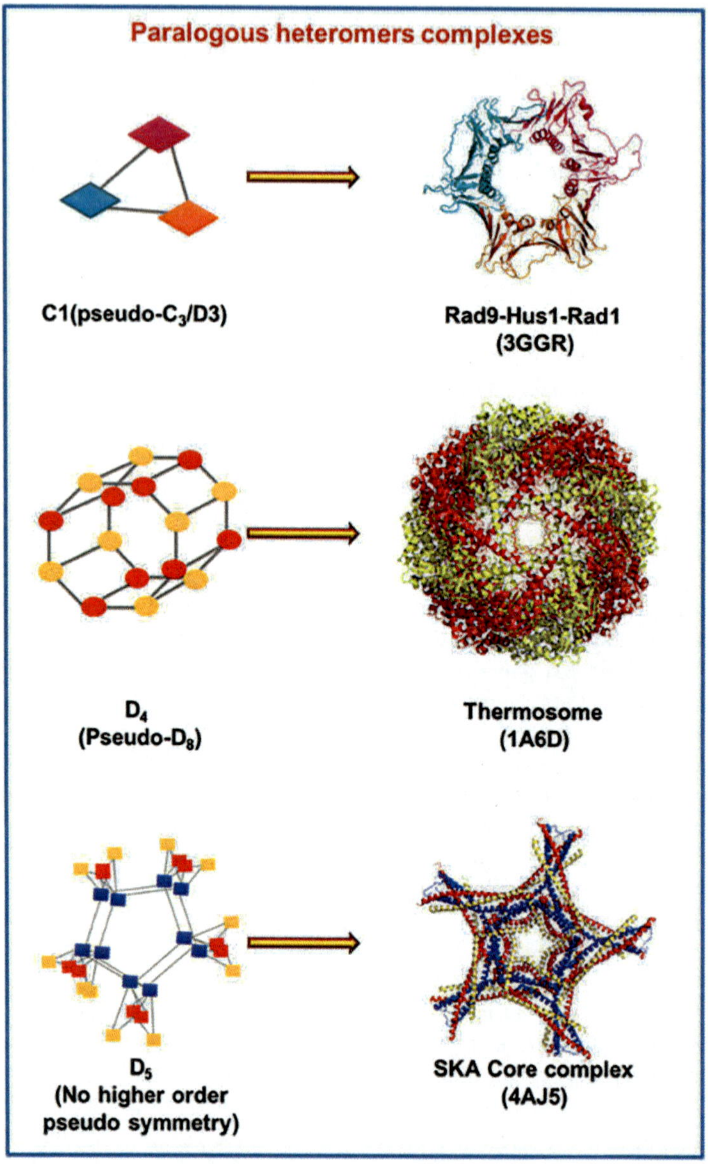

Fig. 4.7 Schematic showing the quaternary structure topology of paralogous heteromer complexes and illustrative proteins with C_1, D_4, and D5 symmetry are displayed on the other side of the arrow. PDB IDs are mentioned in parenthesis

As different positions are occupied by the distinct paralogous subunits, higher order pseudosymmetry does not exist for such protein assemblies.

There are some other heteromeric complexes that comprise *non-paralogous* subunits and have an overall symmetry in structure. For instance, there are two copies of two different subunits in the complex tryptophan synthase, where each of them has *C*2 symmetry, while there are three copies each of three different subunits in the complex formate dehydrogenase having *C*3 symmetry. There are some heteromeric complexes that comprise both non-paralogous and paralogous subunits. The proteasome-proteasome activating nucleotidase (PAN) complex falls under this category, where the α- and β-subunits of the proteasome have similarity with one another, but not to its interactor PAN (Fig. 4.8). The three different subunits are repeated for fourteen times, thus furnishing a D7 symmetry (Yu et al. 2010).

In certain heteromers, some symmetry elements are also mixed leading to uneven stoichiometries of subunits (i.e., disparate number of each type of subunit) (Fig. 4.9). For instance, ATP synthase is a protein complex where two paralogous subunits with pseudo-*C*6 symmetry form a hexamer and connect with a membrane-embedded region having *C*11 symmetry. On a similar note, the p2 base-plate protein complex in lactococcal phage is a trimer. It comprises a core with *C*3 symmetry and a basal hexamer with *C*6 symmetry (Stock et al. 1999).

Apart from this, there are six elements that have *C*3 symmetry locally. There are also some heteromeric complexes which are completely devoid of any kind of symmetry (Fig. 4.10). For instance, in the RNase H2 complex, there are three different subunits that are asymmetrically organized. RNA polymerase II is also a complicated asymmetric heteromer with 10 distinct subunits, despite the fact that some level of pseudosymmetry is brought in by the core, paralogous subunits (Sciara et al. 2010).

4.3.3 Dynamic Nature of Protein Complexes

Analogous to flexibility spectrum of monomeric proteins that can range from extremely rigid to intrinsically disordered, dynamics is an inherent component of protein complexes (Fuxreiter et al. 2014; Marsh et al. 2012; Mittag et al. 2010a; Tompa and Fuxreiter 2008). Such dynamic aspects are more prominent in protein complexes that are related to energy, i.e., protein complexes acting as molecular machines. In such complexes, the different subunits are subjected to coordinated movements for performing a particular function (Alberts 1998). ATP synthase is one such protein complex that acts as a motor. It drives the synthesis of ATP via proton movement and the consequent rotations of its stalk and associated conformational changes (Stock et al. 1999). In the thermosome protein complex, the conformational changes induced by the hydrolysis of ATP assists in protein folding, interactions and release of substrate (Ditzel et al. 1998). Similarly, the protein complex of RNA polymerase II is subject to large structural reorganization during the phases of transcription initiation and elongation (Cramer et al. 2001; Hahn 2004).

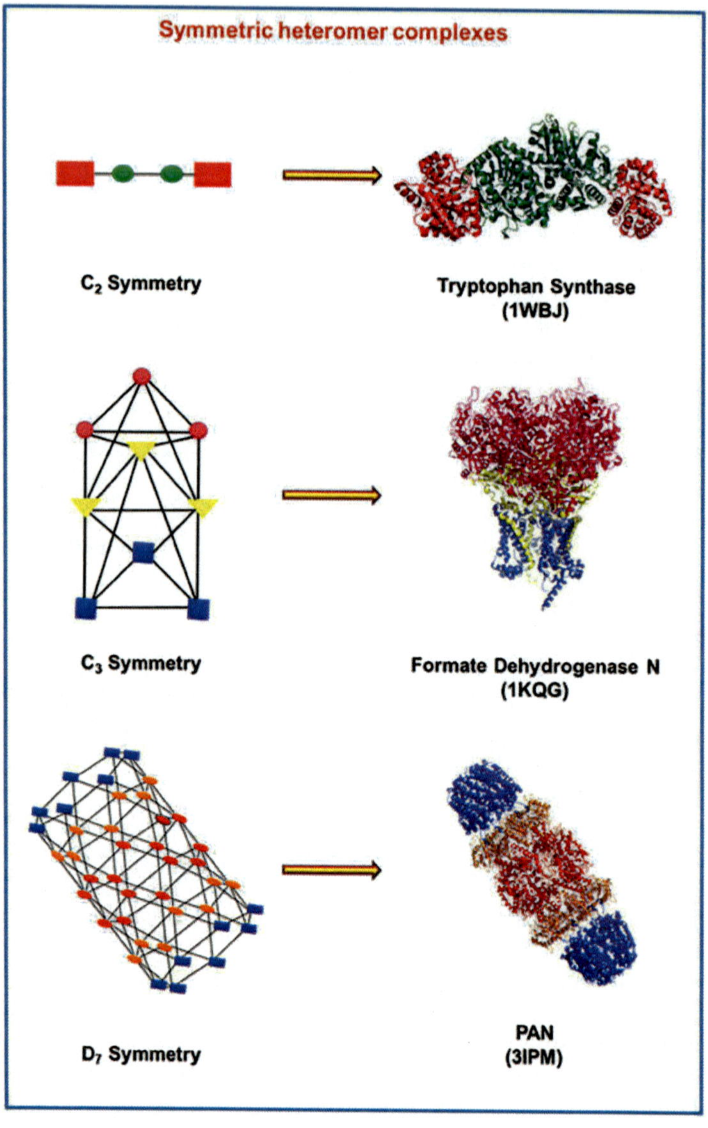

Fig. 4.8 Schematic showing the quaternary structure topology of symmetric heteromer complexes and illustrative proteins with C_2, D_3, and D_7 symmetry are displayed on the other side of the arrows. PDB IDs are mentioned in parenthesis

Apart from such energy-driven protein complexes, there are other protein complexes that also exhibit dynamic properties. Previously, it was arduous to study the dynamics directly, but with the advent of novel NMR techniques, researchers have gained considerable access to the dynamics of protein complexes

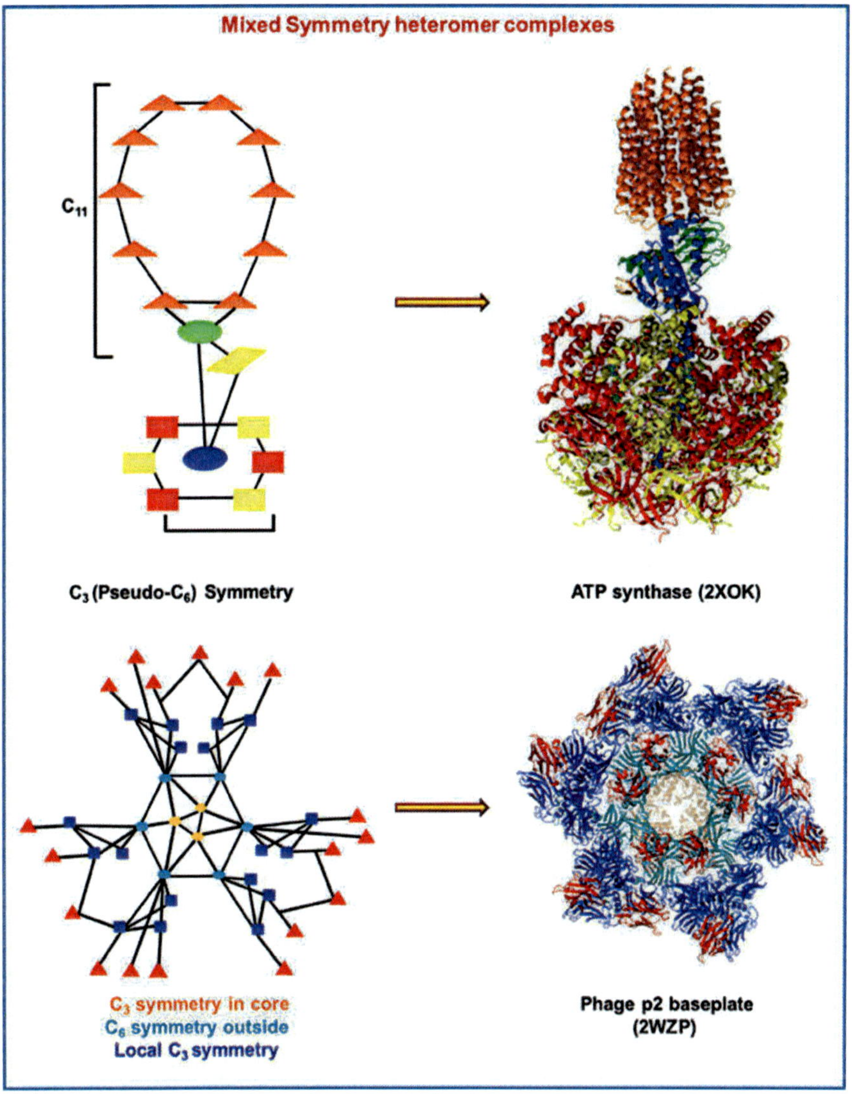

Fig. 4.9 Schematic showing the quaternary structure topology of mixed symmetry heteromer complexes and illustrative proteins with C_2, D_3, and D_7 symmetry are displayed on the other side of the arrows. PDB IDs are mentioned in parenthesis

(Sprangers et al. 2007). For example, the α-subunits in the archaeal proteasome experience fluctuations and equilibrate on a time scale between its closed and open conformations. This equilibration is important for the regulation of the proteolytic chamber (Latham et al. 2014; Religa et al. 2010; Sprangers and Kay 2007). On a similar note, the archaeal exosome core exhibits slow movements that are essential

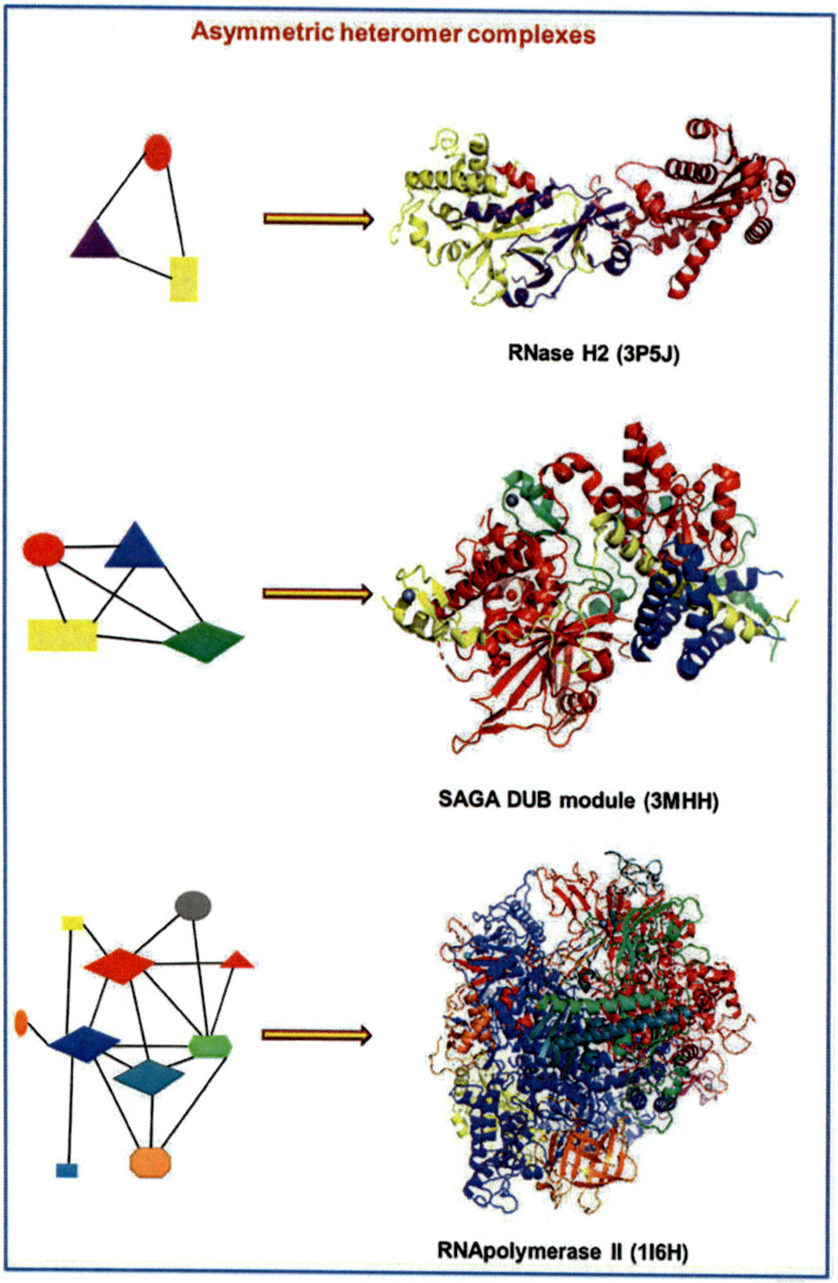

Fig. 4.10 Schematic showing the quaternary structure topology of asymmetry heteromer complexes and illustrative proteins with differential asymmetries are displayed on the other side of the arrows. PDB IDs are mentioned in parenthesis

for associating with the exosomal cap (Audin et al. 2013). It should be noted that the homologous complexes of eukaryotic organisms from archae are usually studied as they have more number of repeated subunits. This is especially advantageous in some experimental techniques like in NMR, where they enhance the simplicity of the spectra and also the strength of the NMR signal. The spectrum of dynamics is broad and its extreme ends represent the partially disordered or extremely dynamic proteins.

Protein complexes having such intrinsically disordered proteins have also been termed as fuzzy complexes (Mittag et al. 2010a; Tompa and Fuxreiter 2008). Despite the fact that existence of disorder in homomeric complexes was attributed to experimental artifacts (Nourse and Mittag 2014), heteromers that are highly dynamic have been reported. A variety of developed modelling techniques that integrate some experimental methods have enabled researchers to obtain structural models of protein complexes in utmost detail (Marsh et al. 2010; Mittag et al. 2010b; Wells et al. 2008). For instance, cystic fibrosis transmembrane conductance regulator (CFTR)—14-3-3β complex has been illustrated in the form of an ensemble model (Bozoky et al. 2013). 14-3-3β has been demonstrated to bind solely with 8 disparate sites via weak interactions rather than a single region on the regulatory domain (Baker et al. 2007). Owing to the phosphorylation of seven of such sites, the binding mechanism involved in this protein complex is considered as a graded and dynamic reaction of phosphorylation.

Additionally, one must take into account the inherent dynamics of the topologies of the quaternary structures. For example, there are several protein complexes that equilibrate between two or more states of self-assembly (such as dimer-tetramer or monomer-dimer) (Deprez et al. 2000; Dey et al. 2010; Jaffe 2005). Heat-shock protein αB-crystallin is a striking example of a protein complex that has unique self-assembling capacity. It forms different quaternary structures that can interconvert and comprise subunits ranging from approximately 10–40 (Aquilina et al. 2003; Haley et al. 1998; Stengel et al. 2010).

4.3.4 Three-Dimensional Complex: A Case Study on Protein Complex Classification

The Protein Data Bank furnishes a lot of structural information on homo- or heteromeric protein complexes. To get better insights into the data, it is essential to hierarchically classify the protein complexes in a manner analogous to the classification of protein domains by CATH or SCOP (Murzin et al. 1995; Orengo et al. 1997). Levy et al. have approached to classify protein complexes by placing them with respect to topological classes, where one considers a point to denote each polypeptide chain and solely takes into account the interface pattern between chains (Levy et al. 2006). The authors then subdivided these classes with respect to structure and sequence of the comprising subunits.

The previous classification methods regarded only structural parts rather than the entire complexes. For example, in SCOP and CATH, proteins were subjected to division into their evolutionary and structural domains, respectively. Subsequently, the proteins were classified in accordance with their structural homology with other domains. Information on domain–domain interfaces have been categorized in several databases since domains have the ability to interact with one another. However, there is an inherent disadvantage of utilizing such an approach for protein complexes as they contain several polypeptide chains with each chain comprising of more than one domain. These properties or features that are dependent on the entire protein complex (such as evolution, symmetry, size, pathway of assembly) cannot be analyzed by considering only the domain pairs that interact. Some manually curated subsets have been employed in certain small-scale studies to address aspects such as oligomer evolution (Brinda and Vishveshwara 2005; Goodsell and Olson 2000), geometric and biochemical properties of protein complexes (Ponstingl et al. 2005), and pathways of assembly in proteins with several subunits (Friedman and Beychok 1979; Levy et al. 2005). However, none of these studies has taken into account all identified structures from entire protein complex's point of view.

Levy et al. first proposed the first hierarchical classification of entire protein complexes whose three-dimensional structures are known. This classification was done on the basis of the graphical representation of their basic structural aspects. This classification furnished the first ever outline of all the protein complexes in the Protein Data Bank (PDB) and enabled the derivation of non-redundant sets in detail. This classification of protein complexes, based on structure utilized a new scheme of visualizing and comparing the protein complexes (Fig. 4.11) (Levy et al. 2006). The authors represented each complex's graph in a simplified form, denoting polypeptide chains by nodes and interfacial chains by edges. Then authors utilized a customized protocol of graph matching to compare and contrast the protein complexes. This protocol considered the graph's topology as it characterized the factors such as sequence and structure similarity between the constituting chains as well as chain-chain interfacial patterns. These properties assisted in classifying the protein complexes in a hierarchical manner. It also yielded a set of protein complexes that are non-redundant such that one can utilize them for the unbiased derivation of statistics. This has been illustrated via drawing on differential levels of classification to address issues on evolution, symmetry, and topology of protein complexes.

It has been demonstrated that multi-meric complexes fall between 33 and 66% of all known structures, taking into account the redundancy factor. From the authors' analysis on topological organization of subunits, it has been revealed that only a small number of organization form in reality as compared to the number obtained theoretically. It has been hypothesized that since majority of the protein complexes comprise less than 4 subunits, they are mostly homomeric. Furthermore, there is a strong proclivity to attain symmetry, even in heteromeric complexes. Lastly, via comparing protein quaternary structure database with "biological units" in the Protein Data Bank, several possible errors related to the assignments at the quaternary structure level have been identified. All these results made on the classification of protein complexes are available both on a webserver and a database (Levy et al. 2006). These studies are aimed to elucidate the evolution–structure relationships of protein complexes in a better and unique fashion.

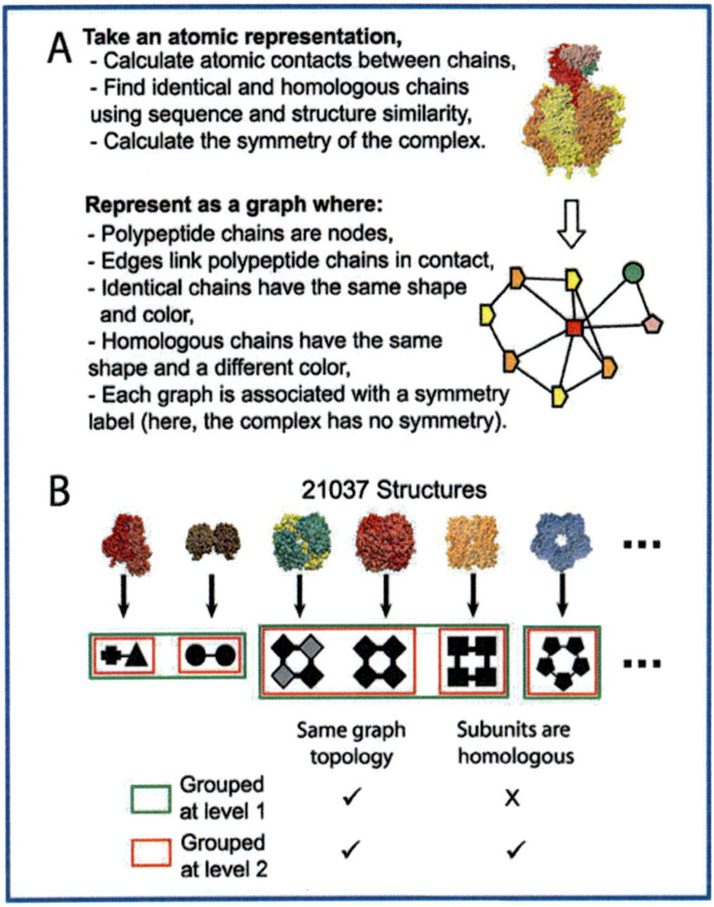

Fig. 4.11 Depiction of protein complexes in the form of graphs; (**a**) constituting polypeptide chains in a graph are displayed as nodes and the interfaces between those polypeptide chains are represented as edges, (**b**) customized graph matching procedure is employed to compare all the protein complexes. Complexes sharing identical graphical topologies are assembled into the first level of hierarchy as denoted by the green box. Additionally, if the subunit structures are related by their SCOP domain architectures, they are assembled at the second level as displayed by red boxes (Adapted from Levy et al. 2006)

4.4 Evolution of Protein Complexes

A considerable interest has been observed among researchers to understand how protein complexes evolve at the level of quaternary structure. Certain important aspects of protein complex evolution have been outlined below. Two mechanisms by which heteromeric complexes evolve have also been discussed in detail in the following sections (Fig. 4.12).

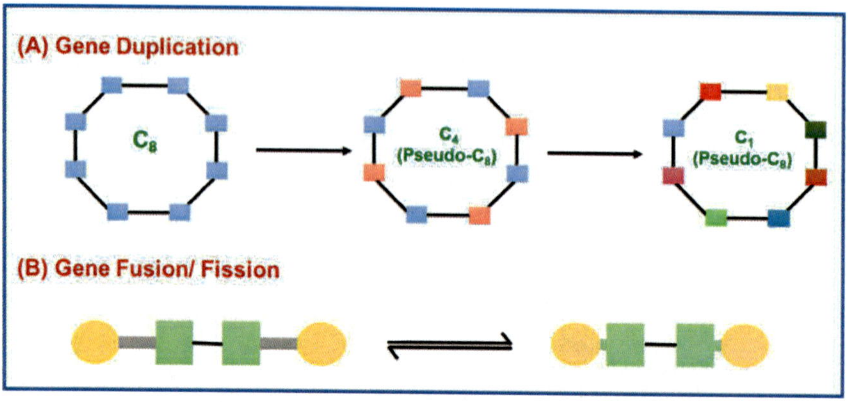

Fig. 4.12 Schematic illustration of mechanisms of heteromer evolution (**a**) duplication of a gene leading to the formation of proteins that assembles to form a homomeric complex which can further lead to the heteromeric complex formation. Multiple duplication events can lead to the formation of distinct paralogs giving rise to paralogous heteromers. (**b**) Gene fusion can lead to the formation of protein complex via a fusion of previously separated subunits within the same complex. Gene fission can also take place which is a rare process

4.4.1 Evolutionary Changes in Self-Assembly State

When one considers the evolution of homomeric protein complexes, it is vital to understand whether the complexes have undergone evolution from a precursor, either with quaternary or monomeric structure. One possible way by which the phylogeny of a homomeric complex may be traced is to find out the homologous proteins that have different quaternary structures (Perica et al. 2012a). For proteins whose sequence identities cross the 90% threshold, the quaternary structure is almost and always conserved. On the other hand, lower sequence identities result in greater divergence. For example, proteins that have sequence identities in the range of 30–40%, the probability of quaternary structure conservation is 70% (Levy et al. 2008). The large-scale search for homologous proteins having known structures has thus enabled research groups to trace the evolutionary history of several homomeric protein complexes.

On an interesting note, researchers have found a strong correlation between homomer symmetry and evolutionary pathway. Strong preference for the steps in evolution that corresponds to symmetric changes for self-assembly has been observed. Therefore, a monomer can either evolve into a cyclic *Cn* or a dimeric *C*2 complex, and a dimeric *C*2 complex can undergo evolution into a dihedral *Dn* complex with variable size; however, a cyclic complex may only undergo dimerization for forming a dihedral complex. Additionally, almost all of the recognized transitions in evolution have been observed to match perfectly with the predicted or determined pathways of assembly. It implies that older interfaces in homomeric complexes are usually bigger than the ones that have been generated recently (Dayhoff et al. 2010; Kim et al. 2006). Therefore, sub-complexes that are energetically favorable and are formed in solution during the process of assembly are more tendentious to correlate with precursors (Marsh and Teichmann 2014a).

Some mechanisms via which proteins waver their self-assembly state have been identified. The change in quaternary structure has been attributed to point mutations, either a direct one located in the interfacial region or an indirect one that exerted its effects in an allosteric manner (Malay et al. 2005; Nishi and Ota 2010; Perica et al. 2012a). Sometimes such mutations result in the exchange of a monomeric tertiary structural element with a new copy of itself, thus generating a "domain-swapped" dimer (Bennett et al. 1995). Generally, such domain swapping has been perceived in 5% of protein families (Huang et al. 2012). On the other hand, 24% of homomeric complexes have been observed comprising intertwined structures, including domain swapping (Mackinnon et al. 2013). Finally, deletions and insertion in the interfacial areas have been demonstrated to modulate the quaternary structure of the protein complex (Akiva et al. 2008; Hashimoto et al. 2010; Hashimoto and Panchenko 2010; Nishi et al. 2011).

4.4.2 Evolution of Protein Complexes Through Gain and Loss of Subunits

Apart from the changes occurring in the self-assembly state, evolution of protein complexes may be brought about via the loss and/or gain of subunits. New subunits are usually acquired via events such as gene duplication (Archibald et al. 1999). When the gene coding for a self-assembling protein in a homomeric complex undergoes duplication, the two identical genes initially generate the same complex using protein molecules from both the gene copies. However, the sequence divergence of the genes may result in the formation of paralogous proteins, whereby there would be a slight difference existing between the two proteins. In such a scenario, these two different proteins may assemble into paralogous heteromeric complexes. In case mutations enhance the energetic favorability of the heteromeric complex with respect to the homomers, then via virtue of evolution, the complex would eventually become heteromeric exclusively (Fig. 4.12a). This process has been observed in protein complexes from eukaryotes, where gene duplication events are common. This phenomenon can result in each subunit of a cyclic ring complex to be coded by a disparate gene. For instance, the eukaryotic proteasome and thermosome are coded by 14 and 8 paralogous genes, respectively.

Despite the fact that paralogous heteromeric protein complexes are fairly widespread, majority of the heteromeric complexes in archae, bacteria, and eukaryotes are not wholly made up of paralogous subunits (Marsh and Teichmann 2014b). In these complexes, it is probable that new subunits were eventually acquired in a sequential manner, because the possibility of the simultaneous assembly of several new subunits is very low. It is challenging to infer the gain of subunits just via the comparison of protein structures owing to the nature of structural data of proteins. If a certain subunit is present in a specific protein complex, but not in the homologous complex in another species, it does not necessarily imply a true difference from the evolutionary perspective. On the other hand, it indicates the bias of the researchers performing the experiment towards proteins that are amenable to structural characterization. To sketch out evolution in a step-wise manner, it is easier to compare the genes of evolutionarily related organisms because of the greater abundance of sequence data as compared to structural data (van der Sluis and Driessen 2006).

For example, a study attempted to trace the evolutionary history of the NADH-ubiquinone oxidoreductase (Complex I) that has 14 distinct subunits in bacteria and 46 in the mammalian mitochondria (Gabaldón et al. 2005). The authors estimated the relative and probable evolutionary ages of the individual subunits in the mammal, and tracked back 6 subunits to the ancestor (bacteria) where they were lost. Interestingly, the authors observed that the new subunits were acquired in a stepwise fashion, contrary to the complex's early evolution, where distinct heteromers were amalgamated to form a chimera (Friedrich 2001; Mathiesen and Hagerhall 2003).

On a similar note, a genomic, comparative method was utilized to identify new and old subunits of protein complexes in humans (Marsh and Teichmann 2014b). Interestingly, a strong preference was observed for the newer subunits to have more flexibility than their evolutionarily ancient counterparts. Researchers have reasoned that this enhanced, inherent flexibility greatly assists in the assembly of protein complexes from different subunits. This implies that there is an increase in subunit flexibility with an increase in the number of subunits contained in a protein complex.

Indeed the literature suggests that one can assess the relative flexibilities of the distinct subunits from the structure of the heteromeric complex that can further aid in constructing evolutionary pathways (Marsh 2013; Marsh and Teichmann 2011; Marsh et al. 2012). Heteromer evolution is strongly related to the strength of conservation of PPIs across various species. Although several experiments have demonstrated that protein interactions are usually weakly conserved, some of the results have been attributed to non-functional PPIs and false positives (Landry et al. 2013). Strikingly, when one takes into accounts the members of a protein complex, the tendency of conservation of PPIs is much greater. For instance, when pairs of interacting human proteins within a complex are considered, 90% of them are co-purified from yeast (Van Dam and Snel 2008).

Furthermore, the process of gene fusion can lead to change in the quaternary structure of a complex (Fig. 4.12b). In this process, two individual genes are fused to form one open reading frame. The mechanism of gene fusion has been established as one of the driving force behind multi-domain protein evolution (Björklund et al. 2005; Buljan et al. 2010; Marsh and Teichmann 2010; Pasek et al. 2006). It is interesting to note that gene fusion usually concerns the pair of genes that code for proteins of the same complex (Enright et al. 1999; Marcotte et al. 1999). Therefore, the number of different subunits in a protein complex usually decreases via gene fusion without affecting the complex's overall assembly. Gene fission is the reverse process of gene fusion, via virtue of which there is a split of single gene into two. Although gene fission can cause a gain in the number of subunits, this process has been found to be rarer than gene fusion (Fong et al. 2007; Kummerfeld and Teichmann 2005).

4.4.3 Adaptive and Non-adaptive Drivers of Quaternary Structure Evolution

It is important to understand why proteins have undergone evolution to assemble into complexes. For heteromeric protein complexes, the answer is clear; the evolution serves the living cell with potential and obvious benefits by amalgamating different polypeptide chains with different functions into one complex. Several benefits of assembly of homomeric complexes have also been postulated. For example, it drives the positive selection of proteins with self-assembling capacity. Although the functional benefits have been discussed in detail elsewhere (Goodsell and Olson 2000; Perica et al. 2012b), some of them are listed here as follows:

Often, the interfaces that are formed as a result of interaction between identical subunit assist in the function of the complex. For instance, ligand binding sites (Heldin 1995) and active sites (Dobson et al. 2004; Navia et al. 1989) can be formed at such interfaces. Homomeric complexes also assist in allostery and acquiring cooperative function. Several examples of symmetric homomers have been identified that involve allostery (Koshland Jr et al. 1966; Monod et al. 1965).

Homomeric complexes also denote an essentially type of coding economy. In other words, it is easier for the cell to assemble a gigantic homomeric complex using numerous copies of one small subunit instead of using a larger protein. This is because less genetic material is required to code the small constituting subunit protein (Crick and Watson 1956). The property of self-assembly also confers the stability since a large homomeric complex has smaller surface-to-volume ratio compared to that of free monomers (Bershtein et al. 2012). Furthermore, as a smaller number of amino acid residues are exposed on the surface, the chances of promiscuous interactions with other protein partners are also minimized (Yang et al. 2012).

In spite of the obvious necessities of self-assembly, Lynch has speculated that all the benefits have not yet been evidenced (Lynch 2013). According to Lynch, since stochastic and non-adaptive processes may also govern evolutionary variation in homomerization in many cases, and one should be careful while attributing an adaptive reason to self-assembly in protein complexes (Lynch 2012). Additionally, self-assembly may be a property intrinsic to many proteins, since self-associating and symmetric interfaces are energetically more favorable (André et al. 2008; Lukatsky et al. 2007). This idea is also supported by the inclinations of protein molecules to crystallize and/or aggregate (Durbin and Feher 1996; Knowles et al. 2014).

Heteromer evolution, particularly the paralogous heteromeric complexes may also have a non-adaptive driving force. In essence, it may denote an evolutionary "ratchet" that is unidirectional in nature (Doolittle 2012; Ruano-Rubio and Fares 2007). In other words, the accumulation of neutral and random mutations post-gene duplication may result in a situation where there is no resultant effect on the function of the protein. Initially, a random sampling between two related proteins may form a functional complex. However, eventual mutations may disrupt the homomeric interactions, while keeping the functional paralogous heteromeric complex intact.

In such a scenario, both the paralogous genes become essential and random mutations are highly improbable to restore the homomeric complex.

A recent study involving ancestral gene resurrection has evidenced that although such paralogous heteromers are easy to be formed as they are highly unlikely to reverse to the homomer (Finnigan et al. 2012). The great abundance of paralogous heteromers in eukaryotes is thus attributed to non-adaptive processes. Non-paralogous heteromers may also be evolutionarily driven by such non-adaptive factors. For example, mutations that are mildly deleterious would usually disrupt individual proteins' stability; but they could be tolerated in cases of weak negative selection, such as in species that have small population size (Lynch 2007). This might consequently result in secondary selection for novel interactions that are capable of rescuing the protein stability in protein complexes (Fernández and Lynch 2011).

4.5 Mechanism of Protein Complex Formation

At the simplest level, the assembly of a protein complex comprises binary interactions involving polypeptide chain pairs. It is important to consider the amount of conformational alterations that occur in the protein's free and bound states while examining these interactions. It is intriguing to note that the factor that most strongly determines the extent of the conformational change is the free state's inherent flexibility (Dobbins et al. 2008; Tobi and Bahar 2005). In spite of exceptions, it has been generally observed that with increasing flexibility of the monomer, there are larger differences in the conformation upon binding (Marsh et al. 2012). However, it is unfortunate that most of the data that is available on monomers is biased toward rigid proteins, since greater flexibility tends to correlate with greater difficulty of crystallization (Marsh 2013; Yu et al. 2000).

In spite of this bias, several flexible proteins have been crystallized in their unassociated states. For instance, cyclin-dependent kinase inhibitor 2 that has moderate flexibility is subjected to considerable conformational rearrangements upon associating with cyclin A (Fig. 4.13) (Cheng et al. 2006). An important and interesting category of protein interactions comprises intrinsically disordered proteins that are commonly found in eukaryotes. When studied in vitro, they have a tendency to be completely or partially disordered (Wright and Dyson 1999). Nevertheless, several intrinsically disordered proteins are involved in the formation of ordered structures inside a protein complex (Dyson and Wright 2002; Hegyi et al. 2007). For instance, a spinophilin contains large amount of intrinsically disordered area. However, when it interacts with protein phosphatase 1, this region experiences a great disorder-to-order transition and undergoes complete folding (Ragusa et al. 2010).

A protein complex with more than two subunits, involves perplexed mechnsim of assembly in contrast to a simple binary protein complex. In this regard, it is important to remember Levinthal's folding paradox, where it has been postulated that the assembly of a protein complex has likelihood to occur via an ordered and

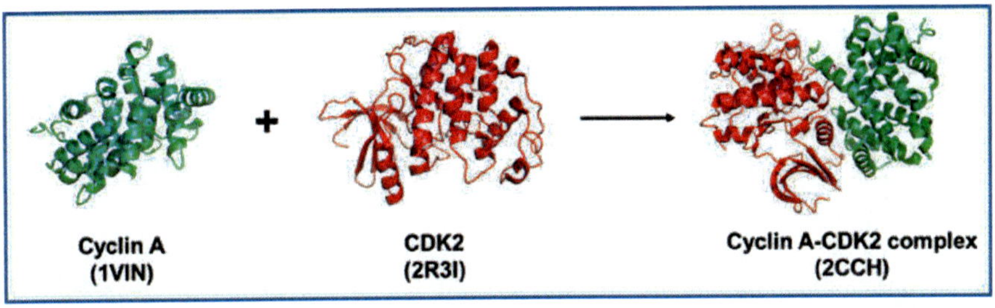

Fig. 4.13 Cartoon representation of proteins: cyclin A (green) and CDK2 (red). Flexible binding between the two proteins is also shown which displays the presence of moderate flexibility in CDK2 in its unbound state. Moderate conformational changes were observed in CDK2 upon binding to cyclin A

energetically favorable pathway, since the probability of the simultaneous association of all subunits in the precise orientation is infinitesimally low. Therefore, scientists have wondered: whether the assembly of protein complexes will follow a single pathway, or whether they take up multiple parallel pathways, similar to the proposed models of protein folding for individual proteins (Dill and Chan 1997).

Techniques such as electrospray mass spectrometry have greatly assisted in our understanding of protein assembly pathways because they have aided in probing the identities of the various sub-complexes formed during assembly and disassembly in the solution-phase (Hernandez and Robinson 2007). These techniques have been further complemented with ion mobility mass spectrometry, for distinguishing the intermediates that are structurally distinct but have equivalent masses (Ruotolo et al. 2008). On the other hand, complementary techniques such as collision-induced dissociation have been employed to observe gas-phase disassembly pathways (Hall et al. 2013). When a homomer series was subjected to electrospray mass spectrometry, it was observed that almost all of the tetrahedral or dihedral complexes dissembled through one sub-complex intermediate: either cyclic or dimeric (Levy et al. 2008). Contrary to this, the researchers did not observe any intermediates of cyclic complexes. This result implied that the assembly had very limited routes and a single, dominant pathway was usually taken up by homomers during their assembly (Fig. 4.14).

For example, a monomer has the capacity to assemble into a dimeric (Fig. 4.14a) or cyclic (Fig. 4.14b) complex. A dimeric complex can undergo further dimerization to generate a tetrameric *D*2 complex (Fig. 4.14c) or undergo trimerization to generate a *D*3 hexamer (Fig. 4.14d). A cyclic complex, on the other hand, can undergo dimerization to generate a *D*3 hexamer (Fig. 4.14e) or undergo tetramerization to generate a tetrahedral complex (Fig. 4.14f). Obviously, the instantaneous formation of a cyclic ring from free monomers is improbable in a single reaction step. Thus, assembly intermediates that are asymmetric and transiently populated are not uncommon. However, it should be noted that symmetric intermediates are more favored energetically and thus are easier to detect by experimentation. On a similar note, characterization of heteromers by electrospray mass

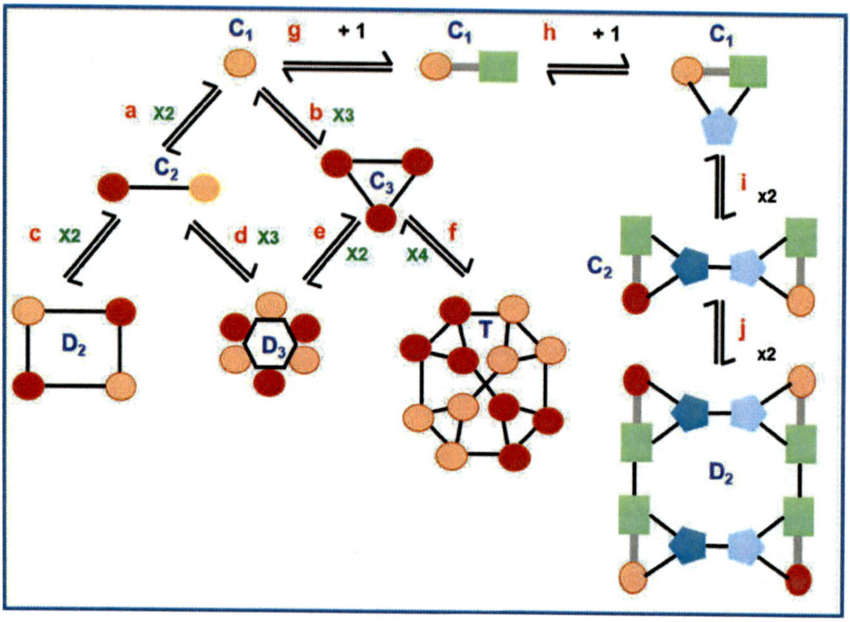

Fig. 4.14 Multiple pathways for the assembly and evolution of protein complex quaternary structures. Protein complexes can assemble and evolve via a variation in their self-assembly state (**a–f**, **i**, **j**) or via an addition of new subunits (**g**, **h**)

spectrometry has demonstrated the presence of intermediate sub-complexes and a single pathway of disassembly in majority of the cases (Hall et al. 2012). Nevertheless, there are certain differences from the homomers. Compared to asymmetric homomeric intermediates, the asymmetric sub-complexes for heteromers are more common as they are equally favored energetically as the symmetric intermediates.

Secondly, parallel disassembly has been evidenced with complicated topologies (Marsh et al. 2013). Studies till date have suggested that simple complexes may usually involve single assembly pathways, while complexes that comprise several subunits are likely to involve multiple pathways (Talkington et al. 2005). One can predict the pathways of assembly of homomers and heteromers with accuracy if the complex's 3D structure is known. This can be done by assuming a correspondence between dissociation energy and intersubunit interface size. This correspondence is strong enough for predictions with respect to the disassembly process of a single complex, where each step involves a limited number of different sub-complexes (Marsh and Teichmann 2014a).

4.5.1 Co-translational Assembly of Protein Complexes

While multi-protein assemblies are widespread, the pathways that govern their formation at the quaternary level are still not very well understood. However, it is alluring to note that several protein complexes studied in both eukaryotes and prokaryotes have been observed to assemble co-translationally. In prokaryotes, the interactions of nascent chains with mature proteins or other nascent chains are encouraged by the polycistronic organization of the mRNA and the related co-translation of other associated cell components. The polycistronic arrangement of operons in prokaryotes assists in integrating all the aspects of synthesis of a heteromeric protein complex, such as (1) regulation of gene expression in a coordinated manner, (2) regulation of the stoichiometry of subunits, and (3) regulation of co-translational interactions (Fig. 4.15).

In a recent study, it has been postulated that the individual subunit levels and their stoichiometry are regulated mainly at the translational level, by discrete mechanisms that are not well known (Li et al. 2014). It is widely believed that PPIs are promoted

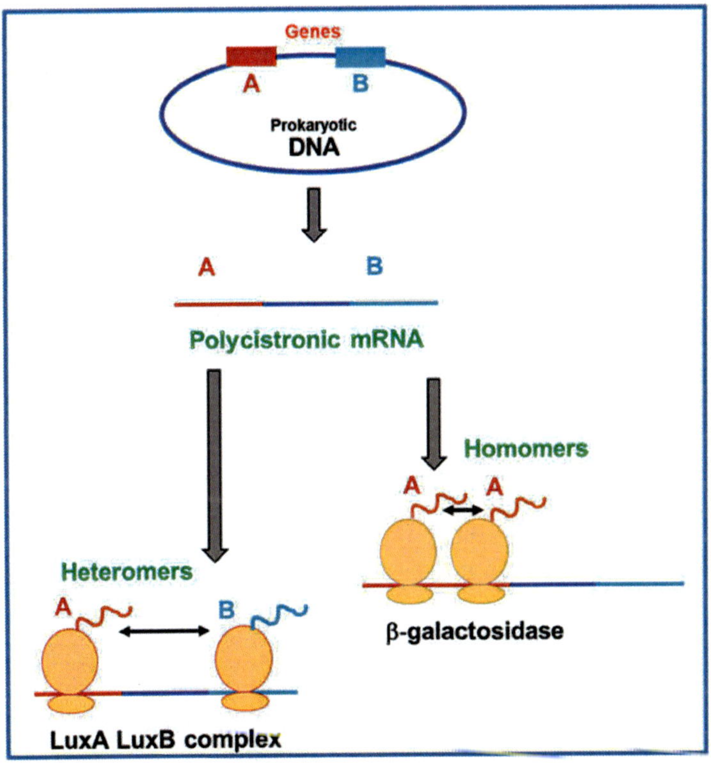

Fig. 4.15 In prokaryotes, genes on polycistronic operons are co-translated on polysomes giving rise to high concentrations of nascent protein and facilitating either post-translational or co-translational assembly of protein complexes

through polycistronic expression and that formation of protein complexes involves co-translational mechanisms because most of the multi-subunit complexes in prokaryotes are encoded by operons. It is also assumed that co-translation is dependent on the availability of subunits as a part of the regulatory mechanism of the assembly. Despite on-going studies, it is still not clear as to how co-translational mechanisms act as quality control systems and warrant the correct assembly of several hundred heterogenous protein complexes in a cell. It is hypothesized that the coupling of the folding process of proteins with that of their co-translational interactions may help in increasing their fidelity to form a protein complex with correct conformation. Therefore, research groups have regarded the co-translational mechanism of protein complex formation to be fundamental to the processes of cellular function (Williams and Dichtl 2018).

A study on the assembly of heterodimeric luciferase complex has demonstrated that the process is mediated co-translationally and is assisted by the synthesis of the subunits in a spatially confined manner from a polycistronic mRNA (Shieh et al. 2015). Researchers speculate whether protein complex assembly is coupled with translation in eukaryotes as well. This is because of different chaperone groups and rarer polycistronic mRNAs in eukaryotes. Shiber et al. performed a comprehensive and mechanistic study on the assembly of protein complexes in the eukaryotic system utilizing ribosome profiling (Shiber et al. 2018). At almost residue-level resolution, the authors determined the interactions among nascent subunits from 12 heteromeric complexes in in vivo conditions in yeast. The study revealed that out of the 12 complexes, 9 assembled in a co-translational manner, while the remaining 3 were governed by specific chaperones. The authors further determined that the co-translational assembly was a unidirectional process that resulted in the interaction of a mature subunit with its nascent cognate subunit, precluding the inclination for aggregation. The study demonstrated that initiation of the association of the subunits in a co-translational manner has direct co-incidence with the event of the nascent interaction domain being fully exposed at the exit point of the ribosomal tunnel. A ribosome-associated heat-shock chaperone, known as Ssb is involved in the assembly process. This chaperone blocks pre-mature interactions between the assembly units via its transient engagement with the interaction domains that are partially synthesized. Prior to the event of association between subunits, Ssb is dislodged from such domains. In a nutshell, co-translational mechanism of protein complex assembly integrates protein synthesis, protein folding, and protein complex formation and is widespread in eukaryotic systems.

4.5.2 Theoretical Background of Protein Complex Assemblies

Sartori et al. have attempted to generate a theoretical basis for assembly of protein complexes utilizing an equilibrium thermodynamic model of self-assembly (Sartori and Leibler 2019). This model reveals four different types of assembly behaviors: diluted protein solution, liquid mixture, "chimeric assembly," and "multifarious assembly." In the latter system, the co-existence of distinct protein complexes has been observed without the formation of chimeric and erroneous structures. The authors have revealed that this regime is satisfied only under two conditions: (1) sufficient heterogeneity in the composition of the complexes and (2) sparse utility of the set of components by the complexes. The authors studied databases on protein complexes and analyzed that protein systems under cellular conditions might have been subject to evolution in such a way that they satisfy both the above-mentioned criteria.

The assembly of protein complexes is highly accurate with respect to composition (Ban et al. 2000). This is an extraordinary phenomenon, because it implies that an assembling complex discriminates its specified components from a multi-component mixture that comprises several hundred other proteins of the proteome. Therefore, if the growing protein complex fails to discriminate from the pool of essentially all cellular proteins, a chimeric structure could be formed that would be of no use to cell function (Garcia-Seisdedos et al. 2019).

If one considers the organization of amino acids into functional proteins as the first stage of generating functional cellular structures, then protein complex assembly can be regarded as the second stage. Thousands of different proteins are encoded by just twenty distinct amino acid residues. Since proteins typically utilize all the 20 residues, their amino acid usage is "dense" instead of "sparse." Additionally, many residues are re-used multiple times within the same protein molecule that makes each protein's compositional heterogeneity low. On the other hand, the assembly of protein complexes is "sparse" as each complex comprises only a few proteins from the entire proteome. However, protein complexes have also been observed to be highly heterogeneous on several occasions, where they have been synthesized from several distinct protein species (Reuveni et al. 2017). The co-existing heterogeneity and sparsity of protein complexes essentially imply that the proteome may not be utilized in a combinatorial manner. The fact that the enormous repertoire of several hundred proteins is combined to form a comparable number of complexes recommends that such "combinatorial expansion" is not generic, but instead linked to certain functions, such as signalling or regulation (Antebi et al. 2017; Stein et al. 2009) (Fig. 4.16). In such cases, proteins are observed to be involved in various complexes, a significant example being the involvement of cyclin-dependent kinases in several cell-cycle regulatory protein complexes (Murray 2004). Proteins can form disordered chimeric structures via promiscuity, a phenomenon by virtue of which they have interactions with several protein partners. Nevertheless, the mechanism of protein complex assembly results in the formation of accurate and completely functional protein complexes (Johansson et al. 2012).

4.5.3 Case Study: Mechanism of Viral Capsid Assembly

When a virus capsid undergoes self-assembly, it not only includes the folding of its subunits and their interactions, but also requires components such as host chaperons and accessory proteins as well as maturation processes via covalent modifications (proteolysis) and other conformational changes (Liljas 1999). Although being

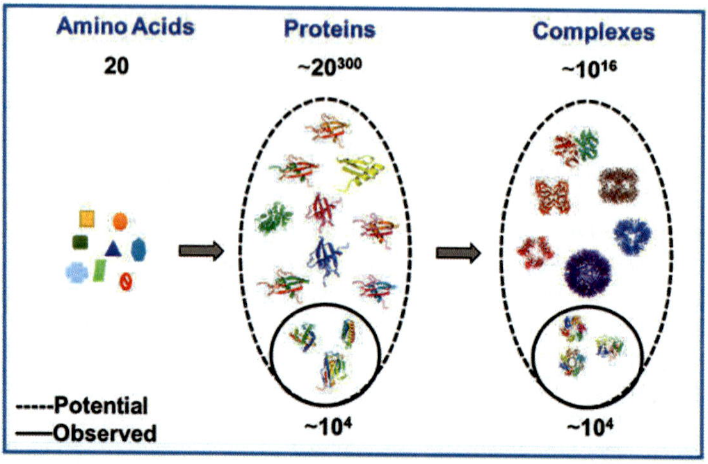

Fig. 4.16 The pattern of expansion from amino acids to proteins to protein complexes. Folding limits the length of a protein which is typically around 300 amino acids. Thus, 20 naturally occurring amino acids have a potency to form 20^{300} proteins, although the repertoire of proteins observed is much smaller, for instance, it is 10^4 in *S. cerevisiae*. This clearly indicates the expansion of proteins from amino acids. In contrast, the combinatorial expansion is not traceable in case of protein complexes where taking the combination on an average four proteins in the formation of protein complex. Since the number of complexes observed is equivalent to the observed number of proteins (Adapted and redrawn from Sartori and Leibler 2019)

complicated, the process is extraordinarily fast. For instance, bacteriophage T4 undergoes an entire cycle of infection, replication, and lysis in just 15 min. Therefore, it is implicative that the time scale of assembly of the viral capsid is in minutes, regardless of the large number of constituting polypeptide chains (Leiman et al. 2003). Despite the fact that the mechanism of capsid assembly subjects the polypeptide chains to great evolutionary pressure, the composition of the interface residues do not vary considerably from that of simpler systems. This observation suggests that the capsid assembly mechanism comprises a series of binary or other low-order steps, similar to that of smaller oligomers (Bahadur and Janin 2008). In the capsid assembly pathway, the oligomeric species serving as intermediates are referred to as *capsomers*. Bahadur et al. proposed that "medium" and "large" sized interfaces are majorly involved in making the capsomers, which can ultimately lead to the formation of the capsid (Bahadur and Janin 2008).

On the other hand, small interfaces are not likely to participate in the assembly process and/or make contributions to the stability of the formed capsid. When bacteriophage P22 and HK97 were studied in vitro, it was found that their capsid assembly involves several oligomeric intermediate levels (Prevelige Jr et al. 1993; Xie and Hendrix 1995). HK97 subunits in solution give rise to pentamers and hexamers that interconvert between them at a slow rate, and the efficiency of the pro-capsid assembly process has been demonstrated to be the highest when the two species are present in a ratio of 1:5. The HK97 assembly process can be explained by the pairwise interfaces that were determined by Bahadur et al. who also discovered "large" fivefold and sixfold interfaces resembling the features of a stable homodimer interface (Bahadur et al. 2003). The authors also recognized "medium" size interfaces that were made between pentamers and hexamers, and also between different hexamer pairs. Such pairwise interfaces were found to contribute to the stability of intermediate oligomeric species and form the capsid by the addition of hexamers in a step-wise manner to the pentamer.

In another instance, Stockley et al. investigated the assembly of bacteriophage MS2 (Stockley et al. 2007). The $T = 3$ capsid of MS2 was observed to dissociate into symmetric dimers. Some of these dimers turned asymmetric when an RNA stem-loop fragment was added. Dimers of hexamers were observed to form symmetric and asymmetric dimers but no pentamer was seen to form. MS was demonstrated to have large twofold and quasi-twofold interfaces, each having a BSA of around 6400 Å^2. The second largest interfaces were found to be the quasi-threefold interfaces in the dimeric hexamers. They were medium in size and their BSA ranged from 800–1100 Å^2 (Fig. 4.17). Properties of the capsid interfaces such as size, composition and other physico-chemical aspects were in line with those observed in stable protein complexes (Janin et al. 2008). The constituting residues of the interfaces have been found to be well-conserved as well (Bahadur and Janin 2008). Therefore, the self-assembly of viral capsids involves the synthesis of various oligomeric intermediates in a step-wise manner and the physico-chemical properties of proteins in the solution state act as the determinant governing the early formation of large interfaces.

4.6 Basis for Formulating Periodic Table of Protein Complexes

The principles of organization underlying the enormous diversity of protein structures at the quaternary level are not yet well understood, as compared to protein folds that have already been studied extensively with respect to their phylogeny and architecture. Ahnert et al. attempted to understand the general principles that underlie protein quaternary structures by considering protein complexes from the viewpoint of their assembly (Ahnert et al. 2015). Several protein complexes involve a spontaneous and ordered assembly pathway that tend to be evolutionarily conserved. Additionally, there is a high correlation between evolutionary and assembly pathways of protein complexes. Therefore, it is implicative that one may utilize information on assembly pathways to gain insight into the types of evolved protein complexes and vice versa. The authors initially examined the basic steps of protein assembly utilizing experiments, existing literature, and a structural analysis of protein complexes on a wide scale. It has been observed that the assembly steps can be majorly categorized into 3 different types, namely—dimerization, cyclization, and heteromeric subunit addition.

Combination of various steps of assembly in different ways in a systematic manner had enumerated a large set of key interface patterns or possible topologies of the quaternary structure. They observed that almost all of the real-life protein

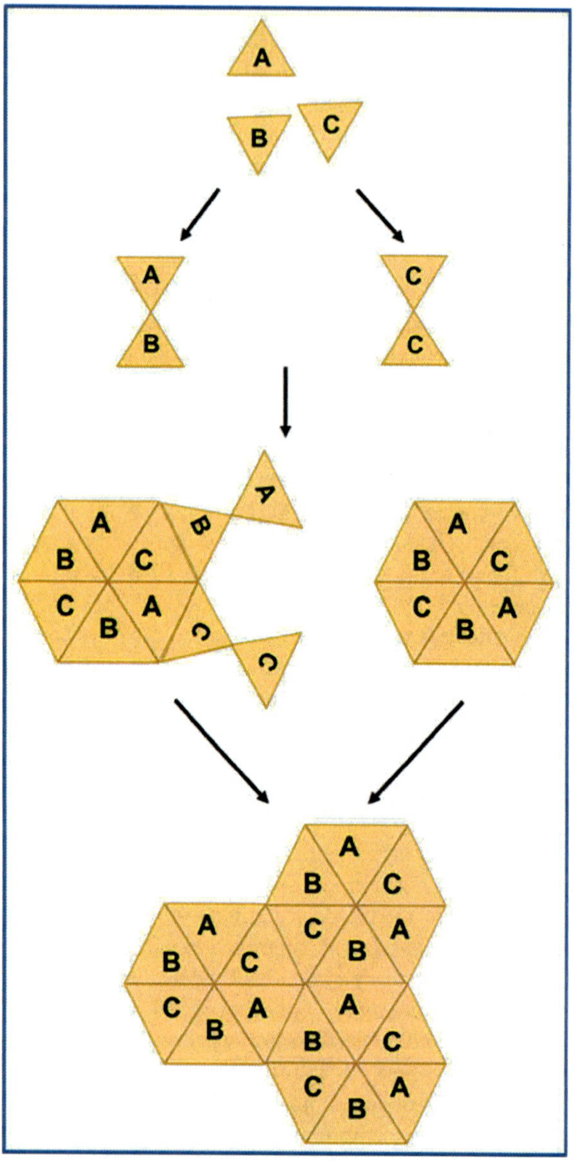

Fig. 4.17 Pathway of self-assembly of T3 bacteriophage MS2 pro-capsid. Pro-capsid composed of 180 identical protein chains that are assembled in $T = 3$ lattice. Three chains named A, B, and C are arranged in cyclic hexameric form. In solution, the subunits A and B, and the identical subunits of C form asymmetric and symmetric dimers, respectively. These dimers aid in the formation of assembly. The largest interface of 6400 $Å^2$ between the symmetric and asymmetric dimers aids in stabilization of the assembly. Medium size interfaces (800–1100 $Å^2$) allow the association of symmetric and asymmetric dimers into dimer hexamers, which further allow the formation of higher order assemblies and step by step addition of hexamers or pair of hexamers

complexes lie within these enumerated topologies. This observation enabled the authors to organize naturally existing protein complexes into a "periodic table" since each heteromer was related to a simpler and symmetric homomer topology. Exceptions were found only in cases of assignment errors in quaternary structure, or circumstances where sequence identical subunits had distinct interactions, thereby introducing asymmetry. Most of these asymmetric complexes were found to fit the periodic table paradigm when their assembly role was taken into consideration. Finally, the authors implemented a periodic table-based model that predicted the expected frequencies of each quaternary structure topology. It has been demonstrated that their model rightly predicted the quaternary structure topologies of electron microscopy and crystal structures that were discovered recently. The results obtained established that this work explained much of protein complex distribution in quaternary structure space and also provided a basic framework to understand their evolution better.

This proposed periodic table has enormous potential to predict and model protein complexes at their quaternary structural levels. Additionally, this periodic table has the capacity to provide information on the topologies of an evolutionary precursor that has most likely given rise to a particular complex. Furthermore, the periodic table can furnish constraints for multi-subunit modelling or docking or both on the basis of the complex symmetry and the subunits' relative arrangements (DiMaio et al. 2011; Esquivel-Rodríguez and Kihara 2012). Moreover, it has been hypothesized that this work has substantial contribution to the field of bio-engineering, as the self-assembly formalism that was presented here enabled the precise specification of essential interfaces that would need engineering to make up a protein complex of a particular topology.

4.7 Protein Complexes for Enhancing Quantitative Proteomics

Protein complexes can be applied for studying phenotypes and phenotype differences, such as differences between human disorders, or between mutants of yeast, etc. Such applications can be usually divided into two classes: (1) non-quantitative and (2) quantitative applications (Srihari et al. 2017). *Non-quantitative* applications utilize protein complexes in a descriptive or qualitative way, such as functionally interpreting a set of proteins in an analysis, or explaining the systematic ground of an observed phenotype. For instance, Dudley et al. utilized phylogenetic profiles of genes from 4700 yeast mutants under various growth conditions to identify gene modules (Dudley et al. 2005). These modules were mapped onto known protein complexes from existing literature to better understand the functions of the genes that were accountable for growth fitness differences in the mutants. Such non-quantitative applications hold significance, but they are not sufficient enough to advance the concept. Since protein complexes mainly provide auxiliary or additional support and the generated outcomes cannot be extrapolated to larger sets. On the other hand, *quantitative* applications utilize protein complexes to make their results more reproducible. For instance, Vandin et al. identified features that could efficiently categorize patients on the basis of cancer subtypes (Vandin et al. 2011).

Protein complexes have become an important source of information for researchers working in the proteomics field. Over time, it has been realized that in order to analyze proteomics data, it is not essential to utilize entire protein interaction networks. Instead, protein complexes that provide information as functional unit are much less complicated and are consequently easier to use (Goh and Wong 2016). Protein complexes essentially serve as the fundamental units of biological function, and are tightly connected to each other, thereby forming a particular network connectivity. Therefore, these protein complexes can be exploited as (non-trivial) subnetworks for explaining the underlying phenotype differences (Fraser and Plotkin 2007; Michaut et al. 2011). Secondly, there are several centralized databases that list out expert-curated protein complexes, whereas there are none for reference subnetworks. Additionally, it is relatively easy to standardize protein complexes because they are usually represented in terms of their constituting proteins that can be validated experimentally and do not require network-mining-based algorithms that can yield variable results. Furthermore, the subcellular localization and functional annotations of protein complexes make it easier to control such complexes as standard for annotations of network-based proteomics. Research groups have already explored the discoursing approaches to utilize protein complexes for quantifying phenotypic differences (Srihari and Ragan 2013; Zhao et al. 2013). The differential expression of protein complexes was gracefully exploited to determine clinical and molecular differences between various cancer subtypes.

4.8 Patented Protein Structural Complexes

In recent years, protein–protein complexes that are linked with various diseases have been taken into consideration as drug targets. These complexes serve as drug targets apart from the classical drug targets that utilize only monomeric proteins. Several of these complexes that have been patented by various organizations are summarized in a tabulated form by Kangueane and Nilofer (2018). Many of these protein complexes have known three-dimensional structures as determined by experimental techniques such as X-ray crystallography. The table summarizes the gene products such as BCAT, KCAT, etc. that are someway involved with diseased conditions. Data on other fields such as patent ID, GenBank ID, year of patent, associated disease, target, and PDB ID have also been reported in the table. GenPept at the National Center for Biotechnology Information (NCBI) was utilized to find records for disease-associated known protein complex structures that also have associated patent rights.

4.9 Conclusions

Although there has been increasing progress on PPIs and protein complexes, there are still certain aspects of them that are not yet well understood. For instance, the coverage of heteromeric complexes is still low as compared to homomeric complexes from the point of view of quaternary structures. Secondly, there is a lot of scope for improvement of understanding protein complexes in the context of living cells and to analyze how they assemble in vivo. Understanding the in vivo functioning of protein complexes gives a holistic idea regarding other aspects such as location of assembly in cells and the contribution of chaperones to the assembly process. Additionally, studying protein complexes has immense potential in the medical field because certain interface mutations have been observed in these complexes accountable for particular human disorders. Thus, it is essential to understand the quaternary structures and other functional roles of protein complexes in detail in order to better predict the effects of mutations on the phenotype of the organism. Further, understanding the assembly and evolutionary pathways of protein complexes can enable researchers to bio-engineer them or develop new drugs to trigger disassembly or inhibit the process of complex assembly.

References

Ahnert SE, Marsh JA, Hernandez H, Robinson CV, Teichmann SA. (2015). Principles of assembly reveal a periodic table of protein complexes. *Science, 350*, aaa2245.

Akiva E, Itzhaki Z, Margalit H. (2008). Built-in loops allow versatility in domain-domain interactions: Lessons from self-interacting domains. *Proceedings of the National Academy of Sciences of the United States of America, 105*, 13292-13297.

Alberts B. (1998). The cell as a collection of protein machines: Preparing the next generation of molecular biologists. *Cell, 92*, 291-294.

André I, Strauss CE, Kaplan DB, Bradley P, Baker DJ. (2008). Emergence of symmetry in homooligomeric biological assemblies. *Proceedings of the National Academy of Sciences, 105*, 16148-16152.

Antebi YE et al. (2017). Combinatorial Signal Perception in the BMP Pathway. *Cell, 170*, 1184-1196 e1124.

Aquilina JA, Benesch JL, Bateman OA, Slingsby C, Robinson CV. (2003). Polydispersity of a mammalian chaperone: mass spectrometry reveals the population of oligomers in αB-crystallin. *Proceedings of the National Academy of Sciences, 100*, 10611-10616.

Archibald JM, Logsdon JM, Doolittle WF. (1999). Recurrent paralogy in the evolution of archaeal chaperonins. *Current biology : CB, 9*, 1053-1056.

Audin MJC, Dorn G, Fromm SA, Reiss K, Schutz S, Vorlander MK, Sprangers R. (2013). The Archaeal Exosome: Identification and Quantification of Site-Specific Motions That Correlate with Cap and RNA Binding. *Angew Chem Int Edit, 52*, 8312-8316.

Bahadur RP, Janin J. (2008). Residue conservation in viral capsid assembly. *Proteins, 71*, 407-414.

Bahadur RP, Chakrabarti P, Rodier F, Janin J. (2003). Dissecting subunit interfaces in homodimeric proteins. *Proteins, 53*, 708-719.

Baker JM, Hudson RP, Kanelis V, Choy WY, Thibodeau PH, Thomas PJ, Forman-Kay JD. (2007). CFTR regulatory region interacts with NBD1 predominantly via multiple transient helices. *Nature structural & molecular biology, 14*, 738-745.

Ban N, Nissen P, Hansen J, Moore PB, Steitz TAJS. (2000). The complete atomic structure of the large ribosomal subunit at 2.4 Å resolution. *Science, 289*, 905-920.

Bennett MJ, Schlunegger MP, Eisenberg D. (1995). 3D domain swapping: a mechanism for oligomer assembly. *Protein science : a publication of the Protein Society, 4*, 2455-2468.

Bershtein S, Mu W, Shakhnovich EI. (2012). Soluble oligomerization provides a beneficial fitness effect on destabilizing mutations. *Proceedings of the National Academy of Sciences of the United States of America, 109*, 4857-4862.

Björklund ÅK, Ekman D, Light S, Frey-Skött J, Elofsson AJ. (2005). Domain rearrangements in protein evolution. *Journal of molecular biology, 353*, 911-923.

Bolanos-Garcia VM (2017) Protein Complexes in the Nucleus: The Control of Chromosome Segregation. In: Macromolecular Protein Complexes. Springer, pp 455-481

Bozoky Z et al. (2013). Regulatory R region of the CFTR chloride channel is a dynamic integrator of phospho-dependent intra-and intermolecular interactions. *Proceedings of the National Academy of Sciences*, *110*, E4427-E4436.
Brinda KV, Vishveshwara S. (2005). Oligomeric protein structure networks: insights into protein-protein interactions. *Bmc Bioinformatics*, *6*, 296.
Buljan M, Frankish A, Bateman A. (2010). Quantifying the mechanisms of domain gain in animal proteins. *Genome biology*, *11*, R74.
Byron O, Lindsay JG (2017) The pyruvate dehydrogenase complex and related assemblies in health and disease. In: Macromolecular Protein Complexes. Springer, pp 523-550
Cao Z, Lindsay JG (2017) The peroxiredoxin family: an unfolding story. In: Macromolecular Protein Complexes. Springer, pp 127-147
Cheng KY et al. (2006). The role of the phospho-CDK2/cyclin A recruitment site in substrate recognition. *The Journal of biological chemistry*, *281*, 23167-23179.
Cramer P, Bushnell DA, Kornberg RD. (2001). Structural basis of transcription: RNA polymerase II at 2.8 angstrom ngstrom resolution. *Science*, *292*, 1863-1876.
Crick FH, Watson JD. (1956). Structure of small viruses. *Nature*, *177*, 473-475.
Dayhoff JE, Shoemaker BA, Bryant SH, Panchenko AR. (2010). Evolution of protein binding modes in homooligomers. *Journal of molecular biology*, *395*, 860-870.
Deprez E, Tauc P, Leh H, Mouscadet JF, Auclair C, Brochon JC. (2000). Oligomeric states of the HIV-1 integrase as measured by time-resolved fluorescence anisotropy. *Biochemistry-Us*, *39*, 9275-9284.
Dey S, Pal A, Chakrabarti P, Janin J. (2010). The subunit interfaces of weakly associated homodimeric proteins. *Journal of molecular biology*, *398*, 146-160.
Dill KA, Chan HS. (1997). From Levinthal to pathways to funnels. *Nature structural biology*, *4*, 10-19.
DiMaio F, Leaver-Fay A, Bradley P, Baker D, Andre I. (2011). Modeling symmetric macromolecular structures in Rosetta3. *PloS one*, *6*, e20450.
Ditzel L, Lowe J, Stock D, Stetter KO, Huber H, Huber R, Steinbacher S. (1998). Crystal structure of the thermosome, the archaeal chaperonin and homolog of CCT. *Cell*, *93*, 125-138.
Dobbins SE, Lesk VI, Sternberg MJE. (2008). Insights into protein flexibility: The relationship between normal modes and conformational change upon protein-protein docking. *Proceedings of the National Academy of Sciences of the United States of America*, *105*, 10390-10395.
Dobson RCJ, Valegard K, Gerrard JA. (2004). The crystal structure of three site-directed mutants of Escherichia coli dihydrodipicolinate synthase: Further evidence for a catalytic triad. *Journal of molecular biology*, *338*, 329-339.
Doolittle WF. (2012). Evolutionary biology: A ratchet for protein complexity. *Nature*, *481*, 270-271.
Dudley AM, Janse DM, Tanay A, Shamir R, Church GM. (2005). A global view of pleiotropy and phenotypically derived gene function in yeast. *Molecular systems biology*, *1*, 2005 0001.
Durbin SD, Feher G. (1996). Protein crystallization. *Annual review of physical chemistry*, *47*, 171-204.
Dyson HJ, Wright PE. (2002). Coupling of folding and binding for unstructured proteins. *Current opinion in structural biology*, *12*, 54-60.
Ellis RJ. (2001). Macromolecular crowding: an important but neglected aspect of the intracellular environment. *Current opinion in structural biology*, *11*, 114-119.
Enright AJ, Iliopoulos I, Kyrpides NC, Ouzounis CA. (1999). Protein interaction maps for complete genomes based on gene fusion events. *Nature*, *402*, 86-90.
Esquivel-Rodríguez J, Kihara D Evaluation of multiple protein docking structures using correctly predicted pairwise subunits. In: BMC bioinformatics, 2012. vol 2. BioMed Central, pp 1-4
Fernández A, Lynch M. (2011). Non-adaptive origins of interactome complexity. *Nature genetics*, *474*, 502-505.
Finnigan GC, Hanson-Smith V, Stevens TH, Thornton JW. (2012). Evolution of increased complexity in a molecular machine. *Nature*, *481*, 360-364.

Fong JH, Geer LY, Panchenko AR, Bryant SH. (2007). Modeling the evolution of protein domain architectures using maximum parsimony. *Journal of molecular biology, 366*, 307-315.
Fraser HB, Plotkin JB. (2007). Using protein complexes to predict phenotypic effects of gene mutation. *Genome biology, 8*, R252.
Friedman FK, Beychok S. (1979). Probes of subunit assembly and reconstitution pathways in multisubunit proteins. *Annual review of biochemistry, 48*, 217-250.
Friedrich T. (2001). Complex I: a chimaera of a redox and conformation-driven proton pump? *Journal of bioenergetics and biomembranes, 33*, 169-177.
Fuxreiter M et al. (2014). Disordered proteinaceous machines. *Chemical reviews, 114*, 6806-6843.
Gabald6n T, Rainey D, Huynen MA. (2005). Tracing the evolution of a large protein complex in the eukaryotes, NADH: ubiquinone oxidoreductase (Complex I). *Journal of molecular biology, 348*, 857-870.
Garcia-Ferrer I, Marrero A, Gomis-Rüth FX, Goulas T (2017) α 2-Macroglobulins: Structure and function. In: Macromolecular Protein Complexes. Springer, pp 149-183
Garcia-Seisdedos H, Villegas JA, Levy ED. (2019). Infinite assembly of folded proteins in evolution, disease, and engineering. *Angewandte Chemie International Edition, 58*, 5514-5531.
Gavin AC et al. (2006). Proteome survey reveals modularity of the yeast cell machinery. *Nature, 440*, 631-636.
Godoy-Ruiz R, Krejcirikova A, Gallagher DT, Tugarinov V. (2011). Solution NMR evidence for symmetry in functionally or crystallographically asymmetric homodimers. *Journal of the American Chemical Society, 133*, 19578-19581.
Goh WWB, Wong L. (2016). Integrating Networks and Proteomics: Moving Forward. *Trends in biotechnology, 34*, 951-959.
Goodsell DS, Olson AJ. (2000). Structural symmetry and protein function. *Annual review of biophysics and biomolecular structure, 29*, 105-153.
Griesenbeck J, Tschochner H, Grohmann D (2017) Structure and function of RNA polymerases and the transcription machineries. In: Macromolecular Protein Complexes. Springer, pp 225-270
Hahn S. (2004). Structure and mechanism of the RNA polymerase II transcription machinery. *Nature structural & molecular biology, 11*, 394-403.
Halebian M, Morris K, Smith C (2017) Structure and assembly of clathrin cages. In: Macromolecular Protein Complexes. Springer, pp 551-567
Haley DA, Horwitz J, Stewart PL. (1998). The small heat-shock protein, αB-crystallin, has a variable quaternary structure. *Journal of molecular biology, 277*, 27-35.
Hall Z, Politis A, Robinson CV. (2012). Structural modeling of heteromeric protein complexes from disassembly pathways and ion mobility-mass spectrometry. *Structure, 20*, 1596-1609.
Hall Z, Hernandez H, Marsh JA, Teichmann SA, Robinson CV. (2013). The role of salt bridges, charge density, and subunit flexibility in determining disassembly routes of protein complexes. *Structure, 21*, 1325-1337.
Hashimoto K, Panchenko AR. (2010). Mechanisms of protein oligomerization, the critical role of insertions and deletions in maintaining different oligomeric states. *Proceedings of the National Academy of Sciences of the United States of America, 107*, 20352-20357.
Hashimoto K, Madej T, Bryant SH, Panchenko AR. (2010). Functional states of homooligomers: insights from the evolution of glycosyltransferases. *Journal of molecular biology, 399*, 196-206.
Havugimana PC et al. (2012). A census of human soluble protein complexes. *Cell, 150*, 1068-1081.
Hegyi H, Schad E, Tompa P. (2007). Structural disorder promotes assembly of protein complexes. *BMC structural biology, 7*, 65.
Heldin CH. (1995). Dimerization of cell surface receptors in signal transduction. *Cell, 80*, 213-223.
Hernandez H, Robinson CV. (2007). Determining the stoichiometry and interactions of macromolecular assemblies from mass spectrometry. *Nature protocols, 2*, 715-726.
Huang YQ, Cao HQ, Liu ZR. (2012). Three-dimensional domain swapping in the protein structure space. *Proteins, 80*, 1610-1619.
Ishii N (2017) GroEL and the GroEL-GroES complex. In: Macromolecular Protein Complexes. Springer, pp 483-504

Jaffe EK. (2005). Morpheeins - a new structural paradigm for allosteric regulation. *Trends Biochem Sci*, *30*, 490-497.

Janin J, Bahadur RP, Chakrabarti P (2008). Protein-protein interaction and quaternary structure. *41*, 133-180.

Jeyaprakash AA, Santamaria A, Jayachandran U, Chan YW, Benda C, Nigg EA, Conti E. (2012). Structural and functional organization of the Ska complex, a key component of the kinetochore-microtubule interface. *Molecular cell*, *46*, 274-286.

Johansson M, Zhang JJ, Ehrenberg M. (2012). Genetic code translation displays a linear trade-off between efficiency and accuracy of tRNA selection. *Proceedings of the National Academy of Sciences of the United States of America*, *109*, 131-136.

Kangueane P, Nilofer C (2018) Patented Protein Structural Complexes in Discovery Platform. In: Protein-Protein and Domain-Domain Interactions. Springer, pp 189-197

Kim WK, Henschel A, Winter C, Schroeder M. (2006). The many faces of protein-protein interactions: A compendium of interface geometry. *PLoS computational biology*, *2*, 1151-1164.

Kleywegt GJ. (1996). Use of non-crystallographic symmetry in protein structure refinement. *Acta Crystallogr D*, *52*, 842-857.

Klotz IM, Langerman NR, Darnall DW. (1970). Quaternary structure of proteins. *Annual review of biochemistry*, *39*, 25-62.

Knowles TP, Vendruscolo M, Dobson CM. (2014). The amyloid state and its association with protein misfolding diseases. *Nature reviews. Molecular cell biology*, *15*, 384-396.

Koshland Jr D, Nemethy G, Filmer D. (1966). Comparison of experimental binding data and theoretical models in proteins containing subunits. *Biochemistry*, *5*, 365-385.

Krogan NJ et al. (2006). Global landscape of protein complexes in the yeast Saccharomyces cerevisiae. *Nature*, *440*, 637-643.

Kühner S et al. (2009). Proteome organization in a genome-reduced bacterium. *Science*, *326*, 1235-1240.

Kummerfeld SK, Teichmann SA. (2005). Relative rates of gene fusion and fission in multi-domain proteins. *Trends in genetics : TIG*, *21*, 25-30.

Landry CR, Levy ED, Abd Rabbo D, Tarassov K, Michnick SW. (2013). Extracting insight from noisy cellular networks. *Cell*, *155*, 983-989.

Latham MP, Sekhar A, Kay LE. (2014). Understanding the mechanism of proteasome 20S core particle gating. *Proceedings of the National Academy of Sciences of the United States of America*, *111*, 5532-5537.

Leiman PG, Kanamaru S, Mesyanzhinov VV, Arisaka F, Rossmann MG. (2003). Structure and morphogenesis of bacteriophage T4. *Cellular and molecular life sciences : CMLS*, *60*, 2356-2370.

Levy Y, Cho SS, Onuchic JN, Wolynes PG. (2005). A survey of flexible protein binding mechanisms and their transition states using native topology based energy landscapes. *Journal of molecular biology*, *346*, 1121-1145.

Levy ED, Pereira-Leal JB, Chothia C, Teichmann SA. (2006). 3D complex: a structural classification of protein complexes. *PLoS computational biology*, *2*, e155.

Levy ED, Boeri Erba E, Robinson CV, Teichmann SA. (2008). Assembly reflects evolution of protein complexes. *Nature*, *453*, 1262-1265.

Levy ED, De S, Teichmann SA. (2012). Cellular crowding imposes global constraints on the chemistry and evolution of proteomes. *Proceedings of the National Academy of Sciences of the United States of America*, *109*, 20461-20466.

Li GW, Burkhardt D, Gross C, Weissman JS. (2014). Quantifying absolute protein synthesis rates reveals principles underlying allocation of cellular resources. *Cell*, *157*, 624-635.

Liljas L. (1999). Virus assembly. *Current opinion in structural biology*, *9*, 129-134.

Lukatsky DB, Shakhnovich BE, Mintseris J, Shakhnovich EI. (2007). Structural similarity enhances interaction propensity of proteins. *Journal of molecular biology*, *365*, 1596-1606.

Lynch M. (2007). The frailty of adaptive hypotheses for the origins of organismal complexity. *Proceedings of the National Academy of Sciences of the United States of America, 104 Suppl 1*, 8597-8604.
Lynch M. (2012). The evolution of multimeric protein assemblages. *Molecular biology and evolution, 29*, 1353-1366.
Lynch M. (2013). Evolutionary diversification of the multimeric states of proteins. *Proceedings of the National Academy of Sciences of the United States of America, 110*, E2821-2828.
Mackinnon SS, Malevanets A, Wodak SJ. (2013). Intertwined associations in structures of homooligomeric proteins. *Structure, 21*, 638-649.
Malay AD, Allen KN, Tolan DR. (2005). Structure of the thermolabile mutant aldolase B, A149P: molecular basis of hereditary fructose intolerance. *Journal of molecular biology, 347*, 135-144.
Marcotte EM, Pellegrini M, Ng HL, Rice DW, Yeates TO, Eisenberg D. (1999). Detecting protein function and protein-protein interactions from genome sequences. *Science, 285*, 751-753.
Marsh JA. (2013). Buried and accessible surface area control intrinsic protein flexibility. *Journal of molecular biology, 425*, 3250-3263.
Marsh JA, Teichmann SA. (2010). How do proteins gain new domains? *Genome biology, 11*, 126.
Marsh JA, Teichmann SA. (2011). Relative Solvent Accessible Surface Area Predicts Protein Conformational Changes upon Binding. *Structure, 19*, 859-867.
Marsh JA, Teichmann SA. (2014a). Parallel dynamics and evolution: Protein conformational fluctuations and assembly reflect evolutionary changes in sequence and structure. *BioEssays : news and reviews in molecular, cellular and developmental biology, 36*, 209-218.
Marsh JA, Teichmann SA. (2014b). Protein flexibility facilitates quaternary structure assembly and evolution. *PLoS biology, 12*, e1001870.
Marsh JA, Teichmann SA. (2015). Structure, dynamics, assembly, and evolution of protein complexes. *Annual review of biochemistry, 84*, 551-575.
Marsh JA, Dancheck B, Ragusa MJ, Allaire M, Forman-Kay JD, Peti W. (2010). Structural diversity in free and bound states of intrinsically disordered protein phosphatase 1 regulators. *Structure, 18*, 1094-1103.
Marsh JA, Teichmann SA, Forman-Kay JD. (2012). Probing the diverse landscape of protein flexibility and binding. *Current opinion in structural biology, 22*, 643-650.
Marsh JA, Hernandez H, Hall Z, Ahnert SE, Perica T, Robinson CV, Teichmann SA. (2013). Protein complexes are under evolutionary selection to assemble via ordered pathways. *Cell, 153*, 461-470.
Mathiesen C, Hagerhall C. (2003). The 'antiporter module' of respiratory chain Complex I includes the MrpC/NuoK subunit - a revision of the modular evolution scheme. *Febs Lett, 549*, 7-13.
McGuffee SR, Elcock AH. (2010). Diffusion, crowding & protein stability in a dynamic molecular model of the bacterial cytoplasm. *PLoS computational biology, 6*, e1000694.
Michaut M, Baryshnikova A, Costanzo M, Myers CL, Andrews BJ, Boone C, Bader GD. (2011). Protein complexes are central in the yeast genetic landscape. *PLoS computational biology, 7*, e1001092.
Mirande M (2017) The aminoacyl-tRNA synthetase complex. In: Macromolecular Protein Complexes. Springer, pp 505-522
Mittag T, Kay LE, Forman-Kay JD. (2010a). Protein dynamics and conformational disorder in molecular recognition. *Journal of molecular recognition : JMR, 23*, 105-116.
Mittag T et al. (2010b). Structure/function implications in a dynamic complex of the intrinsically disordered Sic1 with the Cdc4 subunit of an SCF ubiquitin ligase. *Structure, 18*, 494-506.
Monie TP (2017) The canonical inflammasome: a macromolecular complex driving inflammation. In: Macromolecular Protein Complexes. Springer, pp 43-73
Monod J, Wyman J, Changeux JP. (1965). On the Nature of Allosteric Transitions: A Plausible Model. *Journal of molecular biology, 12*, 88-118.
Murray AW. (2004). Recycling the cell cycle: cyclins revisited. *Cell, 116*, 221-234.

Murzin AG, Brenner SE, Hubbard T, Chothia C. (1995). SCOP: a structural classification of proteins database for the investigation of sequences and structures. *Journal of molecular biology, 247*, 536-540.

Navia MA et al. (1989). Three-dimensional structure of aspartyl protease from human immunodeficiency virus HIV-1. *Nature, 337*, 615-620.

Nguyen HC, Wang W, Xiong Y (2017) Cullin-RING E3 ubiquitin ligases: bridges to destruction. In: Macromolecular Protein Complexes. Springer, pp 323-347

Nishi H, Ota M. (2010). Amino acid substitutions at protein-protein interfaces that modulate the oligomeric state. *Proteins, 78*, 1563-1574.

Nishi H, Koike R, Ota M. (2011). Cover and spacer insertions: Small nonhydrophobic accessories that assist protein oligomerization. *Proteins, 79*, 2372-2379.

Nourse A, Mittag T. (2014). The cytoplasmic domain of the T-cell receptor zeta subunit does not form disordered dimers. *Journal of molecular biology, 426*, 62-70.

Orengo CA, Michie AD, Jones S, Jones DT, Swindells MB, Thornton JM. (1997). CATH - a hierarchic classification of protein domain structures. *Structure, 5*, 1093-1108.

Pané-Farré J, Quin MB, Lewis RJ, Marles-Wright J (2017) Structure and function of the stressosome signalling hub. In: Macromolecular protein complexes. Springer, pp 1-41

Pasek S, Risler JL, Brezellec P. (2006). Gene fusion/fission is a major contributor to evolution of multi-domain bacterial proteins. *Bioinformatics, 22*, 1418-1423.

Pearce FG, Hudson AO, Loomes K, Dobson RC (2017) Dihydrodipicolinate synthase: structure, dynamics, function, and evolution. In: Macromolecular Protein Complexes. Springer, pp 271-289

Perica T, Chothia C, Teichmann SA. (2012a). Evolution of oligomeric state through geometric coupling of protein interfaces. *Proceedings of the National Academy of Sciences, 109*, 8127-8132.

Perica T, Marsh JA, Sousa FL, Natan E, Colwell LJ, Ahnert SE, Teichmann SA (2012b) The emergence of protein complexes: quaternary structure, dynamics and allostery. Portland Press Ltd.,

Perkins JR, Diboun I, Dessailly BH, Lees JG, Orengo C. (2010). Transient protein-protein interactions: structural, functional, and network properties. *Structure, 18*, 1233-1243.

Ponstingl H, Kabir T, Gorse D, Thornton JM. (2005). Morphological aspects of oligomeric protein structures. *Progress in biophysics and molecular biology, 89*, 9-35.

Prevelige Jr PE, Thomas D, King J. (1993). Nucleation and growth phases in the polymerization of coat and scaffolding subunits into icosahedral procapsid shells. *Biophysical journal, 64*, 824-835.

Ragusa MJ, Dancheck B, Critton DA, Nairn AC, Page R, Peti W. (2010). Spinophilin directs protein phosphatase 1 specificity by blocking substrate binding sites. *Nature structural & molecular biology, 17*, 459-464.

Religa TL, Sprangers R, Kay LE. (2010). Dynamic Regulation of Archaeal Proteasome Gate Opening As Studied by TROSY NMR. *Science, 328*, 98-102.

Reuveni S, Ehrenberg M, Paulsson J. (2017). Ribosomes are optimized for autocatalytic production. *Nature, 547*, 293-297.

Ruano-Rubio V, Fares MA. (2007). Testing the neutral fixation of hetero-oligomerism in the archaeal chaperonin CCT. *Molecular biology and evolution, 24*, 1384-1396.

Ruotolo BT, Benesch JLP, Sandercock AM, Hyung SJ, Robinson CV. (2008). Ion mobility-mass spectrometry analysis of large protein complexes. *Nature protocols, 3*, 1139-1152.

Sartori P, Leibler S. (2019). Towards a theory of assembly of protein complexes: lessons from equilibrium statistical physics. *arXiv preprint arXiv:1907.02992.*

Sciara G et al. (2010). Structure of lactococcal phage p2 baseplate and its mechanism of activation. *Proceedings of the National Academy of Sciences, 107*, 6852-6857.

Shiber A et al. (2018). Cotranslational assembly of protein complexes in eukaryotes revealed by ribosome profiling. *Nature, 561*, 268-272.

Shieh YW, Minguez P, Bork P, Auburger JJ, Guilbride DL, Kramer G, Bukau B. (2015). Operon structure and cotranslational subunit association direct protein assembly in bacteria. *Science*, *350*, 678-680.

Sobolevsky AI, Rosconi MP, Gouaux E. (2009). X-ray structure, symmetry and mechanism of an AMPA-subtype glutamate receptor. *Nature*, *462*, 745-756.

Sprangers R, Kay LE. (2007). Quantitative dynamics and binding studies of the 20S proteasome by NMR. *Nature*, *445*, 618-622.

Sprangers R, Velyvis A, Kay LE. (2007). Solution NMR of supramolecular complexes: providing new insights into function. *Nat Methods*, *4*, 697-703.

Srihari S, Ragan MA. (2013). Systematic tracking of dysregulated modules identifies novel genes in cancer. *Bioinformatics*, *29*, 1553-1561.

Srihari S, Yong CH, Wong L (2017) Computational prediction of protein complexes from protein interaction networks. Morgan & Claypool,

Stein A, Pache RA, Bernado P, Pons M, Aloy P. (2009). Dynamic interactions of proteins in complex networks: a more structured view. *The FEBS journal*, *276*, 5390-5405.

Stengel F et al. (2010). Quaternary dynamics and plasticity underlie small heat shock protein chaperone function. *Proceedings of the National Academy of Sciences of the United States of America*, *107*, 2007-2012.

Stock D, Leslie AG, Walker JE. (1999). Molecular architecture of the rotary motor in ATP synthase. *Science*, *286*, 1700-1705.

Stockley PG et al. (2007). A simple, RNA-mediated allosteric switch controls the pathway to formation of a T=3 viral capsid. *Journal of molecular biology*, *369*, 541-552.

Talkington MW, Siuzdak G, Williamson JR. (2005). An assembly landscape for the 30S ribosomal subunit. *Nature*, *438*, 628-632.

Tarassov K et al. (2008). An in vivo map of the yeast protein interactome. *Science*, *320*, 1465-1470.

Tobi D, Bahar I. (2005). Structural changes involved in protein binding correlate with intrinsic motions of proteins in the unbound state. *Proceedings of the National Academy of Sciences of the United States of America*, *102*, 18908-18913.

Tompa P, Fuxreiter M. (2008). Fuzzy complexes: polymorphism and structural disorder in protein-protein interactions. *Trends Biochem Sci*, *33*, 2-8.

Tompa P, Davey NE, Gibson TJ, Babu MM. (2014). A million peptide motifs for the molecular biologist. *Molecular cell*, *55*, 161-169.

Valle M (2017) Pyruvate Carboxylase, Structure and Function. In: Macromolecular Protein Complexes. Springer, pp 291-322

Van Dam TJ, Snel B. (2008). Protein complex evolution does not involve extensive network rewiring. *PLoS computational biology*, *4*, e1000132.

van der Sluis EO, Driessen AJ. (2006). Stepwise evolution of the Sec machinery in Proteobacteria. *Trends in microbiology*, *14*, 105-108.

Vandin F, Upfal E, Raphael BJ. (2011). Algorithms for Detecting Significantly Mutated Pathways in Cancer. *J Comput Biol*, *18*, 507-522.

Venkatakrishnan A, Levy ED, Teichmann SA (2010) Homomeric protein complexes: evolution and assembly. Portland Press Ltd.,

Wells M et al. (2008). Structure of tumor suppressor p53 and its intrinsically disordered N-terminal transactivation domain. *Proceedings of the National Academy of Sciences of the United States of America*, *105*, 5762-5767.

White MR, Garcin ED (2017) D-Glyceraldehyde-3-phosphate dehydrogenase structure and function. In: Macromolecular Protein Complexes. Springer, pp 413-453

Williams NK, Dichtl B. (2018). Co-translational control of protein complex formation: a fundamental pathway of cellular organization? *Biochemical Society transactions*, *46*, 197-206.

Wright PE, Dyson HJ. (1999). Intrinsically unstructured proteins: re-assessing the protein structure-function paradigm. *Journal of molecular biology*, *293*, 321-331.

Xie Z, Hendrix RW. (1995). Assembly in Vitro of Bacteriophage HK97 Proheads. *Journal of molecular biology*, *253*, 74-85.

Xu M, Bai L, Gong Y, Xie W, Hang H, Jiang T. (2009). Structure and functional implications of the human rad9-hus1-rad1 cell cycle checkpoint complex. *Journal of Biological Chemistry, 284*, 20457-20461.

Yang JR, Liao BY, Zhuang SM, Zhang J. (2012). Protein misinteraction avoidance causes highly expressed proteins to evolve slowly. *Proceedings of the National Academy of Sciences of the United States of America, 109*, E831-840.

Yévenes A (2017) The Ferritin Superfamily. In: Macromolecular Protein Complexes. Springer, pp 75-102

Yu L, Reutzel-Edens SM, Mitchell CA. (2000). Crystallization and polymorphism of conformationally flexible molecules: Problems, patterns, and strategies. *Org Process Res Dev, 4*, 396-402.

Yu Y, Smith DM, Kim HM, Rodriguez V, Goldberg AL, Cheng Y. (2010). Interactions of PAN's C-termini with archaeal 20S proteasome and implications for the eukaryotic proteasome–ATPase interactions. *The EMBO journal, 29*, 692-702.

Zhao J, Lee SH, Huss M, Holme P. (2013). The network organization of cancer-associated protein complexes in human tissues. *Scientific reports, 3*, 1583.

5 Experimental Methods for Determination of Protein–Protein Interactions

Abstract

Study of protein–protein interactions in the cell provides better insights into the molecular network of the cell. It is therefore essential to establish the existing interactions by experimental techniques. Such techniques not only help to discover new interactions, but also validate interactions that have been hypothesized. The techniques can be broadly classified as biophysical, biochemical, and genetic. This chapter discusses the techniques under each classification and also outlines their respective working principles, advantages, drawbacks, and recent developments. The future prospects have been delineated based on their current status of applications.

Keywords

Protein–protein interactions · NMR spectroscopy · Light scattering · Cryo-electron microscopy · Mass spectrometry · Fluorescence resonance energy transfer · Co-immunoprecipitation · Synthetic lethality

5.1 Introduction

The identification of PPIs aids researchers to situate proteins in the context of cellular pathways and subsequently assign functional roles to novel proteins (Pelletier and Sidhu 2001). This in turn facilitates the study of cell functioning not only under physiological conditions, but also from pathological perspectives. Further, determination of PPIs on a large scale has the potential to lay down the complex web/network of all protein interactions in the cell to elucidate the "interactome." Understanding the interactome, in turn, forms the basis for systemically analyzing cellular networks, elucidating the proteome organization, and unraveling the dynamic and functional aspects of the cellular organisms (Vidal et al. 2011). Moreover, since anomalous PPIs often form the underlying reason of diseased

K. M. Poluri et al., *Protein-Protein Interactions*,
https://doi.org/10.1007/978-981-16-1594-8_5

conditions in humans, determination of PPIs has become important to design drugs that specifically target such aberrant interactions (Ruffner et al. 2007). For example, researchers have elucidated PPIs responsible for conditions like cancer, autoimmune and infectious diseases. These studies have subsequently aided in the development of therapeutic PPI modulators, some of which have entered the clinical trials or are already in market (Zhou et al. 2016).

As the formation of the protein complexes increases the diversity of protein functionalities, it is not possible to determine the PPIs involved in a protein complex solely by utilizing the genomic information. This makes the research of PPIs relatively difficult. Therefore, techniques based on physics or chemistry need to be applied to determine and analyze PPIs. Further, to understand PPIs completely, the physicochemical properties of proteins need to be taken into account. This makes the research on PPIs interdisciplinary that utilizes and integrates knowledge from fields of biology, chemistry, and physics. Consequently, research methods from different fields are introduced in parallel to understand protein interactions. It therefore becomes essential to understand the principle on which each research method is based on. Understanding the principles not only helps in better accumulation and analysis of collected data, but also aids in developing and technically advancing the equipment (Miura 2018).

Several experimental methods have been developed to elucidate PPIs. These methods are based on principles from either bio-physical, biochemical, or genetic fields (Fig. 5.1). Each of these techniques has its own advantages and pitfalls with respect to specificity and sensitivity. Therefore, research groups utilize different methods for analyzing different aspects of PPIs. Some of them are advantageous for identification of binding partners, while others are more suited to obtain structural data or furnish thermodynamic details or visualize PPIs in living cells in real time. This chapter describes a host of widely used experimental methods for PPIs, briefing the principles, advantages/disadvantages, and PPI case studies of each method.

5.2 Biophysical Techniques

Biophysical techniques are interdisciplinary techniques that obtain information by utilizing principles from the fields of biology and physics. There are several classical biophysical techniques that have found wide utility in the study of PPIs. An array of the techniques applicable to elucidate the structural, thermodynamic, kinetic, and biophysical features of the PPIs have been summarized in Fig. 5.2, and have been discussed in detail in the following sections.

5.2.1 Nuclear Magnetic Resonance (NMR) Spectroscopy

The principle of nuclear magnetic resonance (NMR) phenomenon involves the absorption of electromagnetic radiation of appropriate frequency by magnetically

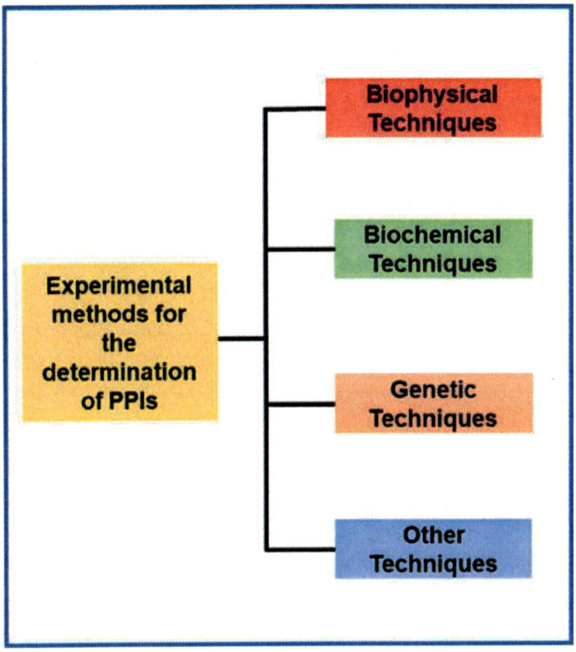

Fig. 5.1 Schematic depicting various experimental methods involved for determination of protein–protein interaction

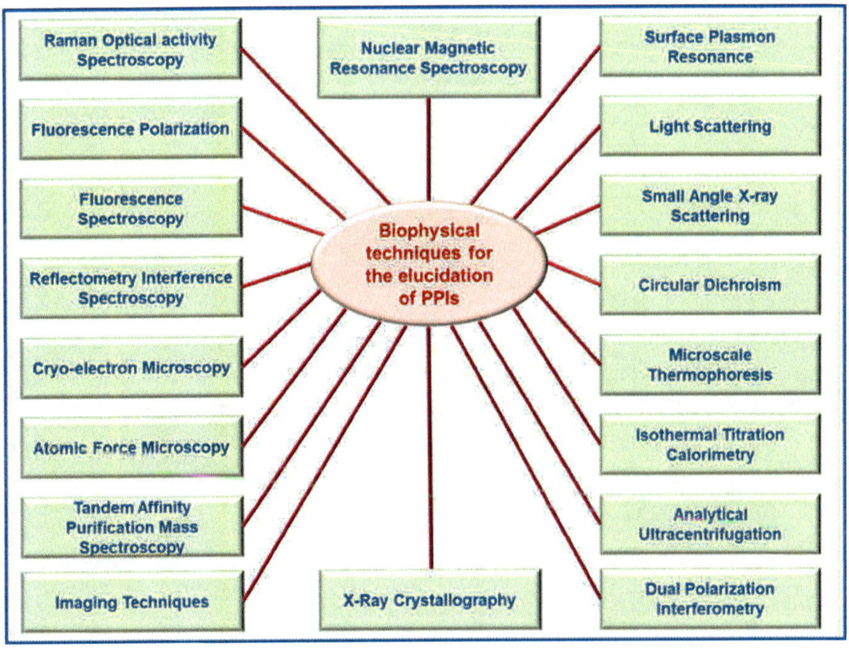

Fig. 5.2 Various types of biophysical techniques involved in detection of PPIs

active nuclei that are oriented by a strong magnetic field. The radiation used is a radio wave and the specific frequency is known as *Larmor frequency* (Fig. 5.3). The *Larmor frequency* is dependent on the chemical environment of each nucleus and depends on factors such as molecular conformations, chemical bonds, dynamic processes, and the local electron density around the nucleus (Thompson et al. 2015). Therefore, even nuclei of the same element can be distinct and give rise to discrete spectral lines (Roberts 1993). The relative positions of the spectral lines are measured in terms of a value known as chemical shift that is convenient for description, treatment, and analysis of macromolecular structures. NMR is most prominently utilized to characterize and determine macromolecular structures in solution. NMR furnishes information on the physical parameters of pKa, ionization states, and hydrogen bonds. It also provides information related to the dynamics of macromolecular backbone and side chains. NMR is advantageous for conducting studies on biomolecules since one is able to perform the experiment under conditions that resemble the physiological environment. It further aids to investigate samples under various conditions, by varying pH/temperature and has gathered popularity even in the field of structure-based drug design (Agarwal et al. 2018; Jaiswal et al. 2018; Moore 1999; Raj et al. 2020).

Both in the solution and solid state, NMR provides information at the atomic level. Therefore, it is also exploited for the identification of contact points between a protein and its cognate ligand or interacting protein partner. The binding sites on a protein can be easily indicated with the aid of chemical shift mapping. This is due to the fact that residues in proximity to the interface region undergo conformational changes upon binding, subsequently affecting the chemical shifts of the concerned nuclei (Gao et al. 2004). In practice, the protein of interest is labeled with isotopes such as ^{15}N or ^{13}C and then its NMR spectrum is acquired. Then, the unlabeled protein partner is simply titrated into the solution and the peak perturbations in the labeled protein's spectra are monitored. The associated structural changes on

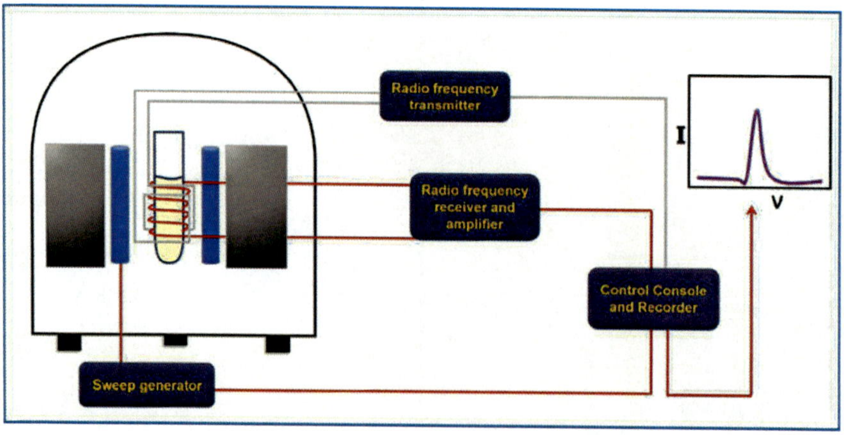

Fig. 5.3 Schematic diagram showing acquisition of NMR spectrum

interaction can then be assessed via the comparison of those two spectra (Liu et al. 2016). The broader spectral changes are usually examined to understand the accompanying changes in the tertiary structure of the protein (Miura 2018). The exact effects that are exerted on the spectrum are dependent on the time scale of the process. Conventionally, PPIs that have an associated K_d of less than 1 μM are said to be in slow exchange (i.e., tight binding) and those having a K_d value greater than 1 mM are said to be in fast exchange (i.e., weak binding) (Gao et al. 2004).

Chemical shift perturbation is the most commonly used NMR technique to determine protein interaction interfaces since it is a simple and sensitive technique. Particularly, this parameter is widely utilized in heteronuclear single-quantum coherence (HSQC) NMR experiments to understand interface regions (Gangele et al. 2020; Garrett et al. 1997; Greene et al. 1998; Gulati et al. 2019; Joshi et al. 2020a; Poluri et al. 2013). However, a host of other NMR spectroscopic parameters to study PPIs are also available (Zuiderweg 2002). They include nuclear Overhauser effect (NOE) which is the transfer of nuclear spin polarization from one population of nuclear spins to another. Its intensity is proportional to the inverse sixth-power of the distance between the two nuclei. The three-dimensional structure of the protein interaction complex involving tight interaction between two proteins can be determined using NOE-derived distance constraints (Dhalluin et al. 2000; Garrett et al. 1999).

Paramagnetic relaxation enhancement (PRE) is another parameter that can furnish PPI-related information by complementing NOE restraints (Clore et al. 2007; Tang et al. 2006). It arises from the magnetic dipolar interactions that exist between an NMR active nucleus and an unpaired electron in a nearby paramagnetic center. Protein complexes can also be structurally characterized with the aid of an NMR parameter known as residual dipolar couplings (RDCs) (Chen and Tjandra 2011; Jensen et al. 2011). This parameter is based on the relative orientations of internuclei vectors. Other NMR derived restraints include pseudo-contact shift, cross saturation, J coupling, etc. A combination of all such restraints aids in the elucidation of a more refined structure of a particular PPI (O'Connell et al. 2009). Apart from the conventional NMR spectroscopy that is performed for soluble proteins, solid-state NMR spectroscopy has also cropped up for studying PPIs involving membrane proteins (Miao and Cross 2013). Similarly, in-cell NMR spectroscopy has progressed to explore PPIs in living cells under physiological conditions (Maldonado et al. 2011).

There are several instances where research groups have employed NMR to delineate PPIs. Williams et al. have applied NMR chemical shift perturbation to map the site on Ras protein that is involved in binding to the cysteine rich domain (CRD) of Raf kinase protein (Williams et al. 2000). Wang et al. have examined the interaction between calcium-bound S100B protein and the third intracellular loop (IC3) of the neurologically important protein dopamine receptor D2 using NMR spectroscopy. The study involved a series of non-destructive experiments and furnished the strength of the protein interaction and the identification of the binding site on the S100B protein (Wang et al. 2019). Jaiswal et al. have delineated the molecular interface between the human SUMO-I protein and *Helicobacter pylori* histone like binding protein Hup using NMR-based chemical shift perturbations and other biophysical techniques (Jaiswal et al. 2019).

D'Silva et al. employed NMR spectroscopy for the formation of complex between MDM2 and p53 proteins (Fig. 5.4). Authors recorded ^{15}N HSQC spectrum for MDM2 alone (Fig. 5.4a) and also in presence of p53 (Fig. 5.4b). The formation of complex was observed as some of cross peaks disappeared upon the formation of complex between MDM2 and p53 (Fig. 5.4b). They also added an inhibitor known as nutlin in the complex and again recorded the ^{15}N HSQC spectrum. Reappearance of cross peaks was observed due to the addition of nutlin that dissociated the complex (Fig. 5.4c) (D'Silva et al. 2005).

5.2.2 Surface Plasmon Resonance (SPR)

Surface plasmon resonance (SPR) is a quantum electromagnetic phenomenon involving a metal–dielectric interface. An SPR experiment utilizes a metal film and a prism attached to the film. The metal film consists of free electrons that are set into collective vibration in the form of a longitudinal wave when light is irradiated on the prism. This is known as a surface plasmon. When a transverse wave corresponding to p-polarized light is irradiated at an angle greater than the threshold value of the critical angle, the surface plasmon resonates with the light's vibration, giving rise to the phenomenon of surface plasmon resonance (SPR). The excitation of the surface plasmon resonance leads to the absorption of light of the resonated wavelength on the metal surface and propagation of an evanescent wave (near field). On the other hand, the wavelength that does not undergo absorption is totally reflected. This leads to the presence of a dark line in the spectrum of the reflected wave and it corresponds to the wavelength absorbed in the metal surface upon excitation of the surface plasmon resonance.

The incidence angle responsible for causing SPR can undergo a shift depending on changes in the metal surface, such as change of protein density. This paves the way to detect PPIs via the measurement of changes in the SPR causing conditions. The incidence angle that causes SPR and is accountable for the dark line in the spectrum is known as the resonance angle. It also aids in the determination of PPIs. In a typical SPR experiment that is designed to determine PPIs, protein is immobilized on one surface of a gold film surface sensor that also contacts a buffer. P-polarized light is then made to irradiate on the opposite surface at incidence angles close to the resonance angle. Only the light irradiated at the resonance angle is absorbed, reflecting out rest of the light irradiated at other incidence angles. Then, the surface of the immobilized protein is brought into contact with a solution containing the candidate protein interactor and the reflected light is assessed in the same way (Fig. 5.5). In case of an interaction occurring between the two proteins, the protein density bound to the surface increases, in turn increasing the refractive index at the sensor surface and causing a subsequent shift in the resonance angle (Miura 2018). The inverse effect is observed during the dissociation process. The resonance

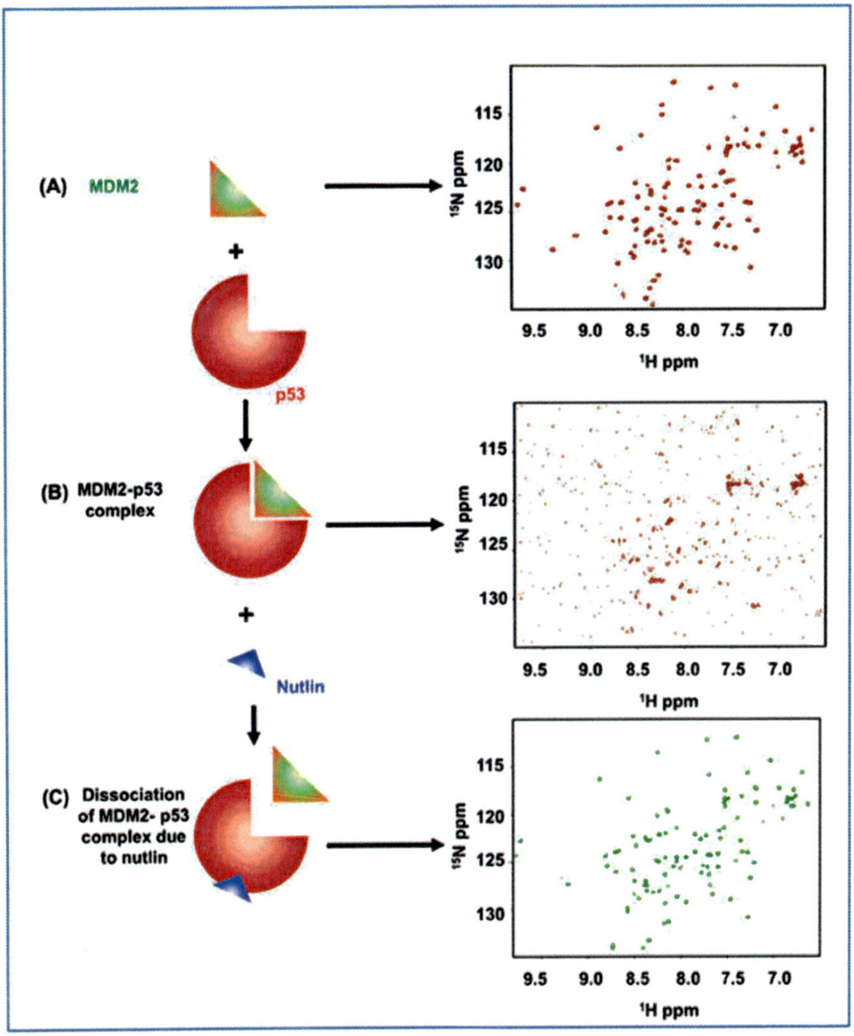

Fig. 5.4 Schematic representation of the effect of an antagonist on the interaction of two proteins. (**a**) represents ^{15}N HSQC spectra of a labeled MDM2 protein. (**b**) displays major cross peaks of MDM2 disappeared upon the formation of MDM2-p53 complex after the addition of p53 protein. (**c**) depicts reappearance of cross peaks in presence of nutlin, a strong inhibitor that dissociated the complex between MDM2-p53 (Adapted from D'Silva et al. 2005)

unit or response unit (RU) is measured to describe the associated change in signal and one RU is correspondent to an angle shift of 10^{-4} degree (Nguyen et al. 2015).

SPR is advantageous to verify PPIs due to a number of reasons (Oesterreicher et al. 2005; Phizicky and Fields 1995; Vernon-Wilson et al. 2000). First, the method is very fast requiring only about 10 min for a typical run. Second, proteins do not require modification such as fluorescent labels or tags. Third, it can analyze

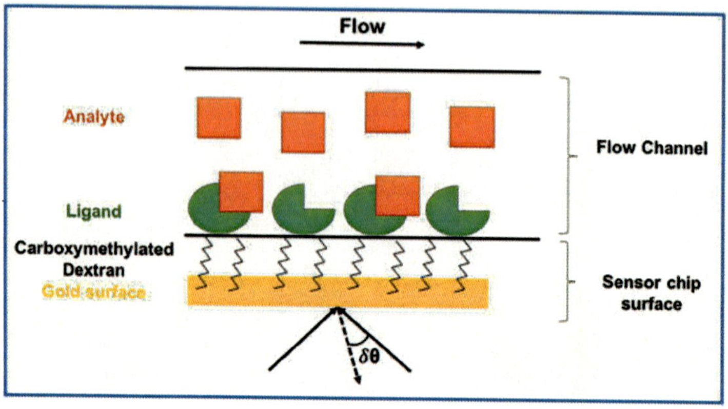

Fig. 5.5 Figure depicting SPR experiment for detection of PPIs

interactions even in complex mixtures and can characterize the interactions in real time. Fourth, it can readily furnish the thermodynamic and kinetic constants of interactions for a wide range of affinities, protein concentrations, molecular weights, and binding rates (Jonsson et al. 1991). Fifth, the technique is sensitive requiring only very small amounts of protein (micro to sub-micrograms). The only potential limitations associated with SPR are non-specific interactions between analyte and sensor surface and the possibility of disturbing the interactions via the need for immobilization of the proteins (Homola 2003).

As the surface-sensitive sensor system is vital to study interacting proteins, various commercially available sensor chips that are robust and reproducible have cropped up. Examples include NTA chip for immobilization of His-tagged proteins, CM5 chip for immobilization via amine, thiol, aldehyde, or carboxyl groups of proteins, SA chip for immobilization of biotinylated proteins, etc. (Rich and Myszka 2000).

Research groups have utilized SPR to monitor PPIs. The quarternary complex of CheY with CheA, CheW, and Tar has been elucidated (Schuster et al. 1993). Similarly, interaction of EGF with EGF receptor (Zhou et al. 1993), monoclonal antibodies with human immunodeficiency virus type 1 core protein p24 (Karlsson et al. 1991), VAMP2 with syntaxin 1A (Calakos et al. 1994), etc. have been demonstrated. In recent times, SPR was exploited by Moscetti et al. to delineate the interactions within the network of the p53 protein (Moscetti et al. 2017) and by Drescher et al. to study the interaction of proteins present in hair cells (Drescher et al. 2009).

5.2.3 Light Scattering

Light scattering refers to a physical phenomenon that involves the interaction of a molecule with light. The encounter of an incident photon with a molecule disturbs the electron orbits of the molecule, producing an induced dipole moment. This, in turn, acts as a source of electromagnetic radiation with energy radiated in all possible directions. If the scattered light is evaluated with respect to its intensity and wavelength, valuable information about the scattering molecule such as binding affinity and stoichiometry of complex interactions can be yielded. Therefore, light scattering is useful for characterizing macromolecular interactions in solution (Some 2013). The advantage of employing light scattering for elucidating PPIs is the fact that it yields absolute measurements without requiring external standards for calibration (Zhou et al. 2016).

Static light scattering (SLS) and dynamic light scattering (DLS) are the most widely used light scattering techniques. In SLS, one collects the time-averaged scattered light intensity at multiple angles and utilizes this data to calculate second virial coefficient that reflects PPIs. In DLS, one records the fluctuations in the scattered light's intensity at a certain angle (Fig. 5.6). This helps in detecting the effect of Brownian motion of the molecules and further yields data about the hydrodynamic radius of molecules and the coefficient of translational diffusion (Murphy 1997). PPIs can be semi-quantitatively characterized by combining a light scattering detection technique with a system for separation, such as size-exclusion chromatography (Kapoor et al. 2010). In order to account for the maintenance of true equilibrium during interaction studies, usually a series of DLS or SLS experiments are performed involving varying compositions. This is known as composition gradient DLS (CG-DLS) or SLS (CG-SLS), and it is employed to determine interaction parameters such as binding affinity and association stoichiometry of PPIs (Attri and Minton 2005; Martos et al. 2010).

Hanlon et al. have employed DLS to study interactions in four protein systems (α-chymotrypsin, soybean trypsin inhibitor, bovine trypsin inhibitor, lysozyme) yielding information about parameters such as equilibrium dissociation constant, stoichiometry of binding, and thermodynamic parameters (Hanlon et al. 2010). Attri et al. studied the self-association properties of insulin using light scattering experiments (both static and dynamic light scattering). They characterized the self-association

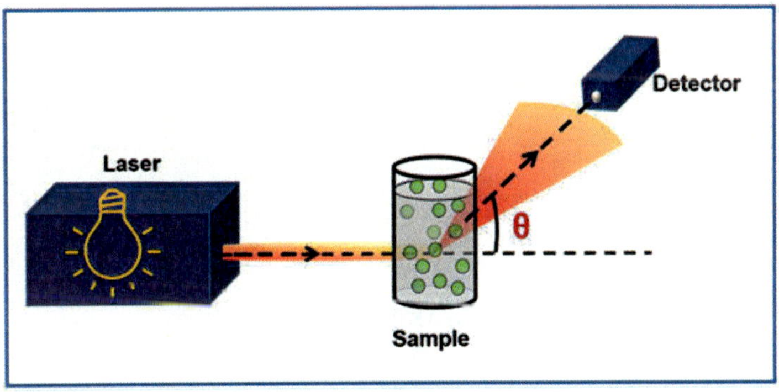

Fig. 5.6 Schematic illustrating the working principle of DLS experiment

phenomenon as a function of the presence of zinc ions and pH of the solution (Attri et al. 2010). They demonstrated that in the presence of zinc, insulin not only exists as a hexamer, but also further interacts with other hexameric units of insulin to form higher-order oligomers (12-mers, 18-mers, etc.). The study also utilized the diffusion and equilibrium constants to determine the molar composition, molecular weight, and translation diffusion constant (Dt) of the insulin oligomers.

5.2.4 Small Angle X-ray Scattering (SAXS)

Scattering is a physical phenomenon in which molecules or atoms, on being exposed to light, re-emit a part of it in different directions with different intensities. In scattering, a particular form of radiation (such as sound, light, or moving particles) is forced to diverge or deviate from a straight trajectory to a different path owing to localized non-uniformities in the material through which it traverses. Small angle scattering (SAS) is one such scattering technique in which collimated radiation is deflected by a small angle (0.1–10°) from its trajectory after interacting with structures that are larger than its own wavelength (Chen and Hennig 2018).

Small angle X-ray scattering (SAXS) is a SAS technique that analyzes the elastic scattering of X-rays at small angles. Based on the range of angles in which a clear scattering signal can be measured, SAXS has the potential to deliver structure-related information for dimensions between 1 and 100 nm (Glatter and Kratky 1982) (Fig. 5.7). For macromolecules such as proteins and protein complexes, SAXS is often preferred over crystallography owing to the non-requirement of protein crystallization and the ability of SAXS to investigate conformational diversity of proteins (Burger et al. 2016). For determining PPIs, SAXS is first conducted independently for two proteins, followed by mixing of the two. The comparison of the acquired structural information is then utilized to judge whether the two proteins have undergone interaction.

Cordeiro et al. studied the interaction between DNA replication and repair regulating $p15^{PAF}$ protein with the proliferating cell nuclear antigen (PCNA) sliding

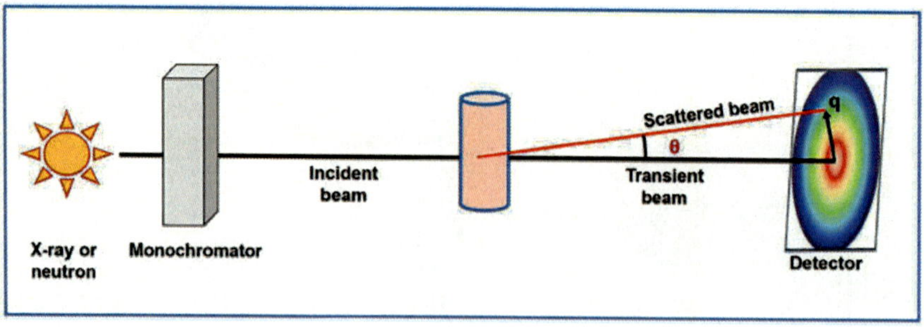

Fig. 5.7 Illustration depicting small angle scattering by sample helps in data acquisition at atomic level

clamp protein. Authors recorded multiple SAXS profiles and elucidated that the PCNA rings penetrated by the N-terminus of p15^{PAF} that subsequently emerges on the back face of PCNA (Cordeiro et al. 2017). Slonimskiy et al. aimed to study the structural conformation of *Synechocystis* FRP (*Syn*FRP) proteins in solution. They engineered a truncated version of *Syn*FRP containing amino acids 8–109 (*Syn*FRP$_{8–109}$) and performed MALLS (multi-angle laser light scattering) and SAXS experiments to establish their common dimeric conformation in the solution (Slonimskiy et al. 2018).

5.2.5 Circular Dichroism (CD)

Circular dichroism or CD refers to the unequal absorption of right-handed and left-handed circularly polarized light. Any beam of light has two time-dependent fields associated with it—the electric field and the magnetic field. When the light is subjected to polarization via suitable filters/prisms, its electric field E oscillates in a single plane in a sinusoidal manner. This sinusoidal wave can be viewed as the resultant of two equal length vectors that trace out circular paths (Fig. 5.8a).

One of the vectors (E_L) rotates counterclockwise, while the other (E_R) rotates clockwise. These two circularly polarized waves have physical existence and are 90° out of phase with respect to one another. The interaction of such circularly polarized light with asymmetric/chiral molecules results in the differential absorption of the right-handed and the left-handed vectors. This consequently causes the light wave

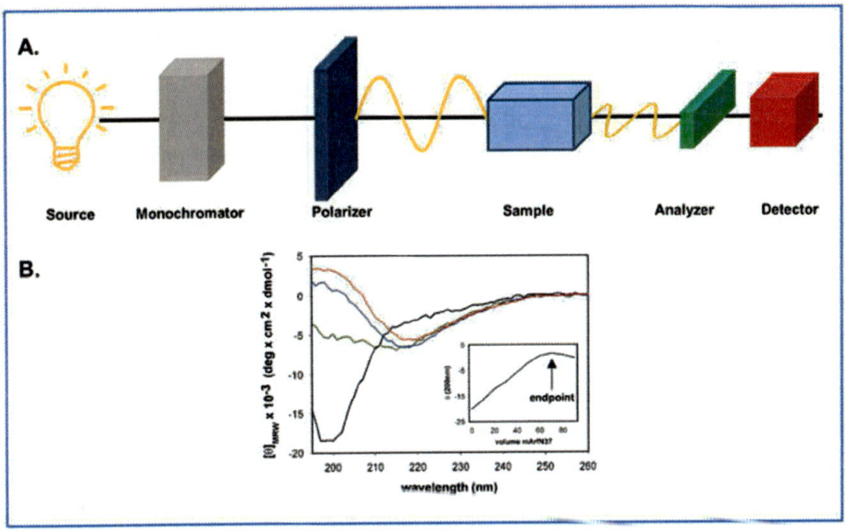

Fig. 5.8 (**a**) Illustration depicting working principle of CD spectroscopy. (**b**) CD spectra showing change in conformation of Hdm2 from random coil to beta-strands on addition of ArfN37 (Adapted from Bothner et al. 2001)

plane to rotate and the resultant of the E_L and E_R vectors to trace out an elliptical path. CD is measured in either of the two units: (1) ΔE or the difference in absorption of the E_L and E_R vectors by a chiral molecule, (2) degrees ellipticity or the angle whose tangent is the ratio of the ellipse's minor to the major axis. The molar ellipticity or [θ] in units of deg cm^2/dmol equates to 3298ΔE (Greenfield 2006).

CD has been proven to be an excellent method for the rapid evaluation of the secondary structure, binding, and folding properties of proteins. The arrangement of the chromophores of polypeptide backbone amides into arrays can split their optical transitions into multiple transitions owing to "exciton" interactions (Sreerama and Woody 2004). Consequently, different structural elements in a protein have characteristic CD spectra. The amide groups of the protein backbone giving rise to the CD spectrum in the far-UV region (190–250 nm) is thus sensitive to the secondary structure of the protein, such as the content of beta-sheets, alpha-helices, and random coil structures present. On the other hand, the CD spectrum of a protein in the near-UV region (250–320 nm), arising from disulfide bonds and aromatic amino acids, is sensitive to some aspects of the protein's tertiary structure (Ranjbar and Gill 2009).

Therefore, the changes in the CD spectrum provides one with information about conformational changes in protein, which can be further extrapolated to detect PPIs since protein complex formation is associated with conformational changes of each of the interacting partners. Research groups who employ CD to determine PPIs initially record the CD spectra of each of the interacting proteins alone to estimate their alpha helix and beta sheet content. Following this, they record the spectrum by mixing the two proteins. The variation in the ratio of the secondary structural elements from that in the original two spectra verifies protein complex formation with an associated change in the secondary structure. The change in the CD spectra upon interaction of two proteins is directly proportional to the quantity of the complex formed. Therefore, CD spectroscopy is deemed as a quantitative technique and can enable us to determine association constants of the interacting partners or the dissociation constant of the complex (Greenfield and Fowler 2002; Honda et al. 1999). Further, CD spectra can be recorded at different temperatures to furnish thermodynamic parameters of the PPIs (Greenfield 2006).

Several programs such as CONTIN, MLR, CDNN, K2D, SELCON that utilize CD to study protein conformational changes accompanying PPIs have cropped up (Zhou et al. 2016). An improved version of CD, known as SRCD (synchrotron radiation circular dichroism) has evolved that utilizes a synchrotron ring as the light source and measures PPIs involving either rigid-body or induced fit mechanisms with improved signal-to-noise ratios (Cowieson et al. 2008; Wallace and Janes 2010). The advantages of using CD to determine PPIs include: (1) requirement of less quantities of proteins, (2) minimal data processing and time requirement, and (3) ability to conduct experiments under varying conditions of pH, temperature, and solvent. However, compared to spectroscopic techniques like NMR, CD furnishes lesser information.

Joseph et al. utilized CD to study the changes in the conformation of α-actinin protein upon interaction with Z-repeats of titin protein. The far-UV CD spectra of the free Z-repeats were typical of random coil elements. However, on titration with

Act-EF34 (last 72 C-terminal amino acids of actinin), there were gains in the helical signals of the titin Z-repeats, denoting the formation of complex (Joseph et al. 2001).

Bothner et al. elucidated via CD spectroscopy that both mArfN37 (N-terminal 37 amino acid residues of mouse Arf protein) and Hdm2 (210–304) are unstructured. However, on mixing mArfN37 and Hdm2 (210–304), they observed a striking transition from random conformations to β-strand secondary structure. This transition was induced on addition of mArfN37 to an Hdm2 fragment or the addition of an Hdm2 fragment to mArfN37 (Bothner et al. 2001) (Fig. 5.8b).

5.2.6 Microscale Thermophoresis (MST)

Microscale thermophoresis (MST) is a powerful tool to determine biomolecular interactions like PPIs. It is based on the basic principle of thermophoresis, which is the focused movement of molecules in a temperature gradient. Thermophoresis is strongly dependent on factors such as charge, size, conformation, hydration shell, etc., and is thus highly sensitive in virtually detecting any change in such molecular properties during interaction events. It enables researchers to precisely quantify molecular events, independent of the nature or size of the investigated sample.

During an MST experiment, an infrared laser is employed to induce a temperature gradient. The directed movement of molecules through this gradient is detected and measured utilizing either intrinsic or covalently attached fluorophores. MST combines the sensitivity of thermophoresis with the precision of molecular fluorescence detection, and thus constitutes a robust, flexible, and quick method to dissect molecular interactions (Jerabek-Willemsen et al. 2014).

In a typical MST experiment to determine PPIs, one of the interacting protein partners is fluorescently labeled either using agents like N-hydroxysuccinimide (NHS) or fusing it to a fluorescent protein such as green fluorescent protein (GFP). Alternatively, label-free MST can also be performed by exploiting the intrinsic UV-fluorescent signal of the aromatic amino acid residues in the protein. An MST instrument heats up a defined sample volume inside a capillary using an infrared (IR) laser and produces a temperature gradient that usually spans 2–6 K. The fluorescence emission of the heat spot within the sample is measured at three time points (before, during, and after the IR laser is switched on) to detect the thermophoretic movement of the fluorescent protein. In the initial stage, fluorescence is recorded at a constant level before heating. On switching on the IR laser, the fluorescence intensity abruptly decreases with the increase in temperature (T-jump). Following this, the fluorescence decays slowly (thermophoresis) until it reaches a steady state. Finally, when the laser is switched off, back diffusion and inverse T-jump will be observed (Zhou et al. 2016). Events like interaction to another protein partner can change the size, charge or hydration shell of the concerned protein, and alter both the thermophoresis process and T-jump. Therefore, by titrating a constant concentration of a fluorescent protein with varying concentrations of its non-fluorescent partner, one can employ MST to determine PPIs (Arbel et al. 2012; Lin et al. 2012; Wienken et al. 2010).

MST offers several advantages over other biophysical methods such as low sample requirement (few microliters of specimen at nano-molar concentrations), fast measurement (determination of binding affinity in 10 min), and easy setting. It also enables one to obtain quantitative data on high-affinity interactions (dissociation constant in the picomolar range). Additionally, MST can be conducted even in complex medium like blood serum (Jerabek-Willemsen et al. 2011).

Lin et al. employed MST to understand the dissociation of Grb2 dimer below a particular threshold concentration. They titrated 55 nM to 100μM of the unlabeled Grb2 protein into a fixed concentration (100 nM) of labeled Grb2 to furnish the dissociation constant (Lin et al. 2012). Uzarska et al. studied the interaction between Ssq1 and Grx5, proteins utilizing MST. They fluorescently labeled Grx5 and titrated increasing amounts of Ssq1 into the solution. Titration with Ssq1 changed the thermophoretic parameters of Grx5, indicating a binary association between the two proteins. The binding constant Ka has been calculated from a theoretical fitting (Uzarska et al. 2013).

5.2.7 Isothermal Titration Calorimetry (ITC)

Isothermal titration calorimetry (ITC) involves the measurement of the energy that is liberated or absorbed by inducing a chemical reaction via mixing two components. A typical ITC experiment is performed by adding one of the reactants stepwise (~10μL per injection) into the other reactant in the reaction cell (~1 mL). With each injection, the energy associated with the corresponding chemical reaction is manifested in the form of heat (q_i) absorbed or released. q_i is proportional to the quantity of one reactant binding to the other in one injection ($v \times \Delta L_i$), where v is the reaction cell's volume and ΔL_i is the increase in the bound ligand's concentration post *i*th injection. It is also proportional to the enthalpy of the binding reaction (ΔH).

ITC instruments in the modern day operate on the principle of heat compensation; therefore, their output response (signal) basically measures the power (microcalories/s) that is required to maintain the temperature difference between the reaction cell and the cell where no reactant is injected (reference cell) (Fig. 5.9). From the obtained signal, the energy change associated with each injection can be estimated by measuring the area under each peak. The peak magnitudes in the signal become progressively smaller until a state of saturation. It denotes that the quantity of uncomplexed reactant in the cell progressively decreases with each injection, until no further unbounded reactant is present. Once the experiment reaches saturation, further injections into the cell produce peaks that correspond to mechanical or dilution effects. They are duly subtracted from the injection peaks before analyzing the ITC output (Leavitt and Freire 2001).

ITC is often employed to understand if there is an interaction between two proteins. It relies on the principle that association of one protein to another will be accompanied with heat change (endothermic or exothermic change). Based on this, a solution of a candidate interactor protein is titrated in small amounts into a solution of the partner protein. If interaction occurs between these two proteins, it can be

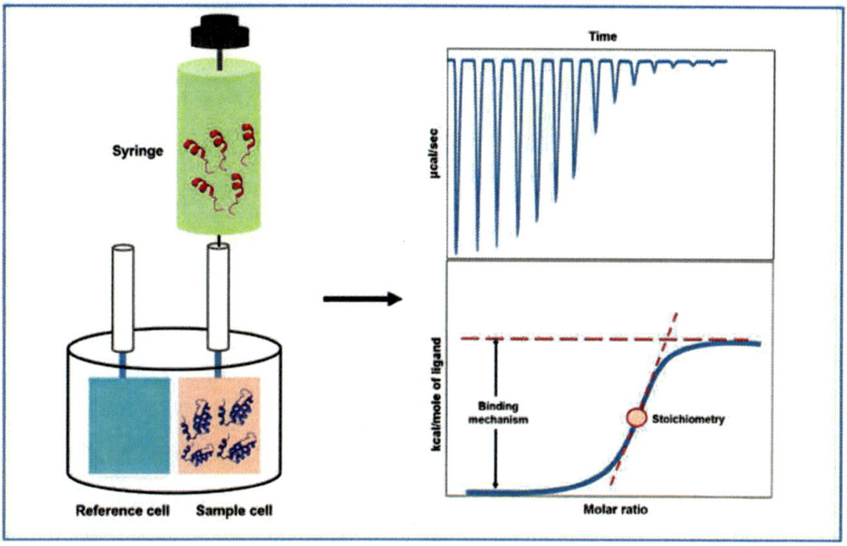

Fig. 5.9 Diagram illustrating working principle of ITC experiment

easily determined by monitoring the change in heat balance. Both interacting proteins (i.e., protein in the syringe and protein in the reaction cell) however must be maintained in the same buffer such that mismatched buffer components do not generate reaction artifacts (Pierce et al. 1999).

ITC has several advantages over other techniques. It is a fast (requires <2 h) and a simple technique that does not require the attachment of proteins to a surface. It is also not affected by constraints of shape, molecular size, or chemical constitution. Further, unlike other techniques that provide us with only the binding affinities, ITC also furnishes thermodynamic parameters such as association constant (Ka), free energy change (ΔG), enthalpy change (ΔH), entropy change (ΔS), heat capacity change (ΔCp), and reaction stoichiometry (*n*) (Velazquez-Campoy et al. 2004) (Fig. 5.9). These data, in turn, enable us to have better insights into the mechanisms underlying PPIs. For instance, the breakdown of data can help to understand the exact contributions to binding via protonation, solvation, structural rearrangements, etc., and data such as reaction stoichiometry can help us validate a protein's oligomeric state (Krachler et al. 2010).

ITC faced one significant limitation in the past where it failed to characterize binding reactions with very high association constants. Beyond a threshold value of the association constants, the characteristic curvature in the signals was lost. However, this problem has recently been mitigated with the aid of displacement ITC (Leavitt and Freire 2001).

Knapp et al. characterized the binding of β-catenin to TCF4 using ITC. Various TCF4 constructs (of different lengths) were titrated into the reaction cell containing β-catenin. The binding isotherm being exothermic was characterized by strong heat

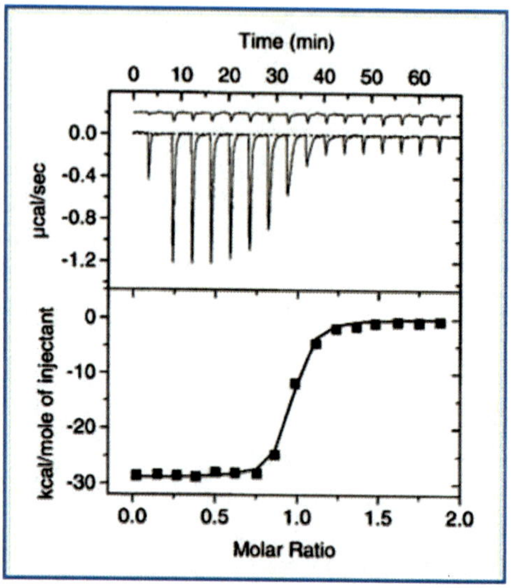

Fig. 5.10 Top panel depicts raw calorimetric data obtained for interaction of TCF4 into buffer (blank) and β-catenin while the lower panel shows the binding curve attained from the integrated heat signals displayed in the above data (Adapted from Knapp et al. 2001)

release at 25 °C. The heat release decreased when β-catenin's binding sites were saturated. Only heat effects due to dilution were observed in the titrations' last few injections (Knapp et al. 2001). Baker et al. elucidated the interaction of turkey ovomucoid third domain (OMTKY3) with porcine pancreatic elastase (PPE) using ITC and characterized the binding energetics (Baker and Murphy 1997) (Fig. 5.10).

5.2.8 Analytical Ultracentrifugation (AUC)

Analytical ultracentrifugation (AUC) is an analytical method that detects macromolecular concentration distribution in real-time, when subjected to centrifugal force. It provides thermodynamic data and first-principle hydrodynamic data including molecular shape, size, mass, stoichiometry of association, energy of association, and thermodynamic non-ideality of the molecules (Laue 2004). The basic principle underlying AUC states the mass redistribution in a gravitational field until the chemical potential energy corresponding to each radial position is exactly balanced by the gravitational potential energy (Fig. 5.11) (Cole et al. 2008). AUC experiments can be broadly categorized into (a) sedimentation velocity (SV) and (b) sedimentation equilibrium (SE).

In SV experiments, one monitors the rate of movement of the boundaries of molecules during redistribution. Such experiments are usually conducted at very high rotor speeds that deplete macromolecules from the meniscus region at the interface and form a concentration boundary that propagates toward the bottom as a function of time (Lebowitz et al. 2002). On the other hand, in a typical SE experiment, one determines the concentration distribution post equilibrium. SE experiments involve relatively lower rotor speeds, where back diffusive forces are

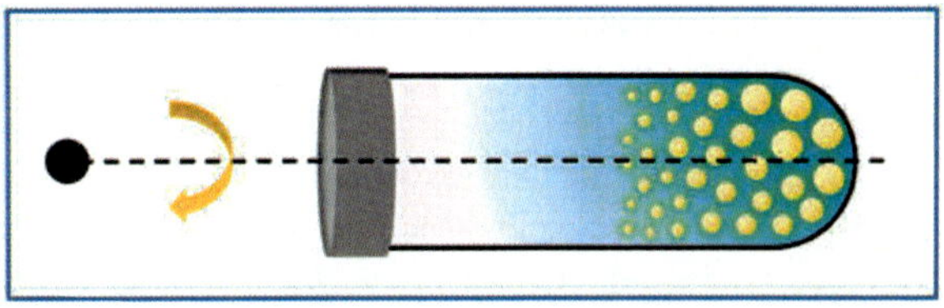

Fig. 5.11 Figure illustrating the mass redistribution in a gravitational field in an AUC experiment

allowed to come into thermodynamic equilibrium with the centrifugal forces. A steady-state concentration distribution is yielded by an SE experiment. Such a distribution serves as a function of mass distribution, buoyant solute mass, and mass action parameters of interacting systems (equilibrium dissociation or association constants) (Harding and Rowe 2010). A technique known as "tracer SE" is a variant of the conventional SE that utilizes a labeled solute component. Tracer SE aids in studying self-associations and hetero-associations with enhanced resolution (Rivas and Minton 2003).

The study of PPIs using sedimentation experiments is based on the key principle that complexes that sediment faster migrate through a solution of components that sediment relatively slower. Consequently, formed complexes that are reversible in nature can re-associate following transient dissociation during the experiment, thereby allowing researchers to characterize transient and weak interactions. Sedimentation experiments also enable the characterization of stoichiometry and binding constants involved in protein complex formation and furnish crucial information on properties of mixed self- and hetero-association. Further, these experiments also provide data on low-resolution structure of protein complexes and conformational changes, since the phenomenon of sedimentation depends on the hydrodynamic features of molecules (Balbo and Schuck 2005).

CD2–CD48 interaction is considered to be responsible for the adhesion of T-cells to antigen presenting cells or other cytotoxic targets. Silkowski et al. studied the affinity of this interaction using AUC technique. The authors concluded that the affinity was weak as the dissociation constant K_d ranged from 60 to 90 μM (Silkowski et al. 1997). On a similar note, Philo et al. assessed the affinity and stoichiometry of the SCF–Kit interaction using AUC (Philo et al. 1996).

AUC offers several advantages over other techniques: (1) it allows to study macromolecules over a broad range of concentrations and solvents (2) it furnishes data that is absolute and independent of standards, since it depends on fundamental properties like mass and gravitation (3) it is non-destructive and enables the study of samples in their biologically relevant solutions (native states) or in free solutions without the prior need for matrix attachment and (4) it allows to study interactions between components that range in size from peptides to multi-protein complexes. The disadvantage on the other hand includes the long running time for SE experiments.

5.2.9 Dual Polarization Interferometry (DPI)

Dual polarization interferometry (DPI) provides high-resolution data on molecular size, density, and mass in real-time. The instrumentation involves a silicon chip consisting of two channels, each of which comprises a sensing waveguide and a reference waveguide. Coherent and focused laser light is allowed to enter the two waveguides to undergo total internal reflection and pass along the waveguide. The reference waveguide provides a constant refractive index, being sandwiched in between two surfaces. This in turn makes the passage of light consistent and exposes the upper side of the sensing waveguide. The light exiting from each waveguide travels through a pin hole before finally combining to form a 2D interference pattern on the detector. The sensing waveguide adsorbs molecules onto its surface and as a consequence, a change in the refractive index of light leads to a change in its evanescent field. This makes the light passing from the top waveguide un-synced with that from the lower guide (reference guide). It is manifested as a shift in the interference pattern (Escorihuela et al. 2015).

DPI has the potential to be used in the study of PPIs. It however requires one of the interacting partners to be immobilized on the waveguide surface. DPI not only furnishes the kinetic parameters and affinity of the interaction, but also provides information on the conformational changes associated with the interaction (Boudjemline et al. 2011).

Biehle et al. studied the interactions between Apolipoprotein E (ApoE) with tissue plasminogen activator (tPA). The authors anchored the activator to amine chips and various isoforms of ApoE were allowed to bind. They observed that interaction of Apolipoprotein E2 with immobilized tPA resulted in considerable conformational changes that were not detected in cases of Apolipoproteins E3 and E4 (Biehle et al. 2004). Balhara et al. utilized DPI to study the interaction between canine plasminogen and enolase from *Streptococcus pyogenes*. They demonstrated that the binding was essential to form stable complexes between both native proteins (Balhara et al. 2014).

5.2.10 Fluorescence Spectroscopy

Fluorescence is one of the two forms of luminescence (emission of light from electronically excited states). Fluorescence involves excited singlet states where the electron in the excited state has spin opposite to that of the electron in the ground state. Consequently, the return of the electron to the ground state is accompanied by the emission of a photon. Fluorescence emission takes place on a time scale of 10^{-8} to 10^{-10} s. It is faster than phosphorescence, the other form of luminescence as it involves the singlet excited state instead of the triplet excited state (Fig. 5.12). The emitted light has lower energy (lower frequency) and thus longer wavelength than the absorbed radiation. Fluorescence data is usually represented as emission spectrum where fluorescence intensity is plotted against wavelength (nm) or wave number (cm^{-1}) (McPhie 2000).

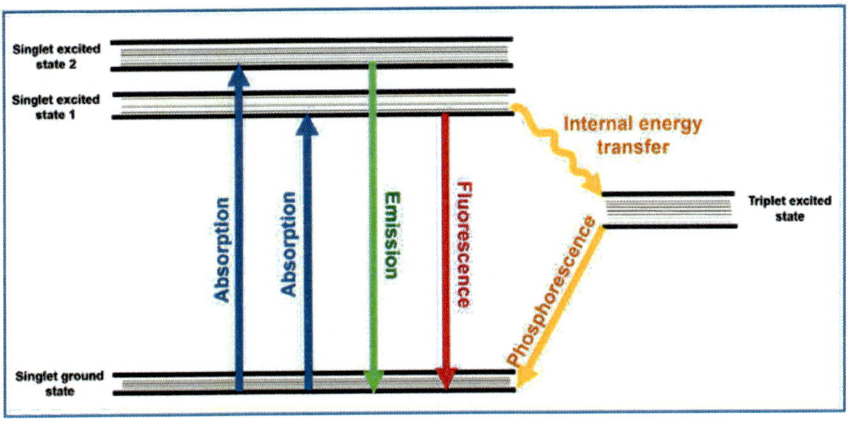

Fig. 5.12 Schematic illustration of underlying principles of the two forms of luminescence—fluorescence and phosphorescence

Tryptophan is the amino acid with the most dominant contribution to a protein's fluorescence spectrum. Monitoring the tryptophan fluorescence paves the way to evaluate PPIs. There are two methods that can be potentially employed to study PPIs. In the first method, changes in the emission spectrum of a protein, either with regard to its fluorescence intensity or the wavelength of its maximum fluorescence is noted on association with its interacting partner. The second method utilizes fluorescent tags to measure PPIs. This is a more sensitive technique to monitor PPIs, considering the tags do not adversely affect the modified protein's function.

Fluorescence spectroscopy, although being a method opted by several research groups to study PPIs, is faced with certain drawbacks. Firstly, interaction between two proteins affects the local environment of only a few tryptophan residues although either of the two interacting partners might have several tryptophan residues. Secondly, if both the interacting partners have tryptophan residues, titration of the second protein into the first might yield a linearly changing background instead of the normal binding curve. In such a case, the background must be duly subtracted to obtain interpretable results. Thirdly, tyrosine fluorescence is less sensitive to environmental changes than tryptophan and thus generates constant background signal (Groemping and Hellmann 2005).

Otto-Bruc et al. utilized this technique to analyze the interaction of cGMP phosphodiesterase g subunit (PDEg) with a subunit of transducin (Ta) in the presence of GDP or GTPgS. The authors titrated PDEg into a solution of Ta-GTPgS leading to an increase in the fluorescence at 320 nm, as compared to that demonstrated by titration of PDEg into buffer alone. The fluorescence intensity did not change on further addition of PDEg when all of Ta-GTPgS were complexed (Otto-Bruc et al. 1993). On a similar note, Mills et al. have elucidated the light-chain kinase/spinach calmodulin interaction by applying fluorescence spectroscopy (Mills et al. 1988).

5.2.11 Fluorescence Polarization (FP)

Fluorescence polarization (FP) is dependent on the molecular movement of fluorophores in solution. When any fluorophore is excited by polarized light, it emits light along different axes in varying intensities. The resulting degree of polarization is inversely related to the rotation of the fluorophore at the molecular level. The molecular rotation is, in turn, greatly dependent on the molecular mass, rotation becomes slower with increase in mass (Du 2015).

FP can be used to study PPIs as it can quantitate the affinity of two protein partners for each other based on the increased rotational time of the formed complex. It can measure dissociation and binding between proteins if one of them is relatively small and tagged with a fluorophore. Commonly used fluorophores for an FP assay include rhodamine, fluorescein, Cy5 (cyanine-5) dyes, or their derivatives, etc.

FP signal (in milli-polarization units, mP) is enhanced via formation of complex and is recorded with the help of a microplate reader. A plot of mP values against multiple concentrations of the labeled molecule generates the binding curve that further furnishes the dissociation constant of the interaction. An FP assay involving truncated motifs that comprise "binding epitopes" or "hotspots" can also be employed to design inhibitors against the given PPI (Lea and Simeonov 2011). FP is an advantageous assay because of its low cost, simple protocol, and high-throughput screening (HTS) capacity (Zhou et al. 2016). However, FP suffers from some pitfalls similar to other fluorescence-based assays, such as quenching, interference from auto-fluorescence and light scattering (Owicki 2000).

Weiel et al. studied the interaction of 30S ribosomal subunits with initiation factor 3 (IF3) by labeling IF3 with fluorescein and utilizing fluorescence polarization. Even after labeling IF3, it was biologically active and interacted with 30S ribosomal subunits. When the 30S ribosomal subunits were present, both the fluorescence lifetime and the fluorescence emission spectrum of fluorescein-labeled IF3 were unchanged. Therefore, the observed increase in fluorescence polarization was attributed to the interaction with 30S ribosomes. The expected change in the polarization was in line with the binding event of a larger molecule (Weiel and Hershey 1981). Kim et al. utilized this method to analyze interactions between T7 polymerase and gene 2.5 protein. For this purpose, they labeled gene 2.5 protein with fluorescein isothiocyanate. Its interaction with T7 polymerase led to decrease in fluorescence and enhancement in anisotropy (Kim et al. 1992).

5.2.12 X-ray Crystallography

X-ray crystallography is a technique that aids in the determination of the 3D structure of molecules, including complex biomolecules like nucleic acids and proteins. This technique elucidates 3D structures of molecules at atomic level. The basic principle underlying this technique is X-ray diffraction. Researchers collect data by diffracting X-rays from a single crystal of the biomolecule. The crystal has a periodic and ordered arrangement of atoms that leads to X-ray scattering and production of a diffraction pattern. Researchers then reconstruct the electron density map from the obtained diffraction data. Finally, from the electron density image, information on several aspects like atoms' mean positions, disorder, chemical bonds etc. is determined (Ryu 2016) (Fig. 5.13).

X-ray crystallography not only helps in the visualization and elucidation of individual protein structures, but also unravels information on PPIs. It determines how proteins interact with other proteins by furnishing data on protein interfaces at the atomic resolution. Although it is a prolific technique and is still considered the "gold standard" for structural analysis of protein complexes with efficient resolution, this technique has drawbacks of its own. Firstly, the number of protein complexes solved by X-ray crystallography is low because of the difficulty of crystallization, production of sufficient sample, and requirement of extensive time and labor (Russell et al. 2004).

Even in the case of small interactomes, where the predicted number of interactions range in tens of thousands, it is not possible to solve all protein complexes experimentally by this technique (Cafarelli et al. 2017). The second major drawback of using this technique to study PPIs is that a crystal does not essentially represent the true protein–protein interface. There are many protein–protein contacts in a crystal that do not occur in nature or in solution. These contacts, also known as "crystal contacts," are not biologically relevant and therefore do not represent the actual "biological contacts." A crystal not only contains such surfeit contacts (biologically insignificant or crystal contacts), but also lacks some of the true biological contacts. This marks a major drawback for X-ray crystallography because differentiating crystal contacts from biological interfaces is extremely difficult. The fundamental challenge in distinguishing these two types of contacts is due to the similarity in their physical basis of formation (Kobe et al. 2008).

Several protein complexes such as the functional complexes of ribosome, RNA polymerase, the GroEL chaperonin, the proteasome, and various other complexes of the cellular transport machinery have already been solved by X-ray crystallography

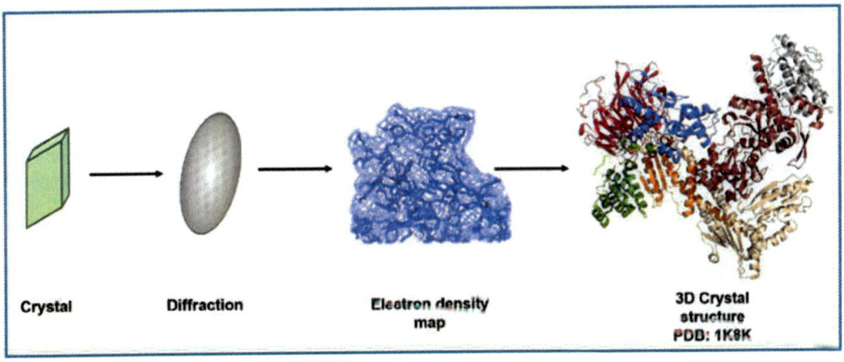

Fig. 5.13 Figure depicting workflow for protein structure determination using X-ray crystallography

(Russell et al. 2004). Robinson et al. determined the crystal structure of bovine Arp2/3 complex at 2.0 Å resolution. Seven proteins are assembled in the complex to function in the initiation of actin polymerization inside eukaryotic cells (Robinson et al. 2001). On a similar note, Liu et al. elucidated the crystal structure of the light harvesting complex of photosystem II in spinach at a resolution of 2.72 Å (Liu et al. 2004).

5.2.13 Reflectometric Interference Spectroscopy (RIFS)

Reflectometric interference spectroscopy (RIFS) is an optical biosensing technique that bears similarity with SPR. Also known as BioLayer Interferometry (BLI), a technique that is very much useful in deciphering biomolecular interactions (Rich and Myszka 2007). RIFS aids to analyze PPIs by interpreting the interference of light that is reflected by multiple surfaces such as the reference layer in the vicinity of the tip of a sensor chip and the surface of protein layer immobilized on the tip.

RIFS utilizes a series of steps to study PPIs. First, the protein of interest is immobilized on the tip of the glass sensor's surface. It is then irradiated with white light from the opposite surface of the sensor (i.e., via epi-illumination) and the spectrum corresponding to the reflected wave is duly recorded. The reflected wave is essentially the wave arising from the interference of the light reflected near the distal surface of the protein and the protein bound to that surface. Next, the sensor is soaked in a buffer containing the candidate interactor protein. In case of an interaction between the two proteins, the density and subsequently the refractive index on the surface are increased. The increase in the refractive index slows down the velocity of the light and makes it to reach the sensor with a slight delay. Therefore, arrival time to the sensor is enhanced in the presence of the protein complex formation. On the other hand, arrival times of reflected light corresponding to the reference layer remain unchanged. The interference pattern of reflected light that is modulated as a result of PPI is interpreted by research groups to examine protein binding. Generally, the tip surface's density that is calculated from the interference spectrum is converted to a more convenient sensogram (plot of values versus time) (Miura 2018).

The RIFS instrument allows flexibility in assays designed for interaction studies. The biomolecular interactions can be studied under various conditions simultaneously with the aid of multiple sensors. It also provides data with high-throughput efficiency. Further, this technique can also be employed to analyze affinity and kinetic parameters of interactions between unprocessed samples (Bhagwat and Kumar 2018).

Kinetics of the interaction between Protein A-human immunoglobulin G under constraints that resemble simple 1:1 binding have been investigated using RIFS. Serial dilutions of IgG were prepared and a microplate was set up for the experiment by aliquoting buffer, biotinylated Protein A, and IgG solutions. A commercial sensor was used to estimate the binding. The raw data was then exported for analysis and determination of constants k(on), k(off), and K_d (Wilson et al. 2010).

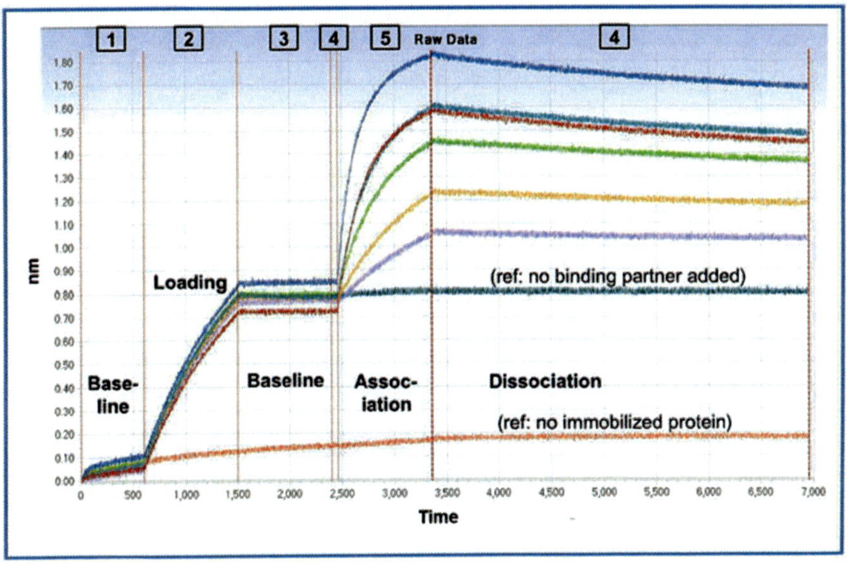

Fig. 5.14 Raw data generated in a RIFS assay to analyze the association of ε-ATP synthase catalytic subunits (Adapted from Shah and Duncan 2014)

Shah et al. utilized this technique to study inhibitory interactions of catalytic complex of *Escherichia coli* ATP synthase with subunit *ε* of the same. The assay furnished the kinetics of the association of subunit *ε* with the catalytic complex (Shah and Duncan 2014) (Fig. 5.14).

5.2.14 Raman Optical Activity Spectroscopy (ROA)

When light is incident on a sample, scattering occurs. The majority of this scattered light retains the same wavelength as that of the incident light. However, some part of the light has a varied wavelength. The scattering of light by a molecule can lead to the absorption of a part of its energy such that the resulting wavelength of the scattered light becomes longer. Conversely, energy from the molecule can be added to the light during scattering to result in a decreased wavelength. This phenomenon is known as Raman scattering. During an experiment, the scattered light from the sample is detected with spectroscope and obtained spectrum is known as a Raman spectrum. The spectrum helps to extract information on the structure and vibration of the molecule (Barron 2015). In Raman optical activity spectroscopy (ROA), spectra are usually acquired via irradiating a sample with monochromatic right and left circularly polarized light. ROA can be extrapolated to study PPIs by the comparison and analysis of the Raman spectra of solutions of individual proteins and their mixtures respectively.

ROA comprises several advantages for studying PPIs. Firstly, it is a rapid method that furnishes reproducible data and requires little to no sample preparation. Secondly, being non-invasive and non-destructive in nature, it aids in the study of biological samples and aqueous solutions. Further, the intensity of the features on the Raman spectra is directly proportional to the concentration of the given species. Lastly, ROA yields very high magnitudes of depth discrimination and spatial resolution.

Kengne-Momo et al. have studied the interaction of anti-*E. coli* immunoglobulin G (IgG) and staphylococcal protein A (SpA) using the Raman spectroscopy. The authors elucidated that the tyrosine residues in the α-helix of SpA are involved in mediating the interaction with IgG. The native protein structures were almost preserved, whereas interaction-mediated changes were seen in the α-helix of SpA. In the adsorbed state, the large modification of the protein structure allowed the authors to identify the tyrosine residues involved in the interaction of IgG and SpA (Kengne-Momo et al. 2012). Li et al. also have employed Raman spectroscopy to elucidate the protein interactions between the brain derived neurotrophic factor (BDNF) and its receptor. The authors concluded that the interaction of BDNF with the receptor generated considerable amount of conformational changes in the receptor (Li et al. 2002).

5.2.15 Cryo-Electron Microscopy (Cryo-EM)

Cryo-Electron Microscopy (Cryo-EM) is a technique that aids in the visualization of a molecule with a transmission electron microscope (TEM). This technique has cropped up with the help of improvements in the performance of the microscope and advances in other aspects like image analysis methods and electron beam detectors. In a cryo-EM experiment, the solution of the protein of interest is first frozen. Next, 2D images of the protein are acquired directly using TEM. Since the resolution is low with only a single image, several similar images, ranging in thousands are acquired to compute and reconstruct the original structure with a better resolution. This is known as single-particle analysis. The formation and structure of a protein complex can also be demonstrated by Cryo-EM analysis (Fig. 5.15).

Cryo-EM has become a preferable method over other techniques because of the need of a small quantity of sample and the lack of need of crystallization. Cryo-EM has become a favorite to several research groups for the study of macromolecular complexes, as is evident from the increasing number of entries in the Electron Microscopy Data Bank or the public repository for electron microscopy density maps of biological systems. On the downside, this technique faces challenges such as high costs of equipment and maintenance and the need for vast computational resources (Nealon et al. 2017).

Further, since cryo-EM sample preparation requires freezing of the sample, a stress factor might be introduced in the sample that might lead to the breakage of PPIs within the multi-protein complexes and consequent dissociation of the complex

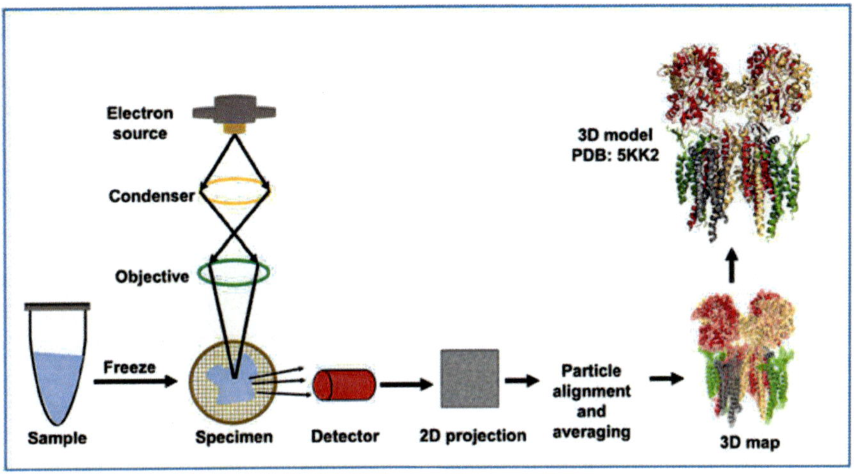

Fig. 5.15 Schematic depicting the working of cryo-electron microscopy and a refined 3D structure of GluA2 receptor-TARP γ2 complex (PDB 5KK2) attained using Cryo-EM

(Taylor and Glaeser 2008). Moreover, structural studies for multi-protein complexes with low or no symmetry still remain a challenge for Cryo-EM. Large complexes are typically unstable or flexible, thereby needing better approaches to deal with the sample heterogeneity and, in turn, more computer power (Costa et al. 2017).

Fronzes et al. employed cryo-EM to study the structure of the bacterial core outer-membrane complex at a resolution of 15 Å. This complex had a molecular weight of 1.1 MDa. It spanned both the bacterial outer and inner membrane and comprised of 14 copies of VirB7, VirB9, and VirB10 proteins (Fronzes et al. 2009). Cryo-EM has also been utilized by Zhao et al. to elucidate the architecture of the GluA2 AMPA receptor-TARP protein complex (Fig. 5.15) (Zhao et al. 2016) and by Gadkari et al. to predict PPIs amongst the coat proteins of dengue virus (Gadkari and Srinivasan 2010).

5.2.16 Atomic Force Microscopy (AFM)

Atomic force microscopy (AFM) is fundamentally an expansion of scanning tunneling microscopy that is suited for live and soft samples such as biological samples (Hansma et al. 1988). AFM has three different modes of operation namely, contact mode, oscillating mode, and force mode and the principle of the technique varies with the mode in use. When AFM is performed in the contact mode, the cantilever of the instrument deflects by the sample surface. During a scan in the *X–Y* dimensional scan, the *Z* position of the piezo is adjusted to maintain the fixed deflection. The resulting AFM image is produced by plotting the piezo's *Z* movement as a function of the *X–Y* position. When AFM is performed in the oscillating mode, a vibration piezo is employed to oscillate the cantilever. The

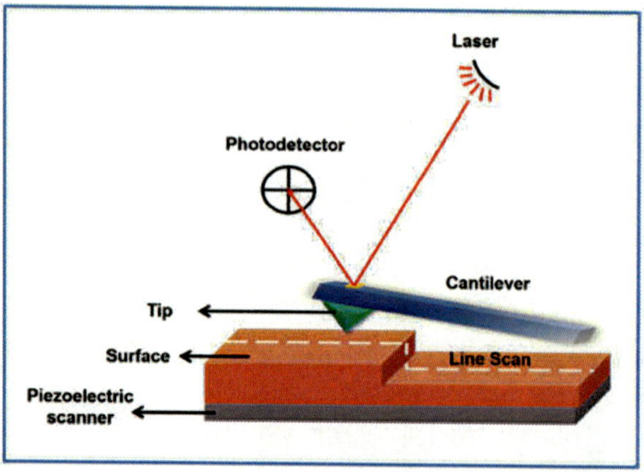

Fig. 5.16 Schematic depiction of working principle of atomic force microscopy

contact of the oscillating cantilever with the sample surface clips the oscillation's amplitude. With the aid of a laser projected on the photodiode, the amplitude of the oscillation is monitored. The amplitude is kept fixed during the scan by utilizing a feedback loop to adjust the piezo's Z position (Fig. 5.16). The image in the oscillating mode is produced in the same manner as contact mode by plotting the Z-movement versus the *X*–*Y* position. In contrast to the contact mode, the oscillating mode exerts lesser lateral forces on the sample. It thereby enhances the lateral resolution of the AFM image and is the most preferred mode for biological applications (Morris et al. 1999). When an AFM experiment is performed in the force mode, the interaction between the sample surface and the AFM tips is recorded in the form of force curves (Zlatanova et al. 2000).

AFM has been applied in several investigations in the biological field. It has been utilized for the quantitative characterization of the thermodynamic and structural properties of protein–protein complexes. Further, it has been employed to determine the association constants and stoichiometries involved in multi-protein complexes. Additionally, AFM has helped researchers to observe the dynamic aspects of biomolecular complexes and to measure the intermolecular forces between different biomolecules. Despite the fact that factors like height of protein can vary in AFM, the volumes of proteins remain unaffected. In fact, several research groups have elucidated that the AFM volumes share a linear dependence with the molecular masses of proteins. As a consequence, volume analysis has cropped up as a robust method to obtain the stoichiometries of protein–protein complexes and determine corresponding association constants (Yang et al. 2003).

AFM has certain advantages when being utilized as a tool for macromolecular complexes characterization. AFM can be performed on samples both in air and under conditions that closely mimic physiological states. Several advances in aspects like cryogenic AFM, solution imaging, carbon nanotube probes, AFM on two-dimensional crystals, high-speed AFM, etc. have improved the operative capabilities of the AFM to investigate biomolecular assemblies. AFM volume analysis, for example, has helped in bridging certain pitfalls of other techniques. It does not require high sample concentrations, thereby overcoming the problems in the assay of proteins that have tight binding constants or low solubility.

However, AFM itself has a few inherent disadvantages. Firstly, volume analysis does not easily apply to proteins having molecular weights less than 20 kDa because of the limitation of AFM resolution. Secondly, the sensitivity, dependence of the measurements on the sample preparation procedures, and the varying experimental conditions make it difficult to estimate the absolute forces involved. Further, there is no universal or standard approach that is utilized for the preparation of the biological samples.

Ratcliff et al. employed AFM to determine the association constant involved in UvrD (*E. coli* DNA helicase II) dimerization. AFM yielded the direct measurement of the thermodynamic properties of this PPI. The authors also analyzed the shifts in the protein oligomeric state distributions under different protein concentrations (Ratcliff and Erie 2001). Ivanov et al. utilized AFM to visualize and detect protein components of the cytochrome CYP11A1-dependent steroid hydroxylase system both in their isolated and complexed forms. The authors determined that proteins Ad and AdR were present as monomers in solutions. The typical heights of the monomeric Ad, AdR, and CYP11A1 images, as measured by AFM, were 1.0 ± 0.2 nm, 1.6 ± 0.2 nm, and 1.8 ± 0.2 nm, respectively. However, the binary AdR/CYP11A1$_{mon}$ and Ad/AdR complexes had heights corresponding to 2.8 ± 0.2 nm and 2.2 ± 0.2 nm respectively, thereby validating the existence of these PPIs in the cytochrome CYP11A1 system (Ivanov et al. 2011).

5.2.17 Tandem Affinity Purification-Mass Spectrometry (TAP-MS)

Mass spectrometry (MS) is a powerful in vitro method for the analysis of macromolecular interactions. MS is based on the principle of producing ions that can be detected depending on their mass-to-charge ratios thereby enabling the identification of polypeptides (Aebersold and Mann 2003). Peptide/protein molecules are converted into ions (gas phase) from their condensed phase by utilizing matrix-assisted laser desorption ionization (MALDI) and electrospray ionization (ESI). Several algorithms have been designed to analyze mass spectra and identify proteins by their sequence. Some of these algorithms determine correlations between experimental and theoretical spectra, whereas others are de-novo algorithms that theoretically interpret the mass spectra. Although MS can potentially characterize interacting proteins, protein complexes need to be first purified prior to their identification and characterization.

For this purpose, tandem affinity purification (TAP) has been developed (Shoemaker and Panchenko 2007). A typical TAP experiment utilizes a TAP tag consisting of a calmodulin binding peptide and two IgG binding domains of *Staphylococcus* protein A, separated by the tobacco etch virus protease cleavage site (Rigaut et al. 1999). The open reading frame (ORF) of the target protein is fused with the TAP tag encoding DNA sequence and is expressed in the yeast system. The expressed target protein is allowed to form native complexes with other interacting

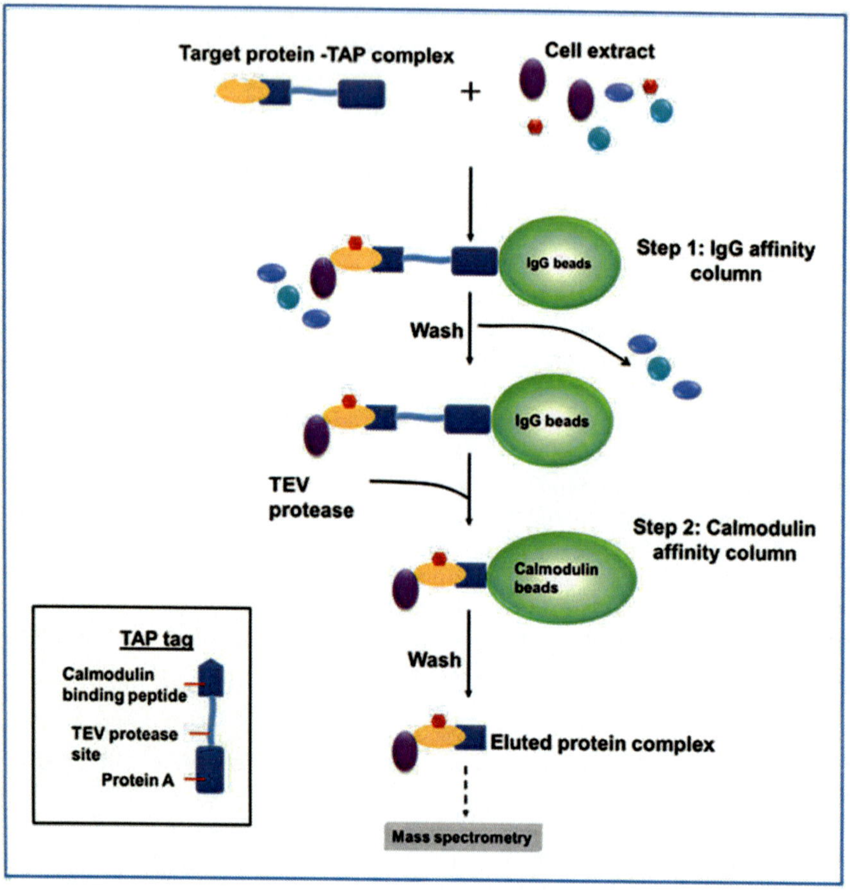

Fig. 5.17 Schematic illustrating the working principle of TAP-MS and structure of TAP tag (inset)

proteins in the yeast. Then, TAP purification is performed in two steps. In the first step, protein A is allowed to bind to an IgG matrix. Following a wash step, a protease is used to cleave protein A from the IgG matrix. In the second step, the eluate of the first step is allowed to incubate with calmodulin-coated beads in the presence of calcium. The target protein complex is then released by performing another wash. Next, the components making up each complex are screened by gel electrophoresis. They are then cleaved with the aid of proteases and subjected to MS for identification (Fig. 5.17). Several TAP tags such as FLAG, Myc, HA, polyhistidine, Strep II, streptavidin-binding peptide (SBP), Streptococcus protein G (ProtG), etc. are available. They help to enhance protein yields in cells other than yeast (Wang et al. 2010). TAP tags comprise two epitope tags and thus require two successive purification steps. Single-step affinity purification (ssAP) can also be exploited that utilizes only one purification tag and consequently a single step (Oeffinger 2012)

TAP-MS has several advantages for the determination of PPIs. Firstly, TAP-tagging enables the identification of a broad variety of protein complexes. It also allows researchers to understand the activeness of multi-meric or monomeric protein complexes existing in vivo. Secondly, TAP significantly decreases non-specific background by employing two-step purification (Zhou et al. 2016). TAP-MS can also yield information on higher-order interactions that are beyond binary complexes (Shoemaker and Panchenko 2007). TAP-MS is however also faced with certain challenges such as the generation of numerous false positives and the inability to determine many true positives that are usually low-abundance complexes or transient associations or specialized interactions dedicated to particular cell types (Snider et al. 2015). Further, TAP only indicates PPIs that are essentially co-complex associations, thus paves the need for further testing to understand which of such indicated interactions are direct physical interactions.

Krogan et al. performed TAP to analyze 4562 tagged proteins of *Saccharomyces cerevisiae*. Each sample preparation was analyzed using MALDI-TOF MS as well as LC-MS to increase the accuracy and coverage (Krogan et al. 2006). Next, machine learning was used to integrate the scores yielded by MS and to aid in the assignment of probabilities to PPIs. The authors concluded that among MS-identified 4087 different proteins, their dataset comprised of 7123 PPIs involving 2708 proteins. Rohila et al. fused 88 cDNAs encoding *Oryza sativa* (rice) protein kinases (PKs) with TAP-tags and expressed them in transgenic rice plants. The PKs and their interacting partners were then purified and identified by TAP-MS. 13 TAP-tagged PKs were found to interact with other rice proteins (Rohila et al. 2009).

5.2.18 Imaging Techniques

Optical imaging has been employed by several research groups to acquire co-localization images that can in turn map, detect, and envision PPIs (Wang et al. 2018). Conventional optical microscopy faces a major challenge known as the Abbe diffraction limit that disables researchers to further improve its resolution. This makes the process of distinguishing objects distinctly in closer proximity (<200–300 nm) using conventional microscopy. Therefore, super-resolution microscopy such as stimulated emission depletion (STED) microscopy and photo-activated localization microscopy/stochastic optical reconstruction microscopy (PALM/STORM) have cropped up with the ability to produce super-resolution (20–100 nm) images (Cui et al. 2019).

Imaging techniques for PPI analysis can be broadly categorized into (a) co-tracking of dual particles, (b) 3D co-localization, and (c) 3D dual-particle tracking. Co-localization analysis furnishes high levels of co-localized signals, implying the vicinity between the two proteins of interest and the likelihood of existing interactions between them. Single-molecule protein proximity index (smPPI) can quantitatively determine the accurate co-localization of several proteins.

5.2.18.1 Co-tracking of Dual Particles

Co-tracking of dual particles helps to track differently labeled proteins simultaneously and image in a time series to furnish separate molecular trajectories (Wang et al. 2015). The spatial separations and particle trajectories can then be evaluated to identify interacting protein molecules. Representative images over the time scale demonstrate the co-diffusion of differently fluorescent-labeled proteins.

5.2.18.2 3D Co-localization

Although 2D dual-color images can aid in the determination of PPIs via the calculation of their co-localization, they may be inaccurate because of the underestimation of the level of co-localization between proteins. This has paved the way for the development of two-color *fluorescence co-localization studies in 3D. It helps in achieving accurate measurements of the* co-localization of labeled proteins at the nanometer scale, thereby enhancing the detection of PPIs in vivo.

5.2.18.3 3D Dual-Particle Tracking

The 3D tracking of two particles utilizes high-resolution spatial localizations and aids to elucidate PPIs in living cells. The 3D co-tracking of fluorescently labeled sub-micrometer structures such as organelles or vesicles residing at great depths within tissues is precise at the nanoscale, thus providing an additional evidence of PPIs in vivo.

Chen et al. combined in vitro and in vivo single-molecule imaging and other techniques to demonstrate how transcription factors assemble on their cognate DNA sites. The authors concluded that enhanceosome assembly is hierarchically ordered. Sox2, being kinetically favored, engages the target DNA first. It then gets involved in a crucial PPI, by an assisted binding to Oct4 protein (Chen et al. 2014).

Lagache et al. utilized three-color structured-illumination microscopy (SIM) imaging to analyze primary hippocampal neurons. They concluded that dendritic abundant proteins PSD95 and Homer were arranged in an asymmetric triangle with the protein Synapsin (Lagache et al. 2018).

5.3 Biochemical Techniques

Biochemical techniques for PPI determination are based on principles from the fields of biology and chemistry. Some classical methods that have been widely used for PPI determination fall under this category are presented in Fig. 5.18 and have been discussed in detail in the following sections.

5.3.1 Forster/Bioluminescence Resonance Energy Transfer (FRET/BRET)

Fluorescence resonance energy transfer (FRET) is a process of non-radiative energy transfer from an excited donor fluorophore to an acceptor fluorophore in close

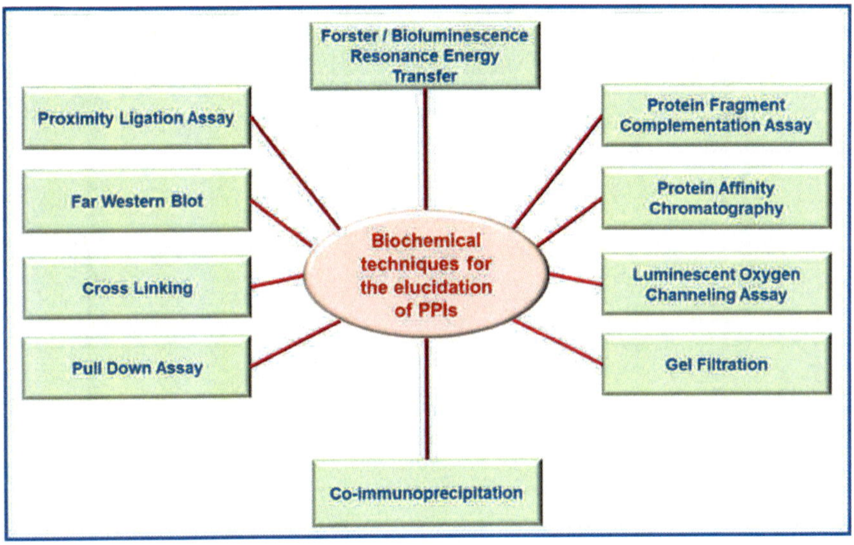

Fig. 5.18 Various types of biochemical techniques used for elucidating the molecular and functional information of PPIs

proximity (1–10 nm) via dipole–dipole interactions (Song et al. 2011). Apart from the close proximity, a donor–acceptor FRET pair needs an overlap between the donor's emission spectrum and the acceptor's excitation spectrum (Fig. 5.19). The widely used FRET pairs come from GFP variants, such as cyan fluorescent protein (CFP)–yellow fluorescent protein (YFP), CyPet–Ypet, blue fluorescent protein (BFP)–GFP, MiCy–mKO, mVenus–mStrawberry, cerulean fluorescent protein (CrFP)–YFP, etc. (Zhou et al. 2016).

In a typical FRET experiment designed to elucidate PPIs, one of the two target proteins is fused with the donor fluorescent chromophore, while the other is fused with the acceptor fluorescent chromophore. Both the fused proteins are subjected to a binding assay after their expression. In case if the two proteins do not interact, the donor fluoresces upon excitation, but the acceptor does not. However, upon interaction of the two proteins, the donor and acceptor come in close proximity. Upon excitation of the donor in this state, non-radiative energy is transferred from the donor to acceptor. In short, the occurrence of PPIs leads to FRET when the donor's fluorescence intensity decreases and that of the acceptor increases. FRET has been extrapolated to in vitro binding assays to design PPI inhibitors via optimized FRET pairs (Song and Liao 2012). It has also been utilized to visualize PPIs in real-time in live cells, with high spatiotemporal resolution in combination with other imaging techniques (Mahajan et al. 1998; Xia et al. 2001).

Bioluminescence resonance energy transfer (BRET) is an analogous method to FRET except for the fact that a bioluminescent protein acts as the donor rather than a fluorescent protein. The luminescent donor protein is usually Renilla luciferase

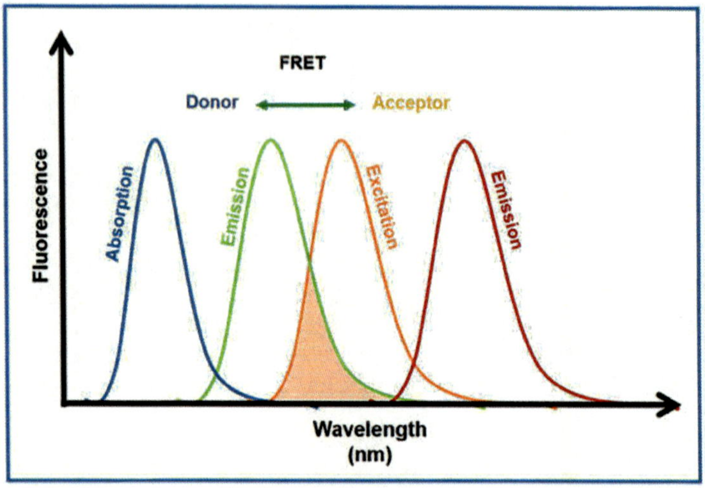

Fig. 5.19 Figure depicting mechanism of fluorescence resonance energy transfer (FRET)

(Rluc), and upon oxidation of its substrate, it emits light to excite the acceptor (Bacart et al. 2008). Improved versions of BRET in recent times have also enabled researchers to investigate PPIs in real-time over an extended timescale (Pfleger et al. 2006).

Although FRET has several advantages, including its capacity to determine dynamic changes in living cells continuously, it faces certain limitations. As FRET is dependent on fluorescent signals for detection of PPIs, fluorescence-based drawbacks such as detector noise, auto-fluorescence, optical noise, spectral bleed-through signals, photo-bleaching, etc. are inherent to this technique (Periasamy 2001). BRET, on the other hand, is free from these problems as it waives the requirement for extrinsic excitation by a light source. Despite such benefits, BRET also has some limitations such as high costs and low quantum yields of luciferin. Further, BRET has the possibility to furnish false-positive re-activity, in the absence of saturation measurements and confirmation of constitutive interactions (Cui et al. 2019).

Wei et al. employed FRET for monitoring PPI between σ^R and RsrA in *Streptomyces coelicolor*. The authors demonstrated that the presence of zinc provided protection to thiols in RsrA and caused the σ^R-RsrA complex to take up a more compact structure (Wei et al. 2014). Mahajan et al. expressed GFP-Bcl-2 and BFP-fused Bax A in the same cell and employed FRET to observe their direct interaction in the mitochondria (Mahajan et al. 1998).

5.3.2 Protein Fragment Complementation Assay

A family of assays, known as protein-fragment complementation assays (PCAs) are designed to detect (PPIs) in vitro, in vivo or in living cells (Michnick et al. 2011). In these assays, the two target proteins (the bait and the prey) are fused to C- or N-terminal fragments of a reporter protein that are complementary in nature and rationally dissected. On interaction of the prey and the bait proteins, the fragments of the reporter protein are brought into close vicinity that leads to the proper folding of the protein, ultimately reconstituting its activity (Michnick et al. 2011) (Fig. 5.20). The reporter proteins that are most commonly used include β-lactamase, dihydrofolate reductase (DHFR), luciferase, beta-galactosidase, yeast cytosine deaminase (yCD), GFP and its variants, ubiquitin, etc. (Zhou et al. 2016). There are several PCA-based techniques and the principle of each of them relies on the detectable effects that are produced by the reporter protein on reconstitution. The effects can vary from survival of cell in selective medium to colorimetric or fluorometric changes.

Split-DHFR PCA utilizes DHFR as the reporter protein. This enzyme catalyzes dihydrofolate reduction into tetrahydrofolate that in turn is essential for cell growth and proliferation. Therefore, the occurrence of PPI between the bait and the prey maintains cell survival in presence of dihydrofolate. In split-ubiquitin PCA, the reassembly of ubiquitin leads to the recruitment of particular ubiquitin specific protease. This protease aids in cleaving the reporter protein tagged to the C-terminal end of ubiquitin. The reporter proteins that detect ubiquitin fragment re-association usually include LexA-VP16, hemagglutinin (HA)-tagged DHFR, Rgpt2, RUra3p, etc. (Zhou et al. 2016).

Biomolecular fluorescence complementation (BiFC) is the most widely used PCA-based technique. Here, the bait and the prey proteins are fused to complementary fragments of protein with fluorescent properties. The interaction between the bait and the prey recovers the fluorescent activity of the protein, that is subsequently analyzed with the aid of flow cytometry or fluorescence microscope (Fujikawa and Kato 2007). Initially, GFP was the first protein to be employed in BiFC assays, but

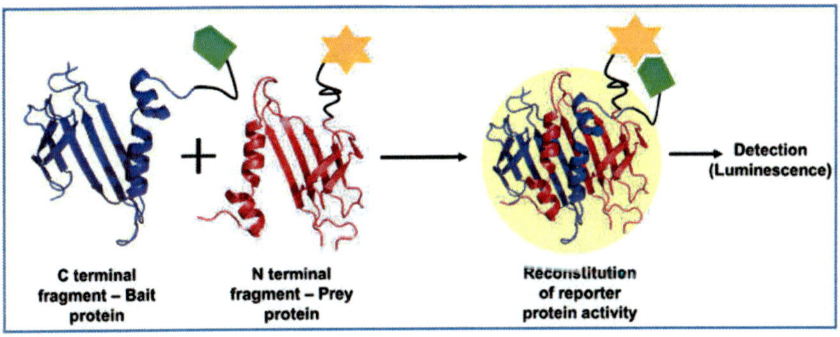

Fig. 5.20 Figure illustrating principle of protein fragment complementation assay (PFCA)

today, over 10 different fluorescent proteins are utilized (Fujii et al. 2018). Another PCA-based assay, known as the split-luciferase assay employs luciferase as the reporter protein to oxidize its substrate luciferin and generate bioluminescence. The substrate has the ability to penetrate cell membranes in a rapid manner, and thus permits in vivo imaging and live-cell assays.

PCAs are posed with several advantages. They have the ability to determine interactions between proteins of any molecular weight or between proteins that are expressed at their endogenous levels. Split-ubiquitin PCA has already been employed to analyze PPIs between membrane proteins and transcription factors that are otherwise impossible to study in other traditional methods such as yeast-two hybrid assay (Dunnwald et al. 1999; Wellhausen and Lehming 1999).

Firstly, BiFC enables the study of weak protein complexes owing to its ability to form irreversible chromophores. The most important advantage that BiFC offers is due to the intrinsic fluorescence of its reporter protein after reconstitution. This waives the requirement of exogenous reagents and also facilitates direct observation in the cells to analyze the subcellular localization of a given PPI. Split-luciferase assay has the ability to detect kinetic parameters of PPIs with a high signal-to-noise ratio. However, PCA-based techniques are also not devoid of limitations. The fused bait and prey proteins might lose a part of their native function, a drawback that also holds true for other techniques involving protein tagging. Secondly, BiFC can yield false positive results because the determination of PPIs is based solely on the proximity of the prey and the bait proteins. Therefore, proteins that do not have direct interaction but stay close within a complex might yield positive results indicating a false PPI (Kodama and Hu 2012). Split luciferase-based PCA also faces a disadvantage. The chemical reaction that needs to generate excited-state luminescent substrates is inefficient as compared to that of fluorescent proteins (Stynen et al. 2012).

Xue et al. performed BiFC to conclude that Phot1 protein that exists as a monomer in the dark conditions undergoes interaction with another Phot1 monomer to form a homo-dimer after being treated with blue light (Xue et al. 2018). Demir et al. elucidated the interaction of CPK21 with SLAH3 in the plasma membrane microdomain after observing co-localized signals from a microdomain marker protein and BiFC fluorescence (Demir et al. 2013).

5.3.3 Protein Affinity Chromatography

Protein affinity chromatography is based on the principle of coupling a protein to a matrix under precise conditions and then selecting proteins from an extract that interact and bind to it. Majority of the proteins in the extract are washed off readily with the help of low concentrations of salt. On the other hand, proteins that are retained in the column are eluted by chaotropic agents or high concentrations of salts (Phizicky and Fields 1995). Candidate target proteins are either coupled covalently or via non-covalent high-affinity interactions. If the protein of interest in the extract is labeled beforehand, the cognate PPI can be determined with utmost sensitivity as other unlabeled polypeptides will be easily separated off.

Protein affinity chromatography has four distinct advantages while being utilized as a method for determination of PPIs. Firstly, it is an exceptionally sensitive method that can determine interactions having very weak binding constants, in the range of 10^{-5} M (Formosa et al. 1991). This falls within 10^{-3} M that is estimated to be the binding constant range for the weakest, physiologically relevant interactions. Secondly, this method allows the testing of all proteins in an extract equally for their possibility of interaction with the coupled protein. Thirdly, one can employ mutant variants of proteins in this technique to understand the exact residues involved in a PPI (Cohen et al. 1993; Truant et al. 1993). Fourth, this method, unlike blotting, detects PPIs that rely on a multi-subunit tethered protein.

However, affinity chromatography also faces certain limitations such as the generation of false positive results and false negative results. False positive results may arise from charge interactions or indirect interactions via a mediator protein. They can also arise owing to the interaction of extract protein with contaminants present within the coupled protein. Moreover, PPIs that are biologically irrelevant may generate false positives. For instance, proteins that are spatially compartmentalized in the living cell may have physical association with each other on coming into contact in the column. On the other hand, false negative results can arise if an interacting protein is unable to exchange with another binding protein. Other challenges of affinity chromatography include the inactivation of protein via coupling. This can, in turn, affect its binding to its cognate partners. However, the most important limitation is the optimization of the concentration of protein bound to the column, which has to be higher than the K_d of the investigated PPI. This leads to the need for a great quantity of protein for PPIs that are weak in nature. On a similar note, the amount of extract that is subjected to the column is also crucial. If a small quantity of the extract is applied, very little protein will be retained in the column to be detected (Sopta et al. 1985). On the other hand, if a very large amount of the protein is subjected to the column, it might compete out with the other minor species that would have otherwise demonstrated interaction with the coupled protein.

Affinity chromatography has been exploited to report a novel interaction between Mex67 (soluble mRNA transport protein) and Nup1 (yeast nuclear pore complex protein) (Belanger 2009). McGaughey et al. investigated the interaction between proteins deoxycytidylate deaminase and thymidylate synthase as they are enzymes involved in catalyzing sequential steps of a reaction. The authors determined the interactions between the enzymes in free and immobilized forms using affinity chromatography (McGaughey et al. 1996).

5.3.4 Cross-Linking

Cross-linking is a technique in which two or more proteins are linked by a cross-linker via the formation of covalent bonds. The cross-linker is a chemical reagent that comprises two or more reactive groups that are conjugated by a linker or spacer region (Chapman 2000). To analyze cross-linked proteins, such as the identification of interacting protein partners, this technique is always accompanied by other downstream techniques such as SDS-PAGE, TAP, or immune-affinity chromatography or MS (Vasilescu et al. 2004). Cross-linkers vary depending on their chemical reactivity and appropriate cross-linkers for the characterization of PPIs are usually selected on an empirical basis.

A homo-bifunctional cross-linker comprising of two NHS esters is the most used cross-linker. With supreme reaction efficiency, it targets primary amine moieties in proteins under physiological conditions. Hetero-bifunctional cross-linkers, on the other hand, usually comprise one sulfhydryl-reactive group and one amino-reactive group. More innovative cross-linkers, such as photo-reactive cross-linkers, contain photo-reactive sites that can be subjected to UV-irradiation to link binding proteins (Weber and Beck-Sickinger 1997). There are several other cross-linkers such as isotope-labeled cross-linkers, fluorogenic cross-linkers, cleavable cross-linkers, trifunctional cross-linkers, etc. (Zhou et al. 2016). These cross-linkers have been utilized to identify and purify cross-linked species from complex reaction mixtures.

In extension of this method, known as label transfer, a latent cross-linking reagent along with a readily detectable tag (radiolabel, biotin, etc.) is conjugated to the target protein through a cleavable linker. Once this modified protein associates with its partner to form the native complex, the cross-linker is activated, followed by the cleavage of the linker arm. This results in the transfer of the tag to the interacting protein partner (Chen et al. 1994).

Cross-linking aids in the characterization of low-affinity PPIs that are otherwise rarely detectable by other methods. Moreover, this method can be employed both in vivo and in vitro to reinforce transient PPIs that are involved in a dynamic process. Some cross-linkers that are membrane-permeable aid in the detection of PPIs in their native niche. Cross-linking, however, has certain weakness that limits its use. Firstly, it has a high probability of capturing non-specific interactions that do not have physical interaction, but are simply in close proximity. Secondly, factors such as low abundance of cross-linked proteins and complexity of downstream processes like tandem mass spectrometry make the identification of cross-linked species a tedious process. Label transfer also faces inherent limitations such as the need of covalent modification of the target protein and reconstitution of the native complex using the modified protein (Liu et al. 2007). Further, the inefficiency of cross-linking chemistry in label transfer also makes it arduous to characterize partner proteins (Lam et al. 2002).

Bleil and Wassarman have utilized cross-linking method for the identification of a 56 kDa protein that binds to ZP3 (Bleil and Wassarman 1990). They employed a hetero-bifunctional, cleavable, and photoactivable cross-linker that comprised of a label on the photo-activated group. The cross-linker was initially conjugated to an isolated protein which was further subjected to incubation in a suitable extract. Photo-activation enabled cross-linking the nearby proteins and cleavage of the cross-linker permitted the transfer of the label to the cross-linked protein. From the experiments, the authors concluded that 90% of the label that is initially present on ZP3 can be transferred to its 56 kDa interactor (Bleil and Wassarman 1990). Sanders et al. employed cross-linking and downstream immunoprecipitation methods to determine the contact of Sec61 protein with tethered proteins involved in translocation (Sanders et al. 1992).

5.3.5 Luminescent Oxygen Channeling Assay (LOCI)

Luminescent oxygen channeling assay is a technique that can determine PPIs by irradiating a mixture of two different kinds of bead-immobilized proteins (Arkin et al. 2012; Ullman et al. 1996; Yasgar et al. 2016). This technique is also known as AlphaScreen where "Alpha" signifies an acronym that stands for "amplified luminescent proximity homogeneous assay." This assay requires acceptor and donor beads that are coated with functional moieties serving as chemiluminescent unit and photosensitizer, respectively.

In a typical LOCI experiment, two different proteins are mixed, where one of them is bound to an olefinic compound (acceptor bead) while the other is bound to phthalocyanine (donor bead). Owing to the short irradiation of phthalocyanine at 680 nm, singlet oxygen is produced that spreads over a constrained distance due to its short half-life. This singlet oxygen induces a cascade of energy transfer reactions that finally generate a luminescent signal if it triggers the thioxene derivatives present in the acceptor beads. If the two proteins interact, the singlet oxygen molecules reach the acceptor beads and a signal corresponding to 612 nm is detected immediately after stopping the 680 nm irradiation. While AlphaScreen utilizes anthracene, thioxene, and rubrene as dyes within the acceptor beads, a LOCI variant known as AlphaLISA employs europium chelate owing to its narrower emission wavelength bandwidth (Zhou et al. 2016).

LOCI is a simple technique and has several technical advantages. First, the generation of around 60,000 singlet oxygen molecules from each donor bead results in an exceptional amplification of the signal. Second, the non-overlapping nature of the emitted light with the irradiation reduces background noise and enhances sensitivity (Miura 2018). Third, having a greater distance of energy transfer (200 nm) than FRET, LOCI facilitates a broader range of analytes than can be studied, including immune-complexes and full-length proteins. Fourth, this method enables researchers to detect PPIs over a broad range of K_d values (picomolar to millimolar). However, LOCI faces the pitfall of the maximum bead capacity of the beads used. At a particular point, known as the "hook point," an excess quantity of the target protein results in oversaturation of the acceptor or donor beads. This in turn blocks the process of association, with a subsequent decrease in the signal (Yegorova et al. 2013).

Zhang et al. employed this technique to design inhibitors for the malicious β-catenin/B-cell lymphoma 9 (BCL9) interaction. The study was based on the principle that the diffusion of the singlet oxygen and the consequent emissions were due to the proximity of the acceptor and donor beads. Therefore, a complete disruption of the PPI by an inhibitor should entirely eliminate the LOCI signal (Zhang et al. 2015). On a similar note, Watson et al. utilized this technique to develop inhibitors against the interaction between AF4 and AF9 proteins (Watson et al. 2013).

5.3.6 Gel Filtration

Gel filtration is a technique that is used for the separation of molecules in a solution on the basis of their mobility. The differences in their mobility in turn arise from the differences in their molecular weights (Bai 2015). In other words, the process of gel filtration acts like a molecular sieve. This method utilizes inert beads made of polymers (polyacrylamide, agarose, etc.) that contain pores of a particular dimension. When a solution containing a mixture of molecules is subjected to the beads in a column, the larger molecules face lesser penetration into the pores and are eluted first. Smaller molecules, on the other hand, are permeable through these beads and are retained for a greater time in the column, resulting in their elution at a later stage. Molecules that fall in the intermediate size range elute on the basis of their molecular radius.

Gel filtration can be utilized in three different ways to estimate binding constants of PPIs (Phizicky and Fields 1995). In the first approach, known as non-equilibrium "small-zone" gel filtration, a solution of a protein and a solution of its interactor are administered to the column in small volume and the resolution is performed in the usual way. The elution profile is then compared with the chromatographs of the individual proteins. In case of formation of a protein complex, the elution is observed earlier than either of the proteins alone. The protein complex can also be purified via fractionating the corresponding peak fraction (Fig. 5.21).

This approach helps to evaluate the binding constant via the measurements of protein concentration that are needed to form the complex. In the second approach, known as Hummel-Dreyer method of equilibrium gel filtration, both the sample and

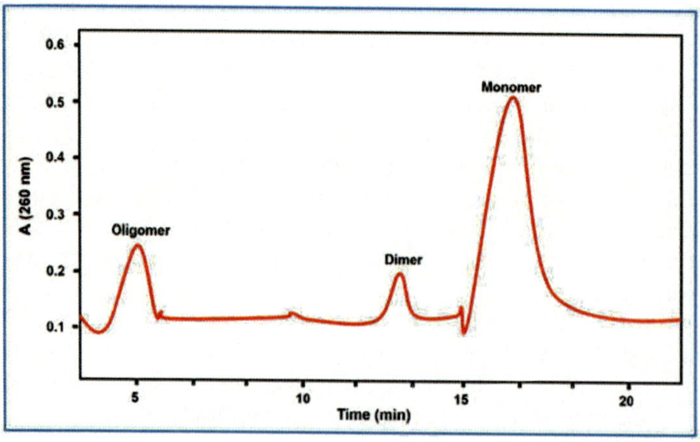

Fig. 5.21 Figure depicting a gel chromatogram showing presence of oligomeric structures

the gel filtration buffer have the ligand (interactor protein) in equal concentrations, but only the sample contains the protein (Hummel and Dreyer 1962). Passing the sample through a column leads to an increase in the ligand concentration at its elution. It is followed by a trough in its concentration profile denoting that some of it has been involved in binding. The binding constant of the cognate PPI can then be calculated by quantitating the concentrations of eluted protein, free ligand, and bound ligand. In the third approach, known as large-zone equilibrium gel filtration, a great sample volume is subjected to the column, followed by the traditional method of buffer elution (Ackers 1970). Owing to the application of a large volume, the eluted protein maintains a constant concentration during the experiment (barring the trailing and leading edges). The elution profiles of these edges are evaluated on the basis of the concentration of the sample administered to the column.

Although gel filtration serves as a convenient technique to study PPIs, the method suffers from a few drawbacks. Firstly, the small-zone approach is not an equilibrium method and the species concentration keeps on changing. This can lead to errors in the calculation of binding constants. For instance, in complexes undergoing rapid equilibrium, the binding constants have the possibility of being greatly underestimated (Gegner and Dahlquist 1991). The Hummel-Dreyer method, in this regard, is flawless. However, it requires a sufficient quantity of protein to perform the experiments and also needs the elution position of the complex to differ from at least one of the interactors. The large zone approach has also found limited utility owing to the requirement of great quantities of protein and rigorous calculation methods of K_d.

Gill et al. have employed gel filtration to assess the association between *E. coli* NusA protein and core RNA polymerase. The values yielded by this method were in line with those generated by fluorescence experiments (Gill et al. 1991). On a similar note, Herberg et al. studied the interaction of cAMP-dependent protein kinase with both PKI and R1 subunit utilizing this technique. The interactions were assessed both in the absence and presence of MgATP (Herberg and Taylor 1993). Recent size exclusion studies on chemokine proteins and their mutants have differentiated their oligomerization properties and their association characteristics (Gangele et al. 2020; Gulati et al. 2018; Joshi et al. 2020b).

5.3.7 Co-immunoprecipitation (Co-IP)

Co-immunoprecipitation (Co-IP) is one of the most utilized techniques aimed to identify PPIs. Based on the fundamental principle of interaction between an antigen and antibody, this method utilizes a bait-specific antibody that enables the precipitation of the bait protein along with its interacting partners from the sample. In essence, it is an extension of affinity chromatography. Co-IP is either followed by Western blot analysis or mass-spectrometry to identify the interactors (Phizicky and Fields 1995) (Fig 5.22).

Depending on the application, both polyclonal and monoclonal antibodies can be used in Co-IP experiments. While monoclonal antibodies recognize single epitopes

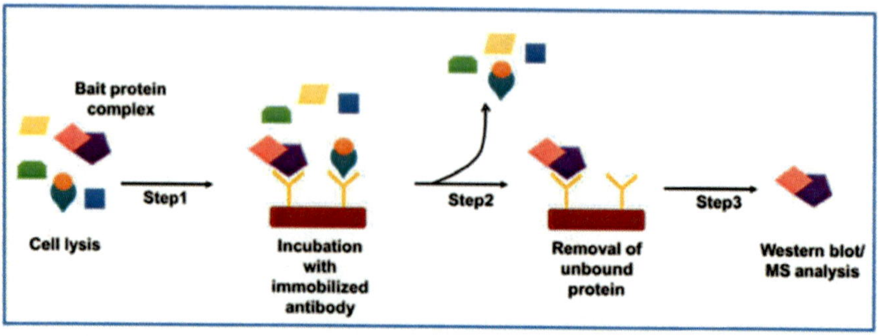

Fig. 5.22 Figure illustrating working principle of co-immunoprecipitation that involves three major steps: (1) Cell lysate including bait protein complex along with non-specific proteins is incubated with immobilized bait-specific antibodies. (2) The bait protein together with its interactors binds with the immobilized antibody and non-specific proteins will be removed. (3) Bait protein complex is further analyzed via Western blot or mass spectrometry (MS)

and thus reduce cross-reactivity and background noise, polyclonal antibodies can recognize multiple epitopes, yielding stronger signals and robust detection. There are several criteria that need to be substantiated before it can be concluded that a Co-IP experiment has yielded authentic results. First, one has to establish that the employed antibody has co-precipitated the protein. This is taken into consideration because of the possibility of precipitation by a contaminating antibody. Second, it has to be validated that the co-precipitated protein (interacting protein) is itself not recognized by the employed antibody. Third, it has to be determined whether the interaction is a direct one or is mediated by another intermediate protein. Fourth, the most arduous criterion is to understand that the result signifies a native PPI. This needs to be substantiated to rule out the possibility that the determined PPI is not a consequence of the lysis of the cell that provides opportunity for compartmentalized and segregated proteins to come in close proximity and be involved in a physical but biologically irrelevant interaction (Phizicky and Fields 1995).

Co-IP has the advantage of reflecting physiological interactions of proteins expressed at endogenous levels since it is carried out in a tissue or whole cell lysates. The technique is non-denaturing in nature and does not induce any artificial effects such as modification or over-expression of the interacting proteins. Elaborated complexes that otherwise face difficulty in vitro assembly can be easily co-precipitated using this technique. Further, the native state of proteins helps to assess the dependence of PPIs on post-translational modifications more realistically (Phizicky and Fields 1995). However, since the concentration of the bait protein is much lower in its native conditions, this technique suffers from limitations of sensitivity (Zhou et al. 2016). Moreover, Co-IP requires sufficient quantity of protein (100–1000 mg) that might not be appropriate for conventional cell-culture assays or tumor biopsies. Lastly, Co-IP does not always denote direct PPIs since co-immunoprecipitating proteins can in essence be a part of larger complexes.

Pelicci et al. utilized Co-IP to study the interaction of SHC with activated EGF receptor (Pelicci et al. 1992). They performed immunoprecipitation to isolate SHC from both EGF stimulated cells and control cells. Subsequent immunoblotting of EGF stimulated samples with the aid of anti-EGFR antibodies proved the co-precipitation of EGFR along with SHC protein. Similar results were also obtained when immunoprecipitation was performed with anti-EGFR antibodies first and then blotted with anti-SHC antibodies. Lin et al. have applied the Co-IP technique to study the interaction between TssB protein and $TssC_{41}$ protein (Lin et al. 2013).

An extension of the co-immunoprecipitation technique, known as LUMIER (LUminescence-based Mammalian IntERactome), is a high-throughput technology that has been developed to study PPIs in mammalian cells. In this technique, a Flag-tagged protein and a Luciferase (LUC)-tagged fusion protein are co-expressed in a transfectable cell line like the HEK-293T cells. Co-immunoprecipitation by utilizing an anti-Flag antibody is used to detect the interaction between the two proteins. Similarly, the LUC-tagged interactor present in the complex was identified by its luciferase activity (Barrios-Rodiles et al. 2017). Barrios-Rodiles et al. utilized collections of Flag-tagged proteins to elucidate protein interaction networks for several Wnt pathway members and TGF-β family receptors (Barrios-Rodiles et al. 2017).

5.4 8 Pull-down Assay

Pull-down assay is also an extension of affinity chromatography that can assist in the determination of PPIs. In the original version of this technique, GST-fused bait protein is expressed in bacterial cells, followed by its immobilization on glutathione beads (Kaelin Jr et al. 1991). The immobilized protein is subsequently utilized to capture its interactors from the cell lysate. Washing steps are incorporated to eliminate non-specific interactions before final elution of the bound proteins. The eluted proteins are then identified with the aid of SDS-PAGE (Fig. 5.23). Often,

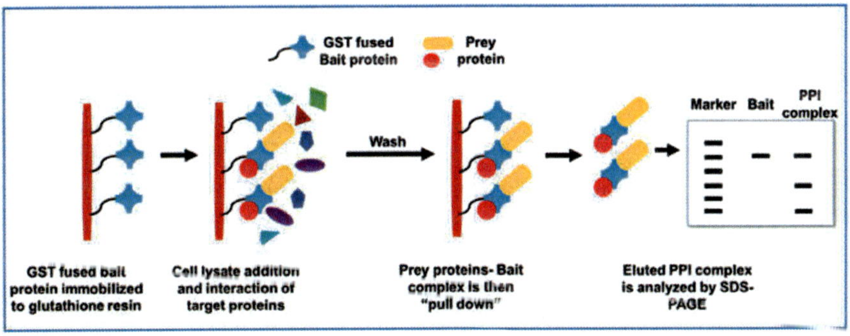

Fig. 5.23 Figure illustrating working principle of pull-down assay. The cell lysate is incubated with immobilized GST fused bait protein. The prey proteins interact with the immobilized tag and eluted after a washing step. The eluted bait–prey complex is analyzed by SDS-PAGE

other fusion tags such as HaloTag, polyhistidine, or biotin are conjugated to the bait protein for their capture by HaloLink resin, Ni-NTA agarose, or streptavidin magnetic beads, respectively (Zhou et al. 2016).

Pull-down assay is one of the most favored techniques by researchers since it can confirm suspected PPIs or even unravel unknown/novel interactions. Variants of pull-down assay have been designed such that the quantity of protein required for a single experiment can be further reduced. For instance, SINgle Bead Affinity Detection (SINBAD) serves as a miniaturized version of the pull-down assay. Since it employs color-coded prey (using quantum dots) and affinity-tagged bait, it enables the direct visualization of interaction events on the bead surface with the aid of fluorescence microscopy. It waives off the steps of elution and gel segregation of isolated protein complexes (Schulte et al. 2008).

In another pull-down assay variant, known as Bead Halo, prey protein with a fluorescent tag and bait protein with an affinity tag are used. Positive interaction between the prey and the bait is denoted with the aid of a fluorescent halo surrounding the beads. This method holds the advantage of detecting interactions of low-affinity, that are otherwise excluded by the washing steps in the conventional pull-down assay (Patel and Rexach 2008). Despite its advantages, pull-down assay can yield false positive or false negative results, as it is an in vitro method and does not provide proteins with their native cellular environment. Therefore, PPIs that are specifically dependent on physiological conditions might not be captured by this technique (Cui et al. 2019).

Patel et al. employed Bead Halo to determine and characterize protein interactions in the machinery that is involved in the nucleus-cytoplasm transport in eukaryotes. The authors unraveled unknown interactions between the importin protein Kap95 and the nucleoporin proteins (nups) Pom34, Nic96, Gle1, Nup84, Ndc1, and She1 that most probably take place during the biogenesis process of the nuclear pore complex (Patel and Rexach 2008). Miura et al. utilized pull-down assay to purify interacting partners of Ash protein from bovine brain. For this purpose, they produced GST–Ash fusion protein and subsequently coupled it to glutathione–agarose to capture interactors of Ash from cytosolic fractions of bovine brain (Miura et al. 1996).

5.4.1 Far Western Blot

Far Western blotting is one of the most popular methods for PPI identification and characterization. In a typical far Western blot experiment, proteins of interest (prey proteins) are first separated electrophoretically. For this separation, the proteins are subjected to treatment with appropriate buffers that not only reduce the existing disulfide bonds, but also linearize it along with addition of uniform negative charges. Following the separation, the proteins are transferred to a PVDF or nitrocellulose membrane with the aid of electroblotting. This membrane is then put in a blocking solution (skim milk/gelatin/biotin) such that the proteins present in this solution "block" the sites on the membrane that are not occupied. The blot is then probed with

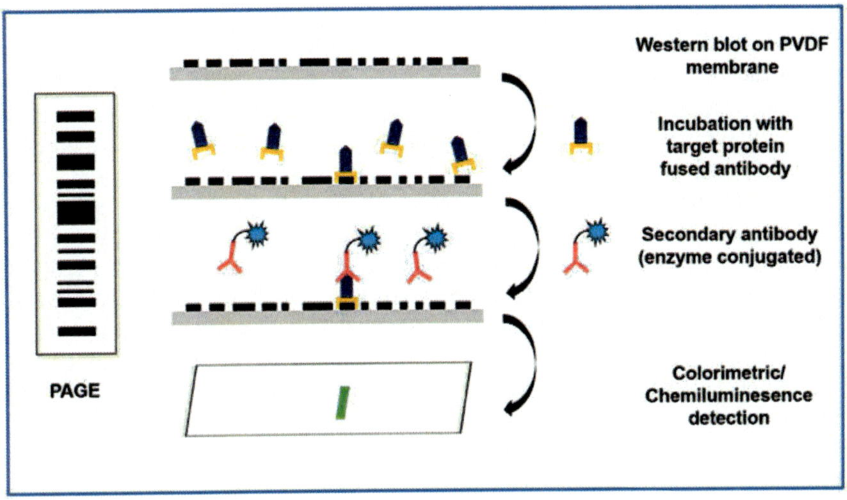

Fig. 5.24 Figure depicting work flow for far Western blot for detection of PPIs. The proteins are separated by electrophoresis and blotted on PVDF membrane by electroblotting and treated with a blocking solution. This blotted membrane is treated with bait fused antibody. The prey–bait–antibody complex is detected by an enzyme-conjugated secondary antibody via colorimetry or chemiluminescence

target proteins that have conjugated fusion tags. The tags are chosen in such a manner that antibodies against them are commercially available. Alternatively, the requirement of tags can be waived off if antibodies are available toward the probing proteins. The antibody employed in the technique is enzyme-conjugated; the enzyme usually being alkaline phosphatase (AP) or horseradish peroxidase (HRP). Following the addition of the antibody and designated incubation time, the enzyme-specific substrate is added such that its conversion to a product confirms if the probing protein has bound to the blotted protein (Fig. 5.24). The commonly utilized detection protocols utilize either colorimetry or chemiluminescence (Chan et al. 2008)

Far Westerns have several technical advantages while being utilized in PPI analysis. First, they are cheap and require simple equipment and/or materials. Second, they involve easy protocols that might be further tailored depending on the requirement. Far Westerns not only help answer several questions regarding PPIs, but also aid in experiments like interaction domain fingerprinting. There are however a few technical disadvantages associated with this technique. First, the fusion tag used may interfere with the structure and/or binding domain of the probing protein and lead to incorrect results even if the blotted protein is displayed correctly on the membrane. Second, this technique is able to detect PPIs only under specific conditions. Therefore, if one does not observe an interaction, it does not necessarily indicate a negative result. This is because the success of the experiment depends largely on the stability of the blotted prey protein that might be tampered by experimental conditions. Third, background noise may arise owing to non-specific interactions that often arise from experimental errors such as incorrect blocking, greater incubation periods, and higher protein concentrations.

Winstone et al. employed dot-blot far Western to investigate the interaction of DmsD with the N-terminal peptide of DmsA in the presence of denaturants (GdnHCl and urea) (Winstone et al. 2006). The authors blotted DmsA-GST fusion protein onto nitrocellulose membrane, followed by its incubation with solutions containing DmsD in varying denaturant concentrations. The extent of DmsD-DmsA interaction was quantified with the aid of an imaging software and compared to that in the absence of denaturants. The authors concluded that complete interaction (100%) took place for urea and GdnHCl concentrations below 2 M and 0.5 M, respectively.

Chan et al. performed Far Western to elucidate the interactome fingerprint of NarJ protein in wild-type *E. coli* and a mutant that lacked the tat operon (Δtat). SDS-PAGE separated membrane fractions from wild-type or Δtat *E. coli* were blotted onto a nitrocellulose membrane that was subsequently blocked overnight with skimmed milk (10%). Then the membrane was incubated with purified NarJ to detect all the plausible interactions of NarJ (Chan et al. 2006).

5.4.2 Proximity Ligation Assay (PLA)

The proximity ligation assay (PLA) is a technique that exploits the property of protein affinity. This assay employs antibodies that have conjugated PLA probes (single-stranded DNA oligonucleotides). When these antibody-probe conjugates bind to two proteins that are in close vicinity (<30 nm) in a given biological system, they enable the ligation of additional DNA molecules with the aid of ligases. This results in subsequent amplification by PCR (polymerase chain reaction) or RCA (rolling circle amplification) mechanism. The resulting amplified DNA molecules then serve as templates to which fluorescently labeled oligonucleotide probes bind. These probes function as markers for an interacting pair of proteins (Fredriksson et al. 2002).

The advantage associated with this technique is the increased specificity of signals that is brought about by the dual recognition criteria of PLA probes to fabricate DNA reporter molecules (Weibrecht et al. 2010). PLA is advantageous over techniques like BiFC, as it does not require tags to identify and/or quantify PPIs. The RCA involved in PLA labels the protein complex by hundreds of fluorophore copies and amplifies the signal to a great extent (Söderberg et al. 2008).

Furthermore, the method is also capable to identify interactions that depend on post-translational modifications (Buntru et al. 2016). It detects endogenous PPIs with respect to the cellular context, as it is independent of overexpression. The technique is however limited by its requirement of antibodies that not only demonstrate high affinity and specificity but are also easily modifiable by DNA oligonucleotides. Moreover, the currently available variants of this assay are unable to furnish information on the binding strengths of PPIs. Such information can only be obtained if one knows about the abundance of both unbound and bound protein molecules (Buntru et al. 2016). Lastly, PLA suffers from the limitation of detecting protein proximity (in nanometer range) rather than physically existing interactions (Leuchowius et al. 2009).

Poulard et al. employed in situ PLA to detect ERα/PI3K interaction endogenously in tumor samples and human breast cancer cells (Poulard et al. 2014). The authors performed PLA by utilizing PI3K- and ERα-specific antibodies. They observed the dimers using both fluorescent detection signals and bright field microscopy. Soderberg et al. utilized proximity ligation in situ assay (P-LISA) to determine the well-known interaction between c-Myc and Max (Söderberg et al. 2006).

5.5 Genetic Techniques

Genetic techniques extract information on PPIs by exploiting the corresponding genes of an organism. These methods usually involve knocking out genes or insertion of foreign genes that are suspected to be responsible for the PPIs under study. Commonly used genetic techniques have been depicted in Fig. 5.25 and have been discussed in detail in the following sections.

5.5.1 Yeast-2-Hybrid (Y2H)

TheYeast-2-Hybrid (Y2H) technique is one of the most popular genetic methods that utilize the yeast GAL4 transcription factor. This transcription factor consists of two domains: an activation domain (AD) and a DNA-binding domain (DBD). These domains however cannot independently function as transcription factors. Only under conditions of close proximity, these domains can reform a functional GAL4 transcription factor (Keegan et al. 1986).

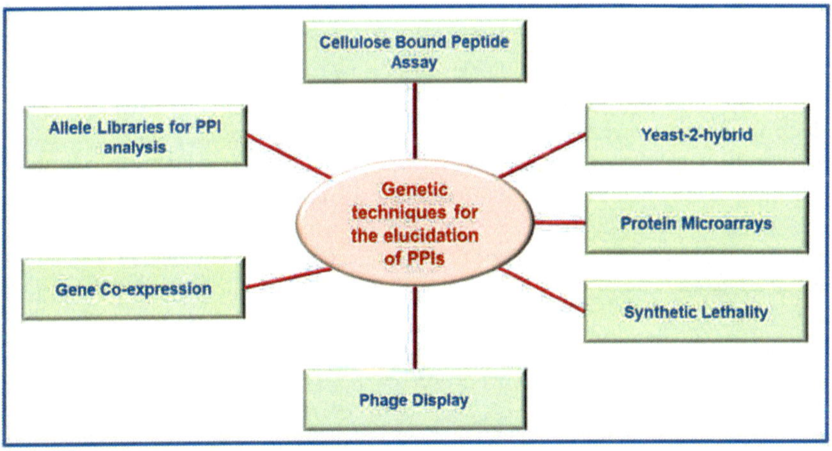

Fig. 5.25 Schematic showing various genetic techniques applicable in detection of PPIs

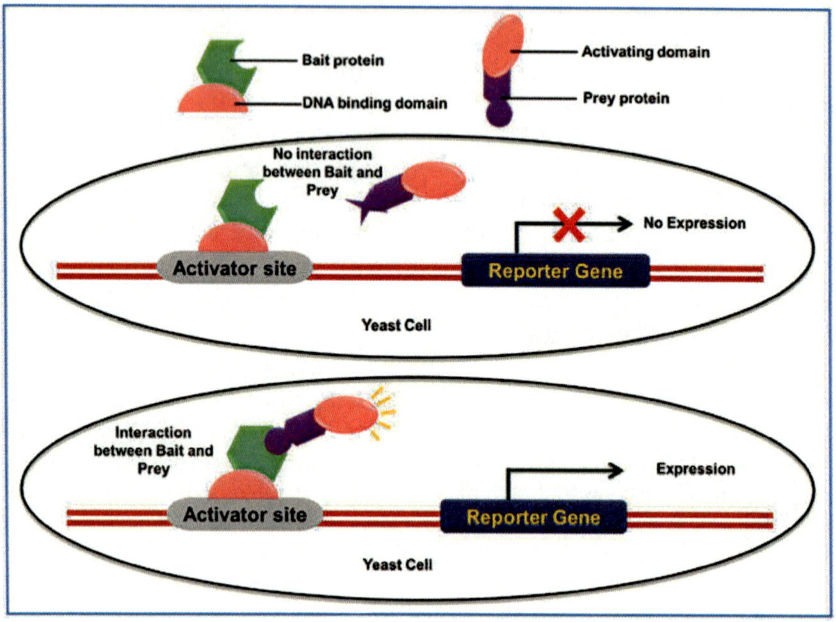

Fig. 5.26 Working principle of yeast-two hybrid system: The GAL4 DNA-binding (DBD) and activating (AD) domains are fused with bait and prey protein, respectively, and transfected into yeast cells. If the bait and prey proteins do not interact, then the reporter gene will not express (above). If the bait and prey proteins interact, it reforms the GAL4 transcription factor and promotes the expression of the reporter gene (below)

In a typical Y2H experiment, the gene sequence of GAL4 DBD, fused to that of the bait protein, is cloned into a plasmid that is transfected into cells. Following expression in the cells, the bait protein binds to the genomic DNA via the fused DBD. Next, a plasmid encoding GAL4 AD-prey fusion protein is transfected into the cells. Only if there is an interaction between the bait and prey protein, the DBD and AD come into close proximity to reconstitute the GAL4 factor. The reporter genes which are located downstream of GAL4 promoter encode proteins that are essential for the cells to proliferate and survive. These proteins are transcribed when the prey interacts with the bait. Thus when there is interaction between the prey and the bait, the cell proliferates in the selective medium but when there is no interaction, the cell does not undergo proliferation. The DNA sequence fused with the AD of GAL4 will be subsequently sequenced to identify the protein interacting with the bait (Fig. 5.26).

Several variants of Y2H have been designed. They include one-hybrid and three-hybrid systems, systems with more than one reporter gene, and systems that detect interactions in prokaryotic and mammalian cells (Shoemaker and Panchenko 2007). The Y2H method has been devised into two major approaches for entire genome screening: (1) matrix-based and (2) library-based. In the former method, a matrix is formed on the basis of prey clones. In each well of the plate a specific prey protein is expressed by a specific clone. Then a series of prey strains is mated with each bait strain and diploids are allowed to form. The diploids that demonstrate interaction of two chimeric proteins are selected depending on the position on the plate and also the expression of a reporter gene. In the latter approach, an undefined prey library consisting of cDNA fragments or ORFs randomly is used to screen each bait. On the basis of their capability to grow on particular substrates the diploid positives are selected. Further, DNA sequencing is used for determination of interacting proteins.

Y2H has gained immense utility in the field of PPIs as it is simple and easy to use. It is an in vivo technique that can be used for wide-range screening of PPIs (Li et al. 2004). Though it is a powerful technique, Y2H still faces certain bottlenecks. Often false positive or negative results are acquired from the Y2H/Y3H system as heterologous genes are expressed in the yeast cells. Y2H is not favorable for proteins present in the subcellular compartments or proteins present in the membranes as it is based on the nuclear translocation of interacting proteins. There are other proteins like transcriptional activators which are unsuited for the Y2H system. These proteins can spontaneously transcribe the reporter genes, independent of the interaction between the bait and prey. Moreover, in case of post-translational modifications that are not found in yeast, Y2H is not effective in detecting the PPIs (Hamdi and Colas 2012).

Durfee et al. employed an enhanced version of Y2H to isolate human cDNAs that code for protein interactors of p110RB (Durfee et al. 1993). One of the clones used in their experiments encoded a novel type 1 protein phosphatase catalytic subunit (PP-1 alpha 2). The authors concluded from their experiments that isoforms of PP-1 alpha isoforms preferentially interact with the hypophosphorylated form of p110RB. Similarly, Chevray et al. utilized the yeast assay system for the identification of interacting partners of Jun (Chevray and Nathans 1992).

5.5.2 Protein Microarrays

Modern day protein microarray technique is a comprehensive high-throughput method. This method helps to analyze several proteins simultaneously by arraying them on a solid surface at high density. There are various types of protein microarrays. Among them, the most widely used method is functional protein microarray. This method involves a large number of protein domains or full-length functional proteins that are immobilized onto chips. It is used to study a proteome entirety in a single experiment and can also be used to study their biochemical properties and assess PPIs (Fig. 5.27) (Hall et al. 2007).

There are several challenges in performing a functional protein microarray. Firstly, it is very difficult to construct a comprehensive library of expression clones. Although microarray fabrication has been possible from proteomes of *Arabidopsis thaliana*, coronavirus, yeast, *E. coli*, and human, it is highly desirable to obtain collections of ORF for other species also (Zhou et al. 2016). The second limitation of this technique is the requirement of high-throughput protein expression and purification. Thirdly, the need for preserving the three-dimensional structure of proteins

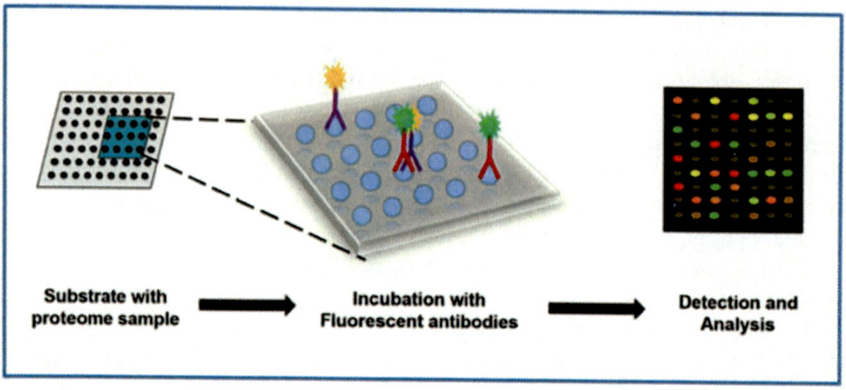

Fig. 5.27 Work flow for protein microarray for detection of PPIs. The proteome samples are immobilized on a chip and incubated with fluorescent target-specific antibodies. The antibodies upon conjugation fluoresce and the intensity of fluorescence are detected and analyzed for determination of protein–protein interaction

makes it essential to develop immobilization methods that do not lead to loss of activity or cause the capture proteins to be incorrectly oriented.

Popescue et al. utilized a microarray comprising of 1133 *A. thaliana* proteins to determine binding partners of Calmodulin (CaM). Screening of the microarrays against several CaMs not only re-identified previously known interactors, but also discovered novel targets (Popescu et al. 2007). Chen et al. utilized a microarray consisting of around 17,000 human ORFs for BAG3 interactome profiling (Chen et al. 2013).

5.5.3 Synthetic Lethality

There are efficient genetic analysis tools for some organisms. In such biological entities, sophisticated methods can be devised to unravel genes that share interactions with other genes. In most of the cases, these novel genes might code for proteins that have physical interaction with proteins encoded by previously identified genes. In rest of the cases, the genetic methods might aid to confirm the interactions between known proteins (Phizicky and Fields 1995). Synthetic lethality is a tool that falls under such classical genetic strategies. Essentially, it serves as an in vivo genetic screen. It elucidates the mechanisms that maintain stability at the phenotypic level despite variations at the genetic level, environmental or mutational effects.

In this technique, deletions or mutations are produced in two genes such that either mutation alone keeps the organism viable, but is lethal when generated together under selective conditions (Bender and Pringle 1991; Ooi et al. 2006). The lethal condition arises usually due to the lack of interaction between the proteins encoded by the two genes. In other words, if dimer AB is required for a certain

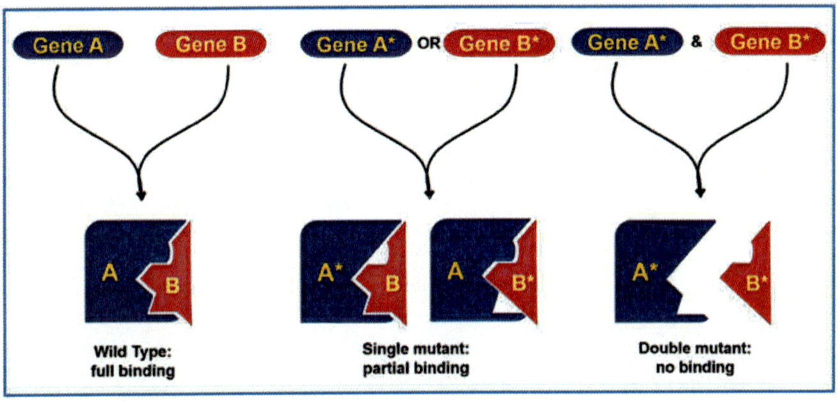

Fig. 5.28 Working principle of synthetic lethality for PPI determination. Here, protein– protein interaction is analyzed by mutation of the genes of protein to be studied. If one of the genes is mutated, partial interaction takes place and keeps the organism viable while when both genes are mutated, a lethal condition arises

cellular function, its loss results in a phenotype that is easily detectable. Mutation alone in A or B produces binding to some extent, but mutation in both A and B yields a double mutant A*B* that completely loses binding and consequent function (Fig. 5.28). The relationships yielded by synthetic lethality do not require the proteins to physically interact and therefore are referred to as "functional interactions" (Rao et al. 2014).

Screens based on synthetic lethality do not yield information on the requirement of the particular PPI. For instance, the proteins encoded by the two genes might comprise the same structure, or one of them could be involved in the regulation of the other, etc. Further, there might be some cases where either of the two mutations can lead to a condition of total inviability for the given organism. As these mutations cause lethality to the organism, they cannot be isolated naturally and requires construction through synthetic approaches (Shoemaker and Panchenko 2007).

Salminen et al. have employed synthetic lethality to elucidate that a mutation based on temperature sensitivity in the SEC4 gene in yeast cells leads to inviability at the permissive temperature when it is combined with temperature-sensitive mutations in other SEC genes (Salminen and Novick 1987). Similarly, Huffaker et al. exploited this technique to conclude that mutations in both a-tubulin and b-tubulin lead to inviable conditions in yeast cells (Huffaker et al. 1987).

5.5.4 Phage Display

Phage display is a relatively new technique that is based on the insertion of foreign DNA into genes of filamentous phage such that a fusion protein is "displayed" on the phage particle surface (Tong et al. 2002). This technique has found utility in several

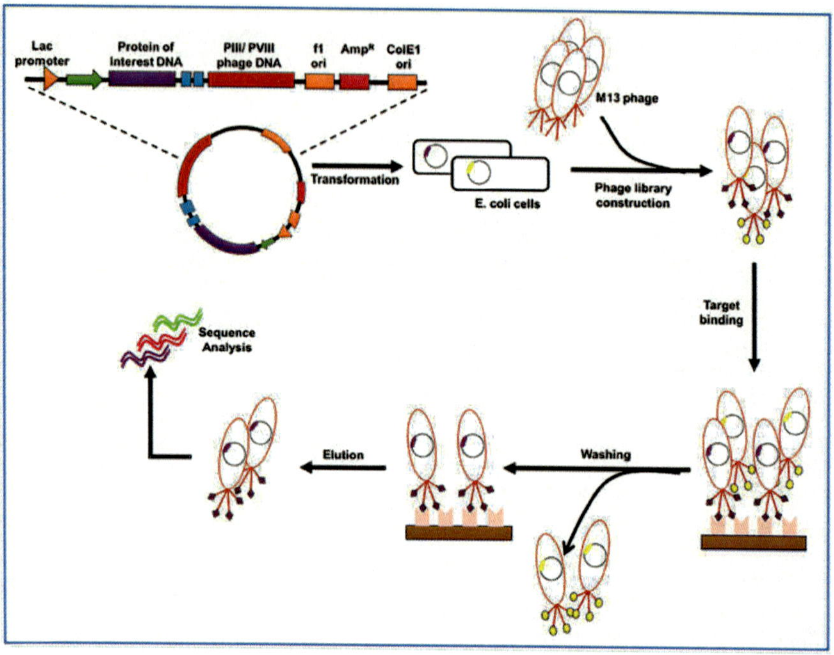

Fig. 5.29 Working principle of phage display for detection of PPIs. The vector containing protein of interest and PIII or PVIII phage element DNA is transformed into *E. coli* cells. The transformed cells are then transfected with M13 phage. The protein of interest/prey is expressed with the coat protein which is incubated with immobilized bait protein and gene sequence was analyzed

applications such as antigen epitope mapping, monoclonal antibody isolation, substrate identification for various proteases, enzyme modulator development as well as determination of PPIs (Pande et al. 2010).

In a typical phage display experiment to elucidate PPIs, a foreign DNA sequence coding for the prey protein is inserted into pVIII or pIII gene of M13 filamentous phage. These genes code for the major or minor coat proteins of the phage, respectively. The gene insertion generates prey protein-coat protein fusions that are displayed on the phage surface. Next, a pool of M13 phages that display different prey proteins are incubated with surface-immobilized bait protein. The phages that display prey proteins sharing interaction with the bait are captured on the surface, while the rest are washed away. Next, the captured preys are eluted and amplified by re-infecting *E. coli* cells. They are enhanced by over 1000-folds via a process known as "panning" that is carried out in successive cycles. The prey protein is finally identified by analyzing the sequence of the chosen phage DNA (Hertveldt et al. 2009) (Fig. 5.29).

Phage display has several advantages that include incorporating the prey protein into a single phage, quick and efficient selection of interacting phages and amplification steps comprising several rounds. This makes it a more sensitive technique compared to existing bioassays. Further, the phages mostly remain functional since the foreign protein comprises a very small fraction of the entire coat protein. Moreover, the phage particles can accommodate up to 50 kDa of the prey protein without any decline in their infective power (Phizicky and Fields 1995).

There are however a few technical disadvantages associated with phage display. For instance, the displayed proteins of M13 phage being secreted via the periplasmic space in the membrane of *E. coli* are often misfolded. Further, PPI screening via phage display requires an ORF library that has large diversity and high quality. Generally, cDNA libraries for phage display are rare, as these libraries can have unpredictable stop codons and/or reading frames that can lead to the generation of out-of-frame phages. Such phage clones encode short peptides that have tendency to outcompete phages encoding correctly displayed proteins (Caberoy et al. 2010). One of the major limitations of this technique is that the phage-coded prey proteins are essentially fusion proteins that might affect their binding with the bait protein. Lastly, phage display is an in vitro approach and thus lacks the advantages of in vivo PPI determination.

Lin et al. have employed phage display and a human brain cDNA library to identify the protein that interacts with kaposin A protein (Lin et al. 2007). Karkkainen et al. have also utilized phage display whereby they inserted genes coding for 296 human SH3 domains into an M13-derived phagemid vector. Using this approach, they objectively identified the SH3 domains that share interaction with proteins of interest (Kärkkäinen et al. 2006).

5.5.5 Cellulose-Bound Peptide Arrays

A peptide screening approach is one of the techniques that aid in the determination of functional sites in a protein that are involved in interaction with other proteins (Werther and Seitz 2008). Such methods therefore furnish a deep insight into known molecular interactions, by rapidly deciphering the chemical nature of these recognition events (Tong et al. 2002). Such information can then be extrapolated for the design of small peptides that can disrupt the interactions in vivo (Frank 2002). SPOT synthesis is one such technique that enables the simultaneous assembly of peptides on a membrane surface. The employed membrane is porous in nature and every peptide is generated on it at a particular spot or distinct site. The cellulose-bound peptide array is then probed for protein interactions in a process similar to Western blot (Frank 1992).

A typical SPOT experiment that aims to investigate protein interaction domains comprises three steps. In the first step, peptides (15- to 20-mer) are generated to identify the binding site. The peptides usually cover the entire sequence with an offset of around 5 amino acid residues. Therefore, a protein that is 1000 amino acid long usually yields peptide fragments, ranging from 200 to 350 in number. In case of limited capacity of SPOT experiments, researchers often narrow down the interaction site to a certain sub-domain utilizing other techniques. In the second step known as "sizing," the core domain responsible for interaction is identified. The binding site shortlisted in the previous mapping step is subjected to further characterization via employing several overlapping peptides. Such peptides have an offset of a single residue and have their sizes reduced stepwise. In the third step termed "analoging," the contribution of single amino acid chains is determined. Traditionally, this step was performed by "glycine or alanine walks" where only one residue in a peptide was exchanged with a glycine or alanine, respectively. However, full replacement studies have now found greater utility owing to yield more information with higher confidence scores. In full replacement methods, each amino acid comprising the core region is replaced by all other amino acids that are genetically encoded.

This technique is advantageous in several ways. First, it is cheap, rapid, and allows analysis on both solution and solid phases. Second, the replacement profiles of the amino acids in the binding site can aid in the determination of cross-reactivity with other proteins. This is especially important for immunological PPIs such as one between pathogenic protein and its cognate host antibody. Third, this method enables one to understand how post-translational modifications contribute to the interaction. This can be done by the synthesis of peptides that have identical sequences but varied modifications at their side chains (Mukhija et al. 1998). Fourth, this method identifies interactions that are conformationally defined. In case, the peptide scanning does not yield a linear epitope, branched double/triple peptides or mimotopes may be employed (Oggero et al. 2004). Fifth, the peptides being assembled on cellulose fibers on the membrane are displayed at high concentrations locally that in turn aid to capture low affinity interactors (Hoffmüller et al. 1999). This technique, however, is not free of limitations. Firstly, the peptide arrays cannot be re-used more than a few times. Secondly, there are drawbacks associated with decreasing the spot size below a threshold (1 mm). Thirdly, the process becomes arduous and costly when the situation demands more number of identical array copies. Fourth, this technique yields complex results when the core residues are not present in a contiguous stretch of sequence.

Oggero et al. employed peptide arrays to determine the epitopes of human granulocyte-macrophage colony-stimulating factor that are recognized by neutralizing and non-neutralizing antibodies (Oggero et al. 2004). Rudiger et al. screened 1633 peptides for their interaction with *E. coli* DnaJ protein for determining the protein's substrate specificity (Rüdiger et al. 2001).

5.5.6 Allele Libraries for PPI Analysis

The residues involved in PPIs may be identified by isolating "Interaction Defective Alleles" (IDAs) from high coverage allele libraries (Werther and Seitz 2008). The IDAs are developed by a method known as reverse two-hybrid, which is technically a variation of the traditional Y2H system (Vidal et al. 1996). In a reverse two-hybrid system known PPIs are weakened or eliminated.

The method starts with a PPI that activates the target reporter gene(s) in yeast followed by perturbing the system to attenuate the reporter(s). These perturbations are usually made via introducing mutations into one or both of the interacting proteins or via the introduction of inhibitory peptides or small molecules (Huang and Schreiber 1997). In the case of a small number of potential perturbations, the status of the reporter genes is usually characterized by replica plating. The greatest technical difficulty associated with this technique is that ORF mutagenesis often yields the stop codons that are present internally. This leads to the generation of truncated proteins within the allele library.

Amberg et al. utilized a set of 35 mutations in actin to map actin–profilin, actin–actin, actin–SH3 domain, and actin–Srv2p interactions. The authors performed the experiments via screening co-transformants for loss of activity in reporter (Amberg et al. 1995). The technique was also used to analyze the interaction between HLH region in Id1 and a bHLH domain in MyoD1. An allele library of the ORF of full-length MyoD1 was utilized in this study. The experiment concluded that most of the alleles failed to grow in the selected media. Many IDAs were isolated for multiple sequence alignment. The results confirmed that majority of the IDAs had mutations clustered in the bHLH region (Gray et al. 2007).

5.5.7 Gene Co-expression

A protein complex that comprises several subunits can only function properly if all of the constituting subunits are present in their correct stoichiometries. This implies that the gene expression profiles of these subunit proteins are regulated to relate them in the perspective of complex formation (Shoemaker and Panchenko 2007). Therefore, researchers can extract information on gene expression levels of proteins that are suspected to be involved in interactions among themselves. One can obtain data on gene expression from cell cycle-related experiments. The similarity in the expression profiles of proteins can then be quantified in the terms of correlation coefficient. The coefficient is calculated between the relative levels of expression or by normalizing the difference between absolute expression levels or via other methods (Kemmeren et al. 2002; Kim et al. 2001; Troyanskaya et al. 2002). Comparison of the quantity distributions of the suspected proteins with that of non-interacting/random pairs of proteins gives one better insight to determine protein interactions. Gene co-expression studies have elucidated the coevolution of the expression levels for pairs of interacting proteins. Therefore, research groups have shifted their attention to the coevolution of gene expression rather than coevolution at the amino acid level (Fraser et al. 2004). The interactions between genes can be interpreted by utilizing DNA microarray conjugated with synthetic lethality.

Jansen et al. utilized gene co-expression studies to study PPIs. They investigated the relationship between mRNA expression levels and PPIs to conclude that subunit components of a given protein complex have similar absolute mRNA levels as well as correlated expression patterns over a time period (Jansen et al. 2002). Permanent protein complexes such as the ribosome and the proteasome were shown to demonstrate the most obvious patterns of co-expression (Jansen et al. 2002). Grigoriev et al. statistically analyzed large-scale data on gene expression and protein interaction in

T7 bacteriophage and yeast and demonstrated that proteins encoded by co-expressed genes are more likely to be involved in interaction (Grigoriev 2001).

5.6 Other Techniques

There are several other techniques that are available apart from the techniques described under biophysical, biochemical, and genetic approaches. These techniques combine several of the features described above, thus providing crucial information about PPIs. These techniques are described in the following sections.

5.6.1 Co-localization

Co-localization is based on the principle of fluorescently labeled proteins and then imaging/visualizing them via fluorescence microscopy to know their location inside cells. If they are co-localized, there is possibility of interactions between the proteins (Bolte and Cordelieres 2006). The biological interaction observed in the cell is then exploited to understand the functional context/cellular function of the interaction. In dual-color images, the co-localization of two components is usually indicated by correlating the pixel intensities. If the co-localized signals yield a high-level coefficient, it is probable that the two target proteins are in proximity to share interactions with each other (Wu et al. 2010).

Several studies have detected spatiotemporal relationships of proteins to infer their interactions and also quantified the overlaps to measure the correlation with respect to space (Costes et al. 2004). The most popular methods indicate interaction by merging the dual-color images and by quantifying the pixels that are overlaid. With the progress in technology such as development in microscopy and computation, co-localization analyses have become quick, efficient, and convenient method for the quantification of PPIs.

Several approaches such as Mander's correlation coefficient, Pearson's correlation coefficient, intensity correlation quotient, and ranked correlations such as Kendall's tau, Spearman's rho, and automatic thresholding method have been used in PPI-based studies (Cui et al. 2019). The co-localization correlation analyses performed with fluorescence microscopy supply quantifiable information with more accurate and refined parameters compared to other traditional techniques (Lagache et al. 2018). The spatial resolution of the images can however vary significantly (Helmuth et al. 2010).

There are three major limitations associated with this technique. First, the spatial co-relation analyses can differentiate between the proteins that have direct interaction and the proteins that form a macromolecular complex. Second, the light microscope suffers limitations with regard to its spatial resolution. This is limited by two factors: the numerical aperture of the objective lens and the wavelength of light. Third, the proximity between two protein molecules usually falls in the nanometer range, which is narrower than visible light's wavelength. Consequently, these proteins appear in the same pixel of the image. This leads to the requirement of high-resolution imaging.

In a fluorescent two-hybrid (F2H) assay, a bait protein that is fluorescently labeled is immobilized. On interaction of the bait with the prey containing a different label, fluorescent signals are co-localized. This, in turn, aids one to visualize and characterize PPIs in real-time from different subcellular compartments in the cell (nucleus, mitochondria, cytoplasm, etc.) (Zhou et al. 2016). Utilizing this technique, Zolghadr et al. studied the interaction between two DNA repair proteins XRCC1 and DNA Ligase III (Zolghadr et al. 2008). Yurlova et al. employed this method to study the interactions between p53 protein and its negative protein regulators, Mdm4 and Mdm2 (Yurlova et al. 2014).

5.6.2 Quantitative Immunoprecipitation Combined with Knockdown (QUICK)

Quantitative immunoprecipitation combined with knockdown (QUICK) is a modern day technique for screening PPIs. QUICK itself comprises the following methods: stable isotope labeling with amino acids in cell culture (SILAC), RNA interference (RNAi), co-immunoprecipitation, and quantitative mass-spectrometry analysis. SILAC uses labeling of proteomes via the incorporation of either a heavy or a normal isotope-substituted amino acid like ^{13}C-labeled lysine (Ong et al. 2002). Owing to the mass difference of the peptides from the heavy isotope and normal isotope labeled samples, a mass spectrometer can be used to differentiate them. The abundance ratio of the corresponding proteins can be accurately determined from the signal intensity ratios of such peptide pairs.

RNAi is a new method that can lead to specific depletion of proteins from living cells (Gunsalus and Piano 2005). QUICK utilizes the SILAC method to metabolically label cells by growing them in medium consisting of either heavy or light stable isotopes of arginine and lysine. Then, it uses RNAi to knock down the protein of interest in one of the two samples. An immobilized antibody is used against this target protein to incubate cell lysates. Mass spectrometry is then used to combine, elute, digest, and analyze the precipitated proteins. There should be greater abundance of the target protein and its interacting partners in the heavy form rather than in the corresponding light form. However, both the light and heavy forms should contain equal amounts of contaminating proteins.

QUICK becomes efficient as it depletes the target enough to return considerable peptide ratios rather than eliminating the target completely. Thus, no false positive results are obtained via this method. However, not all reported interaction partners can be identified by QUICK (Selbach and Mann 2006).

Selbach et al. used QUICK to determine interaction partners of Cbl 9 and β-catenin. The authors utilized a colon carcinoma cell line carrying an inducible hairpin RNA vector (shRNA) to induce β-catenin knockout (Selbach and Mann 2006). Following SILAC labeling, they induced knockdown in the "light cells," but kept the "heavy cells" untreated. A knockdown in the heavy cells was done for another control. After a stipulated time period, the cells were harvested, lysed, and treated with anti-β-catenin antibodies coupled to protein G agarose. The precipitated proteins from the light and heavy cells were combined and subsequently digested. The peptides were then analyzed using LC-MS/MS. As per the expectations, the authors identified β-catenin as a main component of the precipitates, being over eight folds more abundant in the heavy counterpart. The proteins that had abundance ratios of around one were deemed to be non-specific since they were subject to precipitation independent of β-catenin presence. On the other hand, the proteins that had significantly enhanced abundance ratios were denoted to be well-known β-catenin interactors. Ge et al. identified the interacting proteins of 14-3-3ζ using QUICK. Authors have discovered a total of 95 novel interactions by using this method (Ge et al. 2010).

5.6.3 Flow-Induced Dispersion Analysis (FIDA)

Flow-induced dispersion analysis (FIDA) is used to characterize and quantify proteins under natural conditions. FIDA measures the size change of a ligand that occurs on interaction with target proteins. The hydrodynamic radius (size) of an unbound ligand is relatively small under natural conditions but when there is an analyte present, it binds to the analyte leading to an increase in its size. By performing a titration experiment, the K_d of the interaction may be assessed. Analyte concentration is determined by measuring the apparent size of the ligand in an unknown sample. Taylor dispersion analysis (TDA) is performed in capillaries made of fused silica to measure the apparent molecular size. Being an assay that analyzes ligand binding, FIDA has some features which are similar to that of SPR, enzyme-linked immunosorbent assay (ELISA), and biolayer interferometry (BLI) techniques. (Pedersen et al. 2019).

FIDA has certain advantages that prompt research groups to use the technique for characterization of non-covalent interactions that play major roles in PPIs (Jensen and Ostergaard 2010). FIDA is a straightforward assay as it probes a single association event in-solution. The in-built quality control of the assay is possible owing to the absolute measurement of size. Further, complications due to non-specific adsorption of analyte are reduced. This aids in direct measurement in samples like serum and plasma (Pedersen et al. 2019).

The technique also does not delve into surface chemistry. Moreover, it does not rely on secondary antibody interaction and can be easily automated and can be carried out in a shorter time than other prevalent techniques like ELISA. In this method separation of the affinity complex is not essential. This technique can be used for any protein analyte for which there is an existing affinity ligand that has high binding affinity and specificity. The FIDA assay is highly sensitive since it relies on the dissociation constant of the indicator–analyte interaction that often lies in the pico-molar or sub-nanomolar range (Poulsen et al. 2015).

Poulsen et al. injected fluorescently tagged HSA (F-HSA) into a capillary containing varying concentrations of anti-HAS and performed FIDA. The authors analyzed the binding curve and observed two binding events. One of them indicated the interaction corresponding to a small fraction of HAS for concentrations of anti-HAS below 1 nm. On the other hand, the major binding event was observed for concentrations that surpassed 10 nM. The authors reasoned that the heterogenic binding was due to the covalent attachment of fluorescence tags that interfered with the process of epitope recognition. The dissociation constant of this PPI was calculated to be 1×10^{-7} M (Poulsen et al. 2015).

5.7 Conclusions

The majority of the PPI pairs of the human interactome are yet to be explored. Therefore, research has focused on revealing the spatial as well as temporal relationships between interacting proteins. PPI determination studies have gained importance because they divulge information about the cellular processes and also assist research groups to identify better targets for drug design. Biophysical methods such as ITC and AUC, that are based on traditional physical principles, are well-suited for furnishing data on the thermodynamic and hydrodynamic related aspects of PPIs. Biochemical methods such as affinity chromatography have also found wide utility in identifying the binding partners of a target protein. Genomic revolution during the past decade has aided in constructing ORF and cDNA collections at the genome scale. Therefore, genetic methods such as phage display, Y2H, and microarrays have proven to be helpful in PPI network mapping on a large-scale. It should however be noted that all the mentioned techniques have their own drawbacks that can often lead to either false negatives or positives. Therefore, for greater reliability, it is vital to validate results using different techniques. For instance, if a potential PPI is discovered by genetic screens, it should be confirmed using biophysical methods or biochemical assays. Finally, it is essential to note that the wealth of data generated by experimental methods facilitates the need for bioinformatics tools and databases that can analyze raw data and store processed data, respectively.

References

Ackers GK (1970) Analytical gel chromatography of proteins. In: Advances in protein chemistry, vol 24. Elsevier, London, pp 343-446

Aebersold R, Mann M. (2003). Mass spectrometry-based proteomics. Nature, 422, 198-207.

Agarwal N, Nair MS, Mazumder A, Poluri KM (2018) Characterization of nanomaterials using nuclear magnetic resonance spectroscopy. In: Characterization of nanomaterials. Elsevier, London, pp 61-102

Amberg DC, Basart E, Botstein D. (1995). Defining protein interactions with yeast actin in vivo. Nat Struct Biol, 2, 28-35.

Arbel N, Ben-Hail D, Shoshan-Barmatz V. (2012). Mediation of the antiapoptotic activity of Bcl-xL protein upon interaction with VDAC1 protein. J Biol Chem, 287, 23152-23161.

Arkin MR, Glicksman MA, Fu H, Havel JJ, Du Y. (2012). Inhibition of protein-protein interactions: non-cellular assay formats. Assay guidance manual. Bethesda: Eli Lilly

Attri AK, Minton AP. (2005). Composition gradient static light scattering: a new technique for rapid detection and quantitative characterization of reversible macromolecular hetero-associations in solution. Anal Biochem, 346, 132-138.

Attri AK, Fernandez C, Minton AP. (2010). Self-association of Zn-insulin at neutral pH: investigation by concentration gradient-static and dynamic light scattering. Biophys Chem, 148, 23-27.

Bacart J, Corbel C, Jockers R, Bach S, Couturier C. (2008). The BRET technology and its application to screening assays. Biotechnol J, 3, 311-324.

Bai Y (2015) Detecting protein-protein interactions by gel filtration chromatography. In: Protein-protein interactions. Springer, New York, pp 223-232

Baker BM, Murphy KP. (1997). Dissecting the energetics of a protein-protein interaction: the binding of ovomucoid third domain to elastase. J Mol Biol, 268, 557-569.

Balbo A, Schuck P. (2005). Analytical ultracentrifugation in the study of protein self-association and heterogeneous protein-protein interactions. Protein-Protein Interact 14, 253-277.

Balhara V, Deshmukh SS, Kalman L, Kornblatt JA. (2014). The interaction of streptococcal enolase with canine plasminogen: the role of surfaces in complex formation. PLoS One, 9, e88395.

Barrios-Rodiles M, Ellis JD, Blencowe BJ, Wrana JL (2017) LUMIER: a discovery tool for mammalian protein interaction networks. In: Proteomics. Springer, New York, pp 137-148

Barron LD. (2015). The development of biomolecular Raman optical activity spectroscopy. Biomed Spectrosc Imaging, 4, 223-253.

Belanger KD. (2009). Using affinity chromatography to investigate novel protein-protein interactions in an undergraduate cell and molecular biology lab course. CBE Life Sci Educ, 8, 214-225.

Bender A, Pringle JR. (1991). Use of a screen for synthetic lethal and multicopy suppressee mutants to identify two new genes involved in morphogenesis in Saccharomyces cerevisiae. Mol Cell Biol, 11, 1295-1305.

Bhagwat S, Kumar A. (2018). Biolayer interferometry and its applications. J Mol Biol Technol, 2, 106.

Biehle SJ, Carrozzella J, Shukla R, Popplewell J, Swann M, Freeman N, Clark JF. (2004). Apolipoprotein E isoprotein-specific interactions with tissue plasminogen activator. Biochim Biophys Acta, 1689, 244-251.

Bleil JD, Wassarman PM. (1990). Identification of a ZP3-binding protein on acrosome-intact mouse sperm by photoaffinity crosslinking. Proc Natl Acad Sci U S A, 87, 5563-5567.

Bolte S, Cordelieres FP. (2006). A guided tour into subcellular colocalization analysis in light microscopy. J Microsc, 224, 213-232.

Bothner B, Lewis WS, DiGiammarino EL, Weber JD, Bothner SJ, Kriwacki RW. (2001). Defining the molecular basis of Arf and Hdm2 interactions. J Mol Biol, 314, 263-277.

Boudjemline A, Saridakis E, Swann MJ, Govada L, Mavridis IM, Chayen NE. (2011). Use of dual polarization interferometry as a diagnostic tool for protein crystallization. Anal Chem, 83, 7881-7887.

Buntru A, Trepte P, Klockmeier K, Schnoegl S, Wanker EE. (2016). Current approaches toward quantitative mapping of the interactome. Front Genet, 7, 74.

Burger VM, Arenas DJ, Stultz CM. (2016). A structure-free method for quantifying conformational flexibility in proteins. Sci Rep, 6, 29040.

Caberoy NB, Zhou YX, Li W. (2010). Tubby and tubby-like protein 1 are new MerTK ligands for phagocytosis. EMBO J, 29, 3898-3910.

Cafarelli TM, Desbuleux A, Wang Y, Choi SG, De Ridder D, Vidal M. (2017). Mapping, modeling, and characterization of protein-protein interactions on a proteomic scale. Curr Opin Struct Biol, 44, 201-210.

Calakos N, Bennett MK, Peterson KE, Scheller RH. (1994). Protein-protein interactions contributing to the specificity of intracellular vesicular trafficking. Science, 263, 1146-1149.

Chan CS, Howell JM, Workentine ML, Turner RJ. (2006). Twin-arginine translocase may have a role in the chaperone function of NarJ from Escherichia coli. Biochem Biophy Res, 343, 244-251.

Chan CS, Winstone TM, Turner RJ (2008) Investigating protein–protein interactions by far-Westerns. In: Protein–protein interaction. Springer, New York, pp 195-214
Chapman JR (2000) Mass spectrometry of proteins and peptides, vol 146. Springer Science & Business Media, Cham
Chen PC, Hennig J. (2018). The role of small-angle scattering in structure-based screening applications. Biophys Rev, 10, 1295-1310.
Chen K, Tjandra N (2011) The use of residual dipolar coupling in studying proteins by NMR. In: NMR of proteins and small biomolecules. Springer, New York, pp 47-67
Chen Y, Ebright YW, Ebright RH. (1994). Identification of the target of a transcription activator protein by protein-protein photocrosslinking. Science, 265, 90-92.
Chen Y et al. (2013). Bcl2-associated athanogene 3 interactome analysis reveals a new role in modulating proteasome activity. MCP, 12, 2804-2819.
Chen J et al. (2014). Single-molecule dynamics of enhanceosome assembly in embryonic stem cells. Cell, 156, 1274-1285.
Chevray PM, Nathans D. (1992). Protein-interaction cloning in yeast - identification of mammalian proteins that react with the leucine zipper of jun. Proc Natl Acad Sci U S A, 89, 5789-5793.
Clore GM, Tang C, Iwahara J. (2007). Elucidating transient macromolecular interactions using paramagnetic relaxation enhancement. Curr Opin Struct Biol, 17, 603-616.
Cohen BD, Lowy DR, Schiller JT. (1993). The conserved C-terminal domain of the bovine papillomavirus E5 oncoprotein can associate with an alpha-adaptin-like molecule - a possible link between growth-factor receptors and viral transformation. Mol Cell Biol, 13, 6462-6468.
Cole JL, Lary JW, Laue TM. (2008). Analytical ultracentrifugation: sedimentation velocity and sedimentation equilibrium. Methods Cell Biol, 84, 143-179.
Cordeiro TN et al. (2017). Disentangling polydispersity in the PCNA– p15PAF complex, a disordered, transient and multivalent macromolecular assembly. Nucleic Acids Res, 45, 1501-1515.
Costa TR, Ignatiou A, Orlova EV (2017) Structural analysis of protein complexes by cryo electron microscopy. In: Bacterial protein secretion systems. Springer, New York, pp 377-413
Costes SV, Daelemans D, Cho EH, Dobbin Z, Pavlakis G, Lockett S. (2004). Automatic and quantitative measurement of protein-protein colocalization in live cells. Biophys J, 86, 3993-4003.
Cowieson NP, Miles AJ, Robin G, Forwood JK, Kobe B, Martin JL, Wallace BA. (2008). Evaluating protein : protein complex formation using synchrotron radiation circular dichroism spectroscopy. Proteins, 70, 1142-1146.
Cui Y, Zhang X, Yu M, Zhu Y, Xing J, Lin J. (2019). Techniques for detecting protein-protein interactions in living cells: principles, limitations, and recent progress. Science China. Life Sci, 62, 619-632.
D'Silva L, Ozdowy P, Krajewski M, Rothweiler U, Singh M, Holak TA. (2005). Monitoring the effects of antagonists on protein– protein interactions with NMR spectroscopy. J Am Chem Soc, 127, 13220-13226.
Demir F et al. (2013). Arabidopsis nanodomain-delimited ABA signaling pathway regulates the anion channel SLAH3. Proc Natl Acad Sci U S A, 110, 8296-8301.
Dhalluin C et al. (2000). Structural basis of SNT PTB domain interactions with distinct neurotrophic receptors. Mol Cell, 6, 921-929.
Drescher DG, Drescher MJ, Ramakrishnan NA (2009) Surface plasmon resonance (SPR) analysis of binding interactions of proteins in inner-ear sensory epithelia. In: Auditory and vestibular research. Springer, New York, pp 323-343
Du Y (2015) Fluorescence polarization assay to quantify protein-protein interactions in an HTS format. In: Protein-protein interactions. Springer, New York, pp 529-544
Dunnwald M, Varshavsky A, Johnsson N. (1999). Detection of transient in vivo interactions between substrate and transporter during protein translocation into the endoplasmic reticulum. Mol Biol Cell, 10, 329-344.

Durfee T et al. (1993). The retinoblastoma protein associates with the protein phosphatase type 1 catalytic subunit. Genes Dev, 7, 555-569.

Escorihuela J, Gonzalez-Martinez MA, Lopez-Paz JL, Puchades R, Maquieira A, Gimenez-Romero D. (2015). Dual-polarization interferometry: a novel technique to light up the nanomolecular world. Chem Rev, 115, 265-294.

Formosa T, Barry J, Alberts BM, Greenblatt J (1991) Using protein affinity chromatography to probe structure of protein machines. In: Methods in enzymology, vol 208. Elsevier, London, pp 24-45

Frank R. (1992). Spot-synthesis - an easy technique for the positionally addressable, parallel chemical synthesis on a membrane support. Tetrahedron, 48, 9217-9232.

Frank R. (2002). High-density synthetic peptide microarrays: emerging tools for functional genomics and proteomics. Comb Chem, 5, 429-440.

Fraser HB, Hirsh AE, Wall DP, Eisen MB. (2004). Coevolution of gene expression among interacting proteins. Proc Natl Acad Sci U S A, 101, 9033-9038.

Fredriksson S et al. (2002). Protein detection using proximity-dependent DNA ligation assays. Nat Biotechnol, 20, 473-477.

Fronzes R, Schafer E, Wang L, Saibil HR, Orlova EV, Waksman G. (2009). Structure of a type IV secretion system core complex. Science, 323, 266-268.

Fujii Y, Yoshimura A, Kodama Y. (2018). A novel orange-colored bimolecular fluorescence complementation (BiFC) assay using monomeric Kusabira-Orange protein. Biotechniques, 64, 153-161.

Fujikawa Y, Kato N. (2007). Technical advance: split luciferase complementation assay to study protein–protein interactions in Arabidopsis protoplasts. Plant J, 52, 185-195.

Gadkari RA, Srinivasan N. (2010). Prediction of protein-protein interactions in dengue virus coat proteins guided by low resolution cryoEM structures. BMC Struct Biol, 10, 17.

Gangele K, Gulati K, Joshi N, Kumar D, Poluri KM. (2020). Molecular insights into the differential structure-dynamics-stability features of interleukin-8 orthologs: Implications to functional specificity. Int J Biol Macromol, 164, 3221-3234.

Gao G, Williams JG, Campbell SL (2004) Protein-protein interaction analysis by nuclear magnetic resonance spectroscopy. In: Protein-protein interactions. Springer, New York, pp 79-91

Garrett DS, Seok YJ, Peterkofsky A, Clore GM, Gronenborn AM. (1997). Identification by NMR of the binding surface for the histidine-containing phosphocarrier protein HPr on the N-terminal domain of enzyme I of the Escherichia coli phosphotransferase system. Biochemistry, 36, 4393-4398.

Garrett DS, Seok YJ, Peterkofsky A, Gronenborn AM, Clore GM. (1999). Solution structure of the 40,000 M-r phosphoryl transfer complex between the N-terminal domain of enzyme I and HPr. Nat Struct Biol, 6, 166-173.

Ge F, Li W-L, Bi L-J, Tao S-C, Zhang Z-P, Zhang X-E. (2010). Identification of novel 14-3-3ζ interacting proteins by quantitative immunoprecipitation combined with knockdown (QUICK). J Proteome Res, 9, 5848-5858.

Gegner JA, Dahlquist FW. (1991). Signal transduction in bacteria: CheW forms a reversible complex with the protein kinase CheA. Proc Natl Acad Sci U S A, 88, 750-754.

Gill SC, Weitzel SE, von Hippel PH. (1991). Escherichia coli σ70 and NusA proteins: I. Binding interactions with core RNA polymerase in solution and within the transcription complex. J Mol Biol, 220, 307-324.

Glatter O, Kratky O (1982) Small angle X-ray scattering. Academic, New York

Gray PN, Busser KJ, Chappell TG. (2007). A novel approach for generating full-length, high coverage allele libraries for the analysis of protein interactions. Mol Cell Proteomics, 6, 514-526.

Greene MK, Maskos K, Landry SJ. (1998). Role of the J-domain in the cooperation of Hsp40 with Hsp70. Proc Natl Acad Sci U S A, 95, 6108-6113.

Greenfield NJ. (2006). Using circular dichroism spectra to estimate protein secondary structure. Nat Protoc, 1, 2876-2890.

Greenfield NJ, Fowler VM. (2002). Tropomyosin requires an intact N-terminal coiled coil to interact with tropomodulin. Biophys J, 82, 2580-2591.

Grigoriev A. (2001). A relationship between gene expression and protein interactions on the proteome scale: analysis of the bacteriophage T7 and the yeast Saccharomyces cerevisiae. Nucleic Acids Res, 29, 3513-3519.

Groemping Y, Hellmann N. (2005). Spectroscopic methods for the determination of protein interactions. Curr Protoc Protein Sci 20: 28.

Gulati K, Gangele K, Agarwal N, Jamsandekar M, Kumar D, Poluri KM. (2018). Molecular cloning and biophysical characterization of CXCL3 chemokine. Int J Biol Macromol, 107, 575-584.

Gulati K, Gangele K, Kumar D, Poluri KM. (2019). An inter-switch between hydrophobic and charged amino acids generated druggable small molecule binding pocket in chemokine paralog CXCL3. Arch Biochem Biophys, 662, 121-128.

Gunsalus KC, Piano F. (2005). RNAi as a tool to study cell biology: building the genome–phenome bridge. Curr Opin Cell Biol, 17, 3-8.

Hall DA, Ptacek J, Snyder M. (2007). Protein microarray technology. Mech Ageing Dev, 128, 161-167.

Hamdi A, Colas P. (2012). Yeast two-hybrid methods and their applications in drug discovery. Trends Pharmacol Sci, 33, 109-118.

Hanlon AD, Larkin MI, Reddick RM. (2010). Free-solution, label-free protein-protein interactions characterized by dynamic light scattering. Biophys J, 98, 297-304.

Hansma PK, Elings VB, Marti O, Bracker CE. (1988). Scanning tunneling microscopy and atomic force microscopy: application to biology and technology. Science, 242, 209-216.

Harding SE, Rowe AJ (2010) Insight into protein–protein interactions from analytical ultracentrifugation. Portland Press Ltd., London

Helmuth JA, Paul G, Sbalzarini IF. (2010). Beyond co-localization: inferring spatial interactions between sub-cellular structures from microscopy images. BMC Bioinf, 11, 372.

Herberg FW, Taylor SS. (1993). Physiological inhibitors of the catalytic subunit of cAMP-dependent protein kinase: effect of magnesium-ATP on protein-protein interactions. Biochemistry, 32, 14015-14022.

Hertveldt K, Beliën T, Volckaert G (2009) General M13 phage display: M13 phage display in identification and characterization of protein–protein interactions. In: Bacteriophages. Springer, New York, pp 321-339

Hoffmüller U et al. (1999). Interaction of a PDZ protein domain with a synthetic library of all human protein C termini. Angew Chem Int Ed, 38, 2000-2004.

Homola J. (2003). Present and future of surface plasmon resonance biosensors. Anal Bioanal Chem, 377, 528-539.

Honda S, Kobayashi N, Munekata E, Uedaira H. (1999). Fragment reconstitution of a small protein: folding energetics of the reconstituted immunoglobulin binding domain B1 of streptococcal protein G. Biochemistry, 38, 1203-1213.

Huang J, Schreiber SL. (1997). A yeast genetic system for selecting small molecule inhibitors of protein-protein interactions in nanodroplets. Proc Natl Acad Sci U S A, 94, 13396-13401.

Huffaker TC, Hoyt MA, Botstein D. (1987). Genetic analysis of the yeast cytoskeleton. Annu Rev Genet, 21, 259-284.

Hummel JP, Dreyer WJ. (1962). Measurement of protein-binding phenomena by gel filtration. Biochim Biophys Acta, 63, 530-532.

Ivanov YD, Frantsuzov PA, Zollner A, Medvedeva NV, Archakov AI, Reinle W, Bernhardt R. (2011). Atomic force microscopy study of protein-protein interactions in the cytochrome CYP11A1 (P450scc)-containing steroid hydroxylase system. Nanoscale Res Lett, 6, 54.

Jaiswal N, Raikwal N, Pandey H, Agarwal N, Arora A, Poluri KM, Kumar D. (2018). NMR elucidation of monomer–dimer transition and conformational heterogeneity in histone-like DNA binding protein of Helicobacter pylori. Magn Reson Chem, 56, 285-299.

Jaiswal N et al. (2019). Molecular interaction between human SUMO-I and histone like DNA binding protein of Helicobacter pylori (Hup) investigated by NMR and other biophysical tools. Int J Biol Macromol, 123, 446-456.

Jansen R, Greenbaum D, Gerstein M. (2002). Relating whole-genome expression data with protein-protein interactions. Genome Res, 12, 37-46.

Jensen H, Ostergaard J. (2010). Flow induced dispersion analysis quantifies noncovalent interactions in nanoliter samples. J Am Chem Soc, 132, 4070-4071.

Jensen MR, Ortega-Roldan JL, Salmon L, van Nuland N, Blackledge M. (2011). Characterizing weak protein-protein complexes by NMR residual dipolar couplings. Eur Biophys J Biophy, 40, 1371-1381.

Jerabek-Willemsen M, Wienken CJ, Braun D, Baaske P, Duhr S. (2011). Molecular interaction studies using microscale thermophoresis. Assay Drug Dev Technol, 9, 342-353.

Jerabek-Willemsen M, Andre T, Wanner R, Roth HM, Duhr S, Baaske P, Breitsprecher D. (2014). MicroScale thermophoresis: interaction analysis and beyond. J Mol Struct, 1077, 101-113.

Jonsson U et al. (1991). Real-time biospecific interaction analysis using surface-plasmon resonance and a sensor chip technology. Biotechniques, 11, 620.

Joseph C et al. (2001). A structural characterization of the interactions between titin Z-repeats and the α-actinin C-terminal domain. Biochemistry, 40, 4957-4965.

Joshi N, Kumar D, Poluri KM. (2020a). Elucidating the molecular interactions of chemokine CCL2 orthologs with flavonoid baicalin. ACS Omega, 5, 22637-22651.

Joshi N, Nagar N, Gulati K, Gangele K, Mishra A, Kumar D, Poluri KM. (2020b). Dissecting the differential structural and dynamics features of CCL2 chemokine orthologs. Int J Biol Macromol, 156, 239-251.

Kaelin Jr WG, Pallas DC, DeCaprio JA, Kaye FJ, Livingston DM. (1991). Identification of cellular proteins that can interact specifically with the T/ElA-binding region of the retinoblastoma gene product. Cell, 64, 521-532.

Kapoor N, Gupta R, Menon ST, Folta-Stogniew E, Raleigh DP, Sakmar TP. (2010). Nucleobindin 1 is a calcium-regulated guanine nucleotide dissociation inhibitor of Gαi1. J Biol Chem, 285, 31647-31660.

Kärkkäinen S et al. (2006). Identification of preferred protein interactions by phage-display of the human Src homology-3 proteome. EMBO Rep, 7, 186-191.

Karlsson R, Michaelsson A, Mattsson L. (1991). Kinetic analysis of monoclonal antibody-antigen interactions with a new biosensor based analytical system. J Immunol Methods, 145, 229-240.

Keegan L, Gill G, Ptashne M. (1986). Separation of DNA binding from the transcription-activating function of a eukaryotic regulatory protein. Science, 231, 699-704.

Kemmeren P, van Berkum NL, Vilo J, Bijma T, Donders R, Brazma A, Holstege FC. (2002). Protein interaction verification and functional annotation by integrated analysis of genome-scale data. Mol Cell, 9, 1133-1143.

Kengne-Momo R, Daniel P, Lagarde F, Jeyachandran Y, Pilard J, Durand-Thouand M, Thouand G. (2012). Protein interactions investigated by the Raman spectroscopy for biosensor applications. Int J Spectrosc, 2012, 1702-1713.

Kim YT, Tabor S, Churchich JE, Richardson CC. (1992). Interactions of gene 2.5 protein and DNA polymerase of bacteriophage T7. J Biol Chem, 267, 15032-15040.

Kim SK et al. (2001). A gene expression map for Caenorhabditis elegans. Science, 293, 2087-2092.

Knapp S et al. (2001). Thermodynamics of the high-affinity interaction of TCF4 with β-catenin. J Mol Biol, 306, 1179-1189.

Kobe B et al. (2008). Crystallography and protein-protein interactions: biological interfaces and crystal contacts. Biochem Soc Trans, 36, 1438-1441.

Kodama Y, Hu CD. (2012). Bimolecular fluorescence complementation (BiFC): a 5-year update and future perspectives. Biotechniques, 53, 285-298.

Krachler AM, Sharma A, Cauldwell A, Papadakos G, Kleanthous C. (2010). TolA modulates the oligomeric status of YbgF in the bacterial periplasm. J Mol Biol, 403, 270-285.

Krogan NJ et al. (2006). Global landscape of protein complexes in the yeast Saccharomyces cerevisiae. Nature, 440, 637-643.
Lagache T et al. (2018). Mapping molecular assemblies with fluorescence microscopy and object-based spatial statistics. Nat Commun, 9, 698.
Lam YA, Lawson TG, Velayutham M, Zweier JL, Pickart CM. (2002). A proteasomal ATPase subunit recognizes the polyubiquitin degradation signal. Nature, 416, 763-767.
Laue T. (2004). Analytical ultracentrifugation: a powerful 'new' technology in drug discovery. Drug Discov Today, 1, 309-315.
Lea WA, Simeonov A. (2011). Fluorescence polarization assays in small molecule screening. Expert Opin Drug Discovery, 6, 17-32.
Leavitt S, Freire E. (2001). Direct measurement of protein binding energetics by isothermal titration calorimetry. Curr Opin Struct Biol, 11, 560-566.
Lebowitz J, Lewis MS, Schuck P. (2002). Modern analytical ultracentrifugation in protein science: a tutorial review. Protein Sci, 11, 2067-2079.
Leuchowius KJ, Weibrecht I, Landegren U, Gedda L, Soderberg O. (2009). Flow cytometric in situ proximity ligation analyses of protein interactions and post-translational modification of the epidermal growth factor receptor family. Cytom Part A, 75A, 833-839.
Li TS, Talvenheimo J, Zeni L, Rosenfeld R, Stearns G, Arakawa T. (2002). Changes in protein conformation and dynamics upon complex formation of brain-derived neurotrophic factor and its receptor: Investigation by isotope-edited Fourier transform IR spectroscopy. Biopolymers, 67, 10-19.
Li S et al. (2004). A map of the interactome network of the metazoan C. elegans. Science, 303, 540-543.
Lin CW, Tu PF, Hsiao NW, Chang CY, Wan L, Lin YT, Chang HW. (2007). Identification of a novel septin 4 protein binding to human herpesvirus 8 kaposin A protein using a phage display cDNA library. J Virol Methods, 143, 65-72.
Lin CC et al. (2012). Inhibition of basal FGF receptor signaling by dimeric Grb2. Cell, 149, 1514-1524.
Lin JS, Ma LS, Lai EM. (2013). Systematic dissection of the agrobacterium type VI secretion system reveals machinery and secreted components for subcomplex formation. PLoS One, 8, e67647.
Liu Z et al. (2004). Crystal structure of spinach major light-harvesting complex at 2.72 Å resolution. Nat Commun, 428, 287-292.
Liu B, Archer CT, Burdine L, Gillette TG, Kodadek T. (2007). Label transfer chemistry for the characterization of protein-protein interactions. J Am Chem Soc, 129, 12348.
Liu Z, Gong Z, Dong X, Tang C. (2016). Transient protein-protein interactions visualized by solution NMR. BBA Proteins Proteom, 1864, 115-122.
Mahajan NP, Linder K, Berry G, Gordon GW, Heim R, Herman B. (1998). Bcl-2 and Bax interactions in mitochondria probed with green fluorescent protein and fluorescence resonance energy transfer. Nat Biotechnol, 16, 547-552.
Maldonado AY, Burz DS, Shekhtman A. (2011). In-cell NMR spectroscopy. Prog Nucl Magn Reson Spectrosc, 59, 197-212.
Martos A, Alfonso C, Lopez-Navajas P, Ahijado-Guzman R, Mingorance J, Minton AP, Rivas G. (2010). Characterization of self-association and heteroassociation of bacterial cell division proteins FtsZ and ZipA in solution by composition gradient-static light scattering. Biochemistry, 49, 10780-10787.
McGaughey KM, Wheeler LJ, Moore JT, Maley GF, Maley F, Mathews CK. (1996). Protein-protein interactions involving T4 phage-coded deoxycytidylate deaminase and thymidylate synthase. J Biol Chem, 271, 23037-23042.
McPhie P. (2000). Principles of fluorescence spectroscopy, Anal Biochem, 2, 353 354.
Miao YM, Cross TA. (2013). Solid state NMR and protein-protein interactions in membranes. Curr Opin Struct Biol, 23, 919-928.

Michnick SW, Ear PH, Landry C, Malleshaiah MK, Messier V (2011) Protein-fragment complementation assays for large-scale analysis, functional dissection and dynamic studies of protein–protein interactions in living cells. In: Signal transduction protocols. Springer, New York, pp 395-425

Mills JS, Walsh MP, Nemcek K, Johnson JD. (1988). Biologically-active fluorescent derivatives of spinach calmodulin that report calmodulin target protein-binding. Biochemistry, 27, 991-996.

Miura K. (2018). An overview of current methods to confirm protein-protein interactions. Protein Pept Lett, 25, 728-733.

Miura K, Miki H, Shimazaki K, Kawai N, Takenawa T. (1996). Interaction of Ash/Grb-2 via its SH3 domains with neuron-specific p150 and p65. Biochem J, 316 (Pt 2), 639-645.

Moore JM. (1999). NMR techniques for characterization of ligand binding: utility for lead generation and optimization in drug discovery. Biopolymers, 51, 221-243.

Morris VJ, Kirby AR, Gunning AP (1999) Atomic force microscopy for biologists. World Scientific, Singapore

Moscetti I, Cannistraro S, Bizzarri AR. (2017). Surface plasmon resonance sensing of biorecognition interactions within the tumor suppressor p53 network. Sensors, 17, 2680.

Mukhija S, Germeroth L, Schneider-Mergener J, Erni B. (1998). Identification of peptides inhibiting enzyme I of the bacterial phosphotransferase system using combinatorial cellulose-bound peptide libraries. Eur J Biochem, 254, 433-438.

Murphy RM. (1997). Static and dynamic light scattering of biological macromolecules: what can we learn? Curr Opin Biotechnol, 8, 25-30.

Nealon JO, Philomina LS, McGuffin LJ. (2017). Predictive and experimental approaches for elucidating protein-protein interactions and quaternary structures. Int J Mol Sci, 18, 2623.

Nguyen HH, Park J, Kang S, Kim M. (2015). Surface plasmon resonance: a versatile technique for biosensor applications. Sensors, 15, 10481-10510.

O'Connell MR, Gamsjaeger R, Mackay JP. (2009). The structural analysis of protein-protein interactions by NMR spectroscopy. Proteomics, 9, 5224-5232.

Oeffinger M. (2012). Two steps forward—one step back: advances in affinity purification mass spectrometry of macromolecular complexes. J Proteome, 12, 1591-1608.

Oesterreicher S, Blum WF, Schmidt B, Braulke T, Kubler B. (2005). Interaction of insulin-like growth factor II (IGF-II) with multiple plasma proteins: high affinity binding of plasminogen to IGF-II and IGF-binding protein-3. J Biol Chem, 280, 9994-10000.

Oggero M, Frank R, Etcheverrigaray M, Kratje R. (2004). Defining the antigenic structure of human GM-CSF and its implications for receptor interaction and therapeutic treatments. Mol Divers, 8, 257-269.

Ong SE, Blagoev B, Kratchmarova I, Kristensen DB, Steen H, Pandey A, Mann M. (2002). Stable isotope labeling by amino acids in cell culture, SILAC, as a simple and accurate approach to expression proteomics. Mol Cell Proteomics, 1, 376-386.

Ooi SL et al. (2006). Global synthetic-lethality analysis and yeast functional profiling. Tigerpaper, 22, 56-63.

Otto-Bruc A, Antonny B, Minh Vuong T, Chardin P, Chabre M. (1993). Interaction between the retinal cyclic GMP phosphodiesterase inhibitor and transducin. Kinetics and affinity studies. J Biochem, 32, 8636-8645.

Owicki JC. (2000). Fluorescence polarization and anisotropy in high throughput screening: perspectives and primer. J Biomol Screen, 5, 297-306.

Pande J, Szewczyk MM, Grover AK. (2010). Phage display: concept, innovations, applications and future. Biotechnol Adv, 28, 849-858.

Patel SS, Rexach MF. (2008). Discovering novel interactions at the nuclear pore complex using bead halo: a rapid method for detecting molecular interactions of high and low affinity at equilibrium. Mol Cell Proteom, 7, 121-131.

Pedersen ME, Østergaard J, Jensen H (2019) Flow-induced dispersion analysis (FIDA) for protein quantification and characterization. In: Clinical applications of capillary electrophoresis. Springer, New York, pp 109-123

Pelicci G et al. (1992). A novel transforming protein (Shc) with an Sh2 domain is implicated in mitogenic signal transduction. Cell, 70, 93-104.
Pelletier J, Sidhu S. (2001). Mapping protein-protein interactions with combinatorial biology methods. Curr Opin Biotechnol, 12, 340-347.
Periasamy A. (2001). Fluorescence resonance energy transfer microscopy: a mini review. J Biomed Opt, 6, 287-291.
Pfleger KD, Seeber RM, Eidne KA. (2006). Bioluminescence resonance energy transfer (BRET) for the real-time detection of protein-protein interactions. Nat Protoc, 1, 337-345.
Philo JS, Wen J, Wypych J, Schwartz MG, Mendiaz EA, Langley KE. (1996). Human stem cell factor dimer forms a complex with two molecules of the extracellular domain of its receptor, Kit. J Biol Chem, 271, 6895-6902.
Phizicky EM, Fields S. (1995). Protein-protein interactions: methods for detection and analysis. Microbiol Rev, 59, 94-123.
Pierce MM, Raman CS, Nall BT. (1999). Isothermal titration calorimetry of protein-protein interactions. Methods, 19, 213-221.
Poluri KM, Joseph PRB, Sawant KV, Rajarathnam K. (2013). Molecular basis of glycosaminoglycan heparin binding to the chemokine CXCL1 dimer. J Biol Chem, 288, 25143-25153.
Popescu SC et al. (2007). Differential binding of calmodulin-related proteins to their targets revealed through high-density Arabidopsis protein microarrays. Proc Natl Acad Sci U S A, 104, 4730-4735.
Poulard C, Rambaud J, Le Romancer M, Corbo L (2014) Proximity ligation assay to detect and localize the interactions of ERα with PI3-K and Src in breast cancer cells and tumor samples. In: Steroid receptors. Springer, New York, pp 135-143
Poulsen NN, Andersen NZ, Ostergaard J, Zhuang G, Petersen NJ, Jensen H. (2015). Flow induced dispersion analysis rapidly quantifies proteins in human plasma samples. Analyst, 140, 4365-4369.
Raj R, Agarwal N, Raghavan S, Chakraborti T, Poluri KM, Kumar D. (2020). Exquisite binding interaction of 18β-Glycyrrhetinic acid with histone like DNA binding protein of Helicobacter pylori: a computational and experimental study. Int J Biol Macromol, 161, 231-246.
Ranjbar B, Gill P. (2009). Circular dichroism techniques: biomolecular and nanostructural analyses- a review. Chem Biol Drug Des, 74, 101-120.
Rao VS, Srinivas K, Sujini GN, Kumar GN. (2014). Protein-protein interaction detection: methods and analysis. Int J Proteom, 2014, 147648.
Ratcliff GC, Erie DA. (2001). A novel single-molecule study to determine protein-protein association constants. J Am Chem Soc, 123, 5632-5635.
Rich RL, Myszka DG. (2000). Advances in surface plasmon resonance biosensor analysis. Curr Opin Biotechnol, 11, 54-61.
Rich RL, Myszka DG. (2007). Higher-throughput, label-free, real-time molecular interaction analysis. Anal Biochem, 361, 1-6.
Rigaut G, Shevchenko A, Rutz B, Wilm M, Mann M, Seraphin B. (1999). A generic protein purification method for protein complex characterization and proteome exploration. Nat Biotechnol, 17, 1030-1032.
Rivas G, Minton A (2003) Tracer sedimentation equilibrium: a powerful tool for the quantitative characterization of macromolecular self-and hetero-associations in solution. Portland Press Ltd., London
Roberts GCK (1993) NMR of macromolecules. Oxford: Oxford University Press,
Robinson RC, Turbedsky K, Kaiser DA, Marchand JB, Higgs HN, Choe S, Pollard TD. (2001). Crystal structure of Arp2/3 complex. Science, 294, 1679-1684.
Rohila JS et al. (2009). Protein-protein interactions of tandem affinity purified protein kinases from rice. PLoS One, 4, e6685.
Rüdiger S, Schneider-Mergener J, Bukau BJ. (2001). Its substrate specificity characterizes the DnaJ co-chaperone as a scanning factor for the DnaK chaperone. EMBO J, 20, 1042-1050.

Ruffner H, Bauer A, Bouwmeester T. (2007). Human protein-protein interaction networks and the value for drug discovery. Drug Discov Today, 12, 709-716.
Russell RB et al. (2004). A structural perspective on protein-protein interactions. Curr Opin Struct Biol, 14, 313-324.
Ryu W-S (2016) Molecular virology of human pathogenic viruses. Academic, New York
Salminen A, Novick PJ. (1987). A ras-like protein is required for a post-golgi event in yeast secretion. Cell, 49, 527-538.
Sanders SL, Whitfield KM, Vogel JP, Rose MD, Schekman RW. (1992). Sec61p and bip directly facilitate polypeptide translocation into the Er. Cell, 69, 353-365.
Schulte R, Talamas J, Doucet C, Hetzer MW. (2008). Single bead affinity detection (SINBAD) for the analysis of protein-protein interactions. PLoS One, 3, e2061.
Schuster SC, Swanson RV, Alex LA, Bourret RB, Simon MI. (1993). Assembly and function of a quaternary signal transduction complex monitored by surface plasmon resonance. Nature, 365, 343-347.
Selbach M, Mann M. (2006). Protein interaction screening by quantitative immunoprecipitation combined with knockdown (QUICK). Nat Methods, 3, 981-983.
Shah NB, Duncan TM. (2014). Bio-layer interferometry for measuring kinetics of protein-protein interactions and allosteric ligand effects. JoVE, 2014 e51383.
Shoemaker BA, Panchenko AR. (2007). Deciphering protein–protein interactions. Part I. Experimental techniques and databases PLoS Comput Biol, 3, e42.
Silkowski H, Davis SJ, Barclay AN, Rowe AJ, Harding SE, Byron O. (1997). Characterisation of the low affinity interaction between rat cell adhesion molecules CD2 and CD48 by analytical ultracentrifugation. Eur Biophys J Biophy, 25, 455-462.
Slonimskiy YB et al. (2018). Functional interaction of low-homology FRPs from different cyanobacteria with Synechocystis OCP. Bioenergetics, 1859, 382-393.
Snider J, Kotlyar M, Saraon P, Yao Z, Jurisica I, Stagljar I. (2015). Fundamentals of protein interaction network mapping. Mol Syst Biol, 11, 848.
Söderberg O et al. (2006). Direct observation of individual endogenous protein complexes in situ by proximity ligation. Nat Methods, 3, 995-1000.
Söderberg O, Leuchowius K-J, Gullberg M, Jarvius M, Weibrecht I, Larsson L-G, Landegren U. (2008). Characterizing proteins and their interactions in cells and tissues using the in situ proximity ligation assay. Methods, 45, 227-232.
Some D. (2013). Light-scattering-based analysis of biomolecular interactions. Biophys Rev, 5, 147-158.
Song Y, Liao JY. (2012). An in vitro forster resonance energy transfer-based high-throughput screening assay for inhibitors of protein-protein interactions in SUMOylation pathway. Assay Drug Dev Technol, 10, 336-343.
Song Y, Madahar V, Liao JY. (2011). Development of FRET assay into quantitative and high-throughput screening technology platforms for protein-protein interactions. Ann Biomed Eng, 39, 1224-1234.
Sopta M, Carthew RW, Greenblatt J. (1985). Isolation of three proteins that bind to mammalian RNA polymerase II. J Biol Chem, 260, 10353-10360.
Sreerama N, Woody RW (2004) Computation and analysis of protein circular dichroism spectra. In: Methods in enzymology, 383. Elsevier, London, pp 318-351
Stynen B, Tournu H, Tavernier J, Van Dijck P. (2012). Diversity in genetic in vivo methods for protein-protein interaction studies: from the yeast two-hybrid system to the mammalian split-luciferase system. MMBR, 76, 331-382.
Tang C, Iwahara J, Clore GM. (2006). Visualization of transient encounter complexes in protein-protein association. Nature, 444, 383-386.
Taylor KA, Glaeser RM. (2008). Retrospective on the early development of cryoelectron microscopy of macromolecules and a prospective on opportunities for the future. J Struct Biol, 163, 214-223.

Thompson PM, Beck MR, Campbell SL (2015) Protein-protein interaction analysis by nuclear magnetic resonance spectroscopy. In: Protein-protein interactions. Springer, New York, pp 267-279

Tong AH et al. (2002). A combined experimental and computational strategy to define protein interaction networks for peptide recognition modules. Science, 295, 321-324.

Troyanskaya OG, Garber ME, Brown PO, Botstein D, Altman RB. (2002). Nonparametric methods for identifying differentially expressed genes in microarray data. Bioinformatics, 18, 1454-1461.

Truant R, Xiao H, Ingles CJ, Greenblatt J. (1993). Direct interaction between the transcriptional activation domain of human P53 and the tata box-binding protein. J Biol Chem, 268, 2284-2287.

Ullman EF et al. (1996). Luminescent oxygen channeling assay (LOCI): sensitive, broadly applicable homogeneous immunoassay method. Clin Chem, 42, 1518-1526.

Uzarska MA, Dutkiewicz R, Freibert SA, Lill R, Muhlenhoff U. (2013). The mitochondrial Hsp70 chaperone Ssq1 facilitates Fe/S cluster transfer from Isu1 to Grx5 by complex formation. Mol Biol Cell, 24, 1830-1841.

Vasilescu J, Guo XC, Kast J. (2004). Identification of protein-protein interactions using in vivo cross-linking and mass spectrometry. Proteomics, 4, 3845-3854.

Velazquez-Campoy A, Leavitt SA, Freire E (2004) Characterization of protein-protein interactions by isothermal titration calorimetry. In: Protein-protein interactions. Springer, New York, pp 35-54

Vernon-Wilson EF, Kee WJ, Willis AC, Barclay AN, Simmons DL, Brown MH. (2000). CD47 is a ligand for rat macrophage membrane signal regulatory protein SIRP (OX41) and human SIRPα 1. Eur J Immunol, 30, 2130-2137.

Vidal M, Brachmann RK, Fattaey A, Harlow E, Boeke JD. (1996). Reverse two-hybrid and one-hybrid systems to detect dissociation of protein-protein and DNA-protein interactions. Proc Natl Acad Sci U S A, 93, 10315-10320.

Vidal M, Cusick ME, Barabasi AL. (2011). Interactome networks and human disease. Cell, 144, 986-998.

Wallace BA, Janes RW. (2010). Synchrotron radiation circular dichroism (SRCD) spectroscopy: an enhanced method for examining protein conformations and protein interactions. Biochem Soc Trans, 38, 861-873.

Wang L et al. (2010). Cultivation of green algae Chlorella sp. in different wastewaters from municipal wastewater treatment plant. Appl Biochem Biotechnol, 162, 1174-1186.

Wang L et al. (2015). Spatiotemporal dynamics of the BRI1 receptor and its regulation by membrane microdomains in living arabidopsis cells. Mol Plant, 8, 1334-1349.

Wang L, Xue Y, Xing J, Song K, Lin J. (2018). Exploring the spatiotemporal organization of membrane proteins in living plant cells. Annu Rev Plant Biol, 69, 525-551.

Wang Y, Tadayon R, Shaw GS (2019) Monitoring interactions between S100B and the dopamine D2 receptor using NMR spectroscopy. In: Calcium-binding proteins of the EF-hand superfamily. Springer, New York, pp 311-324

Watson VG, Drake KM, Peng Y, Napper AD. (2013). Development of a high-throughput screening-compatible assay for the discovery of inhibitors of the AF4-AF9 interaction using alphascreen technology. Assay Drug Dev Technol, 11, 253-268.

Weber PJ, Beck-Sickinger AG. (1997). Comparison of the photochemical behavior of four different photoactivatable probes. J Pept Res, 49, 375-383.

Wei Z-H, Chen H, Zhang C, Ye B-C. (2014). FRET-based system for probing protein-protein interactions between σ R and RsrA from streptomyces coelicolor in response to the redox environment. PLoS One, 9, e92330.

Weibrecht I et al. (2010). Proximity ligation assays: a recent addition to the proteomics toolbox. Expert Rev Proteomics, 7, 401-409.

Weiel J, Hershey JW. (1981). Fluorescence polarization studies of the interaction of Escherichia coli protein synthesis initiation factor 3 with 30S ribosomal subunits. Biochemistry, 20, 5859-5865.

Wellhausen A, Lehming N. (1999). Analysis of the in vivo interaction between a basic repressor and an acidic activator. FEBS Lett, 453, 299-304.

Werther M, Seitz H (2008) Protein-protein interaction. Springer, New York

Wienken CJ, Baaske P, Rothbauer U, Braun D, Duhr S. (2010). Protein-binding assays in biological liquids using microscale thermophoresis. Nat Commun, 1, 100.

Williams JG, Drugan JK, Yi GS, Clark GJ, Der CJ, Campbell SL. (2000). Elucidation of binding determinants and functional consequences of Ras/Raf-cysteine-rich domain interactions. J Biol Chem, 275, 22172-22179.

Wilson JL, Scott IM, McMurry JL. (2010). Optical biosensing: kinetics of protein A-IGG binding using biolayer interferometry. Biochem Mol Biol Educ, 38, 400-407.

Winstone TL, Workentine ML, Sarfo KJ, Binding AJ, Haslam BD, Turner RJ. (2006). Physical nature of signal peptide binding to DmsD. Arch Biochem Biophys, 455, 89-97.

Wu Y, Eghbali M, Ou J, Lu R, Toro L, Stefani E. (2010). Quantitative determination of spatial protein-protein correlations in fluorescence confocal microscopy. Biophys J, 98, 493-504.

Xia Z, Zhou Q, Lin J, Liu Y. (2001). Stable SNARE complex prior to evoked synaptic vesicle fusion revealed by fluorescence resonance energy transfer. J Biol Chem, 276, 1766-1771.

Xue Y et al. (2018). Arabidopsis blue light receptor phototropin 1 undergoes blue light-induced activation in membrane microdomains. Mol Plant, 11, 846-859.

Yang Y, Wang H, Erie DA. (2003). Quantitative characterization of biomolecular assemblies and interactions using atomic force microscopy. Methods, 29, 175-187.

Yasgar A, Jadhav A, Simeonov A, Coussens NP (2016) AlphaScreen-based assays: ultra-high-throughput screening for small-molecule inhibitors of challenging enzymes and protein-protein interactions. In: High throughput screening. Springer, New York, pp 77-98

Yegorova S, Chavaroche AE, Rodriguez MC, Minond D, Cudic M. (2013). Development of an AlphaScreen assay for discovery of inhibitors of low-affinity glycan-lectin interactions. Anal Biochem, 439, 123-131.

Yurlova L et al. (2014). The fluorescent two-hybrid assay to screen for protein–protein interaction inhibitors in live cells: targeting the interaction of p53 with Mdm2 and Mdm4. J Biomol Screen, 19, 516-525.

Zhang M, Wisniewski JA, Ji H. (2015). AlphaScreen selectivity assay for β-catenin/B-cell lymphoma 9 inhibitors. Anal Biochem, 469, 43-53.

Zhao Y, Chen SS, Yoshioka C, Baconguis I, Gouaux E. (2016). Architecture of fully occupied GluA2 AMPA receptor-TARP complex elucidated by cryo-EM. Nature, 536, 108.

Zhou M, Felder S, Rubinstein M, Hurwitz DR, Ullrich A, Lax I, Schlessinger J. (1993). Real-time measurements of kinetics of EGF binding to soluble EGF receptor monomers and dimers support the dimerization model for receptor activation. Biochemistry, 32, 8193-8198.

Zhou M, Li Q, Wang RX. (2016). Current experimental methods for characterizing protein-protein interactions. ChemMedChem, 11, 738-756.

Zlatanova J, Lindsay SM, Leuba SH. (2000). Single molecule force spectroscopy in biology using the atomic force microscope. Prog Biophys Mol Biol, 74, 37-61.

Zolghadr K et al. (2008). A fluorescent two-hybrid assay for direct visualization of protein interactions in living cells. MCP, 7, 2279-2287.

Zuiderweg ERP. (2002). Mapping protein-protein interactions in solution by NMR spectroscopy. Biochemistry, 41, 1-7.

Prediction, Analysis, Visualization, and Storage of Protein–Protein Interactions Using Computational Approaches

6

Abstract

For many years, protein–protein interactions have been determined solely by virtue of experiments. The experimental techniques employed to elucidate these interactions, however, have faced many inherent limitations. The experimental techniques are time-consuming and not well-suited for creating protein networks that comprise a voluminous amount of data. Consequently, researchers have looked forward to utilize computational techniques to determine such interactions. This chapter will discuss the currently employed computational methods and other web resources like databases and analysis tools. It will also describe how protein interactions are mapped into a network to provide better insight into various pathways that operate at the cellular level. Furthermore, the chapter also throws light on the computational techniques of protein–protein docking and states its implications in further experimental research design.

Keywords

Computational prediction · Evolution · Gene · Protein · PPI analysis · PPI storage databases · PPI networks

6.1 Computational Prediction Methods of PPIs

It is quintessential to identify PPIs to gain a comprehensive understanding of how proteins perform their jobs in a combined manner in a cell in order to accomplish their cellular tasks. PPIs play an imperative role in variety of cellular functions and are also involved in numerous diseases. Hence, PPIs are considered as the crucial drug targets for various diseases. Gaining knowledge about natural PPIs can also aid in designing the artificial PPIs, which can be further employed for fabrication of therapeutics.

K. M. Poluri et al., *Protein-Protein Interactions*,
https://doi.org/10.1007/978-981-16-1594-8_6

Numerous experimental techniques such as X-ray crystallography (Kobe et al. 2008), co-immunoprecipitation (Lin and Lai 2017), NMR spectroscopy (Vinogradova and Qin 2011; Zuiderweg 2002), fluorescence resonance energy transfer (Kenworthy 2001), surface plasmon resonance (Nikolovska-Coleska 2015), electron microscopy (Dudkina et al. 2010), etc. are being used to unravel the PPIs. Experimental approaches for discerning the PPIs exhibit several discrepancies such as: (1) Experimental approaches are highly labor-intensive and time-consuming; (2) All the experimental strategies may not be applicable to all the proteins under consideration (Piehler 2005; Rao et al. 2014); (3) Elucidation/recognition of weak and transient interactions is extremely complicated using experimental approaches (Ngounou Wetie et al. 2013); (4) Handling large volumes of PPI data manually is arduous (Rao et al. 2014); (5) Limitation in detection of number of PPIs at a given point by experimental methods makes it cumbersome to construct PPI network web. All these pitfalls associated with experimental means of disclosing PPIs opened up the way to employ computational techniques for predicting, analyzing, viewing, and storing PPIs.

PPIs identified for different model organisms such as *Homo sapiens*, *Mus musculus*, *Saccharomyces cerevisiae*, *Schizosaccharomyces pombe*, *Escherichia coli*, *Drosophila melanogaster* have been stored in various databases. Efforts have already been made in identifying PPIs however, there is a large chunk of PPIs that have not yet been identified. For instance, statistical analysis reported the presence of around 6,50,000 PPIs, but only around 41,327 interactions have been identified in humans as depicted by HPRD (human protein reference database) (Keshava Prasad et al. 2009). Even for other organisms such as the most well-studied yeast and the *Caenorhabditis elegans*, an important model organism, there is a huge difference between the number of PPIs identified and the PPIs that are still undiscovered (Ding and Kihara 2018). Hence, there is a strong demand of the computational tools for the detection of networks of PPIs in the organism.

Numerous computational methods have been emerged in the last few decades to overcome the limitations of experimental methods for gaining comprehensive knowledge on the PPIs and to complement the experimentally obtained protein interaction outcomes. These methods are divided into two categories, namely: (1) homology-based and (2) template-free machine learning (ML) methods (Xue et al. 2015).

Homology-based methods employ the usage of query protein homologs to discern its interacting partner and hence, they return more accurate and reliable outcomes. A query for which homolog data is not available, the template-free machine learning approach is opted. Supervised learning is a common machine learning method that assembles a predictor by employing differential training datasets. Training datasets represent the data available for both the interacting protein partners and the non-interacting protein cognate pairs. The predictor is then used to deduce the interactions for the new pair of proteins. A conventional ML approach requires the characteristics such as sequence, structure, affinity or any other physiochemical data to represent the protein. It also requires a learning algorithm such as neural networks, support vector machine (SVM), random forests,

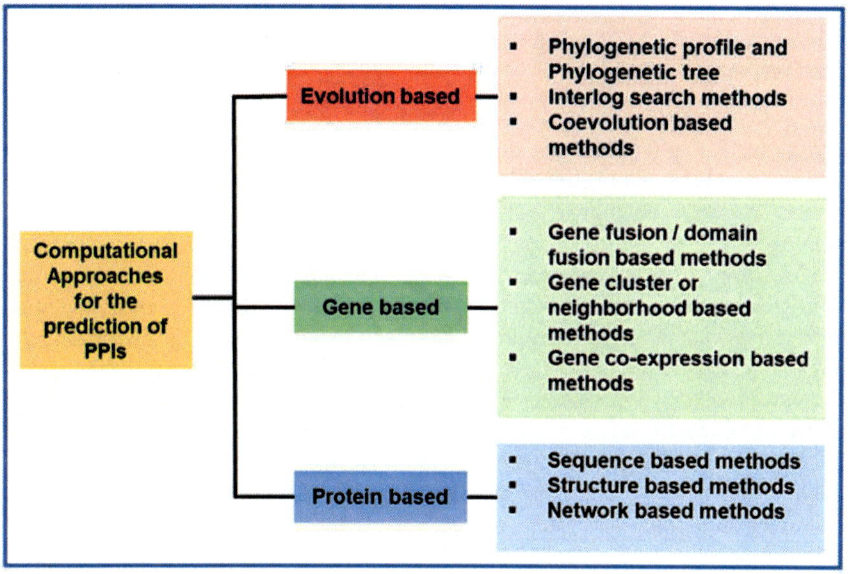

Fig. 6.1 Schematic depicting the classification of computational approaches for the prediction of PPIs

or decision trees that can obtain the present information about the PPIs and further use it to reveal unobserved PPIs (Keskin et al. 2016).

Computational approaches used to deduce PPIs are broadly divided into three classes: (1) evolution-based, (2) gene-based, and (3) protein-based methodologies (Fig. 6.1). They have been classified based on the information they employ to predict PPIs (Keskin et al. 2016).

6.1.1 Evolution-Based Methods

Evolution-based methods consider the incidents happened during the evolutionary history of a protein for the prediction of interacting protein pairs. Tracking the evolution of a protein helps in acquiring the details about the occurrence of changes in the sequence and structure of protein over the course of time, which also aids in obtaining lot of information about its interactions with other proteins. Some of the methods based on concept of evolution of proteins include: (1) phylogenetic profile and phylogenetic tree, (2) interlog-based methods, and (3) coevolution-based methods.

6.1.1.1 Phylogenetic Profile and Phylogenetic Tree

Phylogenetic tree provides details about the historical events happened during the course of evolution of protein. Proteins with the potency of interaction share similarity in their phylogenetic trees. Phylogeny-based methods to predict PPIs are classified into two, namely: phylogenetic profile and phylogenetic tree topology-based methods.

Phylogenetic Profile Method

This is a simple means of predicting PPIs via usage of comparative genomics. This method involves phylogenetic profiling of co-occurrence and co-absence of orthologous proteins in the organisms (Pellegrini et al. 1999). If there is an involvement of two proteins to carry out a biological function, then to perform that function the organism need to possess both the proteins. In contrast, if there is no need of that function for an organism, then both the proteins should be absent, as the presence of only one of the protein is meaningless for the organism.

Steps involved in predicting PPIs using phylogenetic profile method are as follows: (1) Identification of orthologs for all the proteins in queried genome. (2) Construction of a phylogenetic profile for all the proteins in queried genome. This involves binary codes 0 and 1. Values 1 and 0 are assigned if there is presence or absence of orthologous protein, respectively, in the reference genome. Number of reference genomes used for evaluation will be the dimension of the profile. (3) Profile of proteins will be compared, and the pair of proteins that are sharing similar profiles are predicted as interacting pairs. The choice of reference genomes is very crucial in this process of making protein profiles (Sun et al. 2007). Further, the performance of this method is also affected by the threshold value (E-value) used for sequence similarity searches in detection of protein orthologs (Sun et al. 2005). To improve the performance of profiles, the real value vectors of alignment score are being employed rather than binary values for the construction of profiles (Lin et al. 2013).

Phylogenetic Tree Method

Phylogenetic tree method relies on the fact that the interacting protein partners exhibit similar phylogenetic trees among the various species (Goh et al. 2000; Goh and Cohen 2002; Pazos and Valencia 2001). The major reason behind this similarity is that the interacting proteins suffer same evolutionary pressures to maintain their complex functionalities. Algorithm has been developed that can measure the similarity in the trees topologies.

"Mirror tree method" is being used to calculate similarity between the phylogenetic trees as the correlation between the distance matrices of the corresponding trees. Various steps included in this method are: (1) Construction of multiple sequence alignment for the query proteins against the reference organisms, (2) Construction of phylogenetic tree for the proteins, (3) Distance matrices are made for each tree by calculating the pairwise distances between the orthologous proteins, (4) Correlation coefficient between the two distance matrices is calculated. Then correlation coefficient value is compared with the cut-off value for the interacting and the non-interacting protein partners. Coefficient value above the cut-off value denotes that the query proteins are interacting with each other (Pazos and Valencia 2001).

Several modifications have been done to further improve or extend the usage of mirror tree method. Ramani and Marcotte extended the usage of this method to predict the specific interacting protein pairs from the two interacting families of proteins such as ligands and their corresponding receptor proteins (Ramani and Marcotte 2003). Sato et al. improved the algorithm to enhance PPI prediction accuracy by the removal of background similarity, which arises as a result of overall evolutionary distances of organism from individual protein's distance matrices (Sato et al. 2005). Sato et al. further demonstrated the use of partial coefficient which can reduce the organism level similarity at the background while considering the similarity of trees for query proteins in predicting PPIs. They also observed considerable improvement in the performance of partial correlation coefficient for predicting PPIs than the original form of mirror tree methodology (Sato et al. 2006).

Juan et al. further reduced the noise in the mirror tree methodology by considering the network of similarities among all the protein pairs simultaneously instead of considering similarity between the query protein pairs. This is due to the observation that the protein coevolves with many other interacting proteins (Juan et al. 2008b). Herman et al. assessed the performance of mirror tree methods using different set of organisms for constructing tree. The authors found that the outcome of mirror tree method is based on the particular set of organisms chosen for predicting the particular type of interactions. It is recommended that while predicting PPIs, it is better to employ different sets of organism for the optimal performance of mirror tree method (Herman et al. 2011).

Major lead of mirror tree method over phylogenetic profile method for predicting PPIs is that there is no need of having fully sequenced genomes of organisms as trees of interested proteins form the basis of mirror tree method. Further, phylogenetic profiling method does not provide information about the housekeeping genes, as these genes are always present in all the organisms. Partners to those proteins that are unique to particular organism cannot also be predicted using phylogenetic-based methods (Keskin et al. 2016). Moreover, gene duplication events also result in noisy phylogenetic profiles of protein. Functionally important residues are often evolutionary conserved. These residues often form the binding sites for the partner proteins, hence revealing such conserved binding motifs in the protein aids in finding the binding partners. Methods employing such conserved residues information have been designed. For instance, an evolutionary trace method for prediction of PPIs takes into account both divergence in the phylogenetic tree of a protein family and conservation of residues that forms important binding sites during the interaction of protein partners (Lua et al. 2014).

6.1.1.2 Interlog Search Methods

Conservation of PPIs among species forms the basis of interlog search methods. This implies that there is a transfer of PPIs between the species. The conserved PPIs across the species are known as interlogs. Hence, PPIs can be predicted in an organism using the interactions that are known for orthologous genes in other organisms. Interlogs of higher-order eukaryotes are predicted mostly using model organisms. Several automated methods have been developed to search interlogs.

Folador et al. used a strategy employing freely available tools and public databases to map interlogs (Folador et al. 2014). Sahu et al. exploited the combination of interlog search method and motif-based method to unravel PPIs network between *Pseudomonas syringae pv. tomato* DC3000 and *Arabidopsis thaliana* at genomic scale. Underpinning such PPI network between host and pathogen is a highly informative resource for scientific community for the characterization of host–pathogen interactions (Sahu et al. 2014).

BIPS: BIANA (Biologic Interactions and Network Analysis) Interolog Prediction Server is a web-based tool to unravel PPIs on the basis of interlog details contained in various PPI databases. BIANA is a framework to integrate multiple PPI databases. BIANA also provides freedom to the user to use information like gene ontology, orthologous genes cluster, and also several other features including number of transmembrane helices for the prediction of PPIs (Garcia-Garcia et al. 2012). Specificity and sensitivity of the server has been determined employing the known interactomes. Server's specificity varies from 72% to 98% and its sensitivity ranges from 1% to 59%. These values depend on the sequence identity cut-off values employed for the calculation of sequence similarities (Garcia-Garcia et al. 2012).

Interlog search first scans for the conservation of two interacting protein pairs in different species. If the two interacting protein pairs in one species are also conserved in other species, then it is predicted that they are also interacting in other species. It has to be noted that evolutionary conservation is majorly observed among obligate interactions than among transient interactions; hence, the interlog search method is more applicable to obligate interactions than to transient interactions. Several other approaches like functional conservation and domain pair conservation analysis can be used to assess the predicted interlogs. Interlog search method provides reliable outcomes, although this prediction protocol is not applicable for large-scale proteomes (Keskin et al. 2016).

Mathews et al. have used interlog search method to obtain PPI network in *Caenorhabditis elegans* using protein interaction map available for *Saccharomyces cerevisiae*. The authors used two hybrid yeast interaction maps which together contain 1195 interactions. From these interactions, they obtained 257 potential interlogs involving 282 proteins (Matthews et al. 2001). Huang et al. designed database known as POINT (the prediction of interactome database) which integrates the PPI data from various publically available databases. PPI datasets for mouse, worm, yeast, and fruit fly from DIP (database of interacting proteins) have been extracted and used for the prediction of human PPIs. POINT also takes into account the subcellular localization data of proteins from GO (gene ontology) database and the mRNA expression data from cell cycle microarray database to draw conclusions regarding the functional relevance of predicted interacting protein partners in human (Huang et al. 2004).

Lee et al. made an effort to expand orthologous protein pairs from 18 eukaryotic species and combined them with experimentally derived PPIs. They utilized the expanded PPI dataset to predict host–pathogen interactions. Interactions between several human proteins and *Plasmodium falciparum* calmodulin were predicted which may contribute in maintaining the calcium levels in the host cell (Lee et al. 2008).

Geisler-Lee et al. predicted PPIs in *Arabidopsis thaliana* using the homologs from human, *C. elegans*, fruit fly, and yeast. From 3617 proteins in *A. thaliana*, they found 1159, 5913, and 12,907 as high, medium, and low confidence interacting protein pairs. They have also predicted several interactions that were previously unknown in known complexes and pathways. The authors have presented these interactions in a web interface known as Arabidopsis Interactions Viewer (AIV), which will further aid researchers in unraveling the complex signaling pathways in *Arabidopsis* and related families, thus serving this information as preliminary data for the experimentalists (Geisler-Lee et al. 2007).

Gu et al. developed an interactome database for rice known as PRIN (Predicted Rice Interactome Network) to discern PPIs in *Oryza sativa*. PPI data for PRIN is based on the interlogs data from six model organisms, namely: worm, yeast, human, *E. coli*, fruit fly, and *A. thaliana*. For 5049 rice proteins, they found 76,585 non-redundant interactions among rice proteins. They observed that topology of protein interaction network of *O. sativa* and yeast resembles in comparison with the network topologies of other five organisms. For the validation of these interactions, they have also employed the gene ontology, gene expression, and subcellular localization data (Gu et al. 2011).

Dutkowski et al. presented a framework which considers phylogenies of protein families for mapping PPIs between homologous proteins. This framework is based on their previously developed CAPPI (Conserved Ancestral Protein–Protein Interactions) framework that compares PPI networks across species. Duplication and divergence models form the basis for CAPPI. CAPPI is used to infer dependencies between PPIs and to perform evolutionary analysis on the present PPI data (Dutkowski and Tiuryn 2007). CAPPI is capable of comparing many PPI networks at the same time. This algorithm has also revealed numerous functional modules with higher specificity (Dutkowski and Tiuryn 2007). They used this framework to decipher novel PPIs in seven eukaryotes, namely: *Rattus norvegicus, Mus musculus, C. elegans, S. cerevisiae, Drosophila melanogaster, A. thaliana, and Homo sapiens*. They also verified these interactions using function-based gene ontology data. The outcomes from CAPPI proved the ability of CAPPI to return higher amount of known interactions in a blind test and basis for the experimental verification for all the available integrated data. They have also analyzed specific previously validated PPIs in yeast as well as in human and observed that PPIs can be attained solely on the basis of data available for other species without the need of experimental data for the species of interest (Dutkowski and Tiuryn 2007).

6.1.1.3 Coevolution-Based Methods

Proteins interact with each other through amino acids localized at specific sites (also known as patches) present on the exposed protein surfaces. These amino acids exhibit various features like interface residue propensity, accessible surface area, protrusion, hydrophobicity, and planarity. In addition to these features that are observed in all the interacting surfaces, they also contain information about the specific binding. Specificity is dependent on the amino acids constituting the interaction site. An amino acid substitution in one protein interface will be compensated by complementary amino acid substitution in the interface of the other interacting protein partner. The outcome of these substitutions is the coevolution of interacting protein partners.

The term "coevolution" implies the simultaneous evolution or the co-adaptation of the interacting protein partners. The substitution of an amino acid at the protein interface will drive a coordinated amino acid substitution in the interface of its binding protein partner. These changes are also known as compensatory changes to maintain the optimum structural and functional integrity. Coevolution was observed among various receptor and ligand pairs like insulin and interleukins. The main aim of this coordinated evolution among these proteins was to maintain their specific interactions and thereby maintaining their function. Methods have been developed that infer coevolution of proteins based on multiple sequence alignment (MSA) of proteins. Coevolution methods and other genome-based methods for PPIs prediction are jointly known as context-based methods (Argos 1988; Chothia and Janin 1975; Hakes et al. 2007; Hoskins et al. 2006; Janin and Chothia 1990; Janin et al. 1988; Jones and Thornton 1995; Jones and Thornton 1996; Singh 2019).

Coevolutionary analysis has been used by numerous studies for the prediction of PPIs (Aytuna et al. 2005; Craig and Liao 2007; Hakes et al. 2007; Kann et al. 2007; Pazos and Valencia 2001; Sato et al. 2005; Tillier and Charlebois 2009). Coevolutionary analysis can be performed in two ways, namely: (1) Correlated mutation analysis (CMA) and (2) Correlated evolutionary rates.

Correlated mutation analysis is based on determining the amino acids that underwent correlated mutations for the prediction of proteins in contact. Residues in close proximity in proteins are essential for maintaining the structure and stability in the proteins. If one of those residues gets mutated, then it will adversely affect the stability of protein locally unless other residues also get mutated to preserve the interaction and thereby the stability and function of the protein (Bloom et al. 2007; Worth et al. 2009). Such residue-based analysis of correlated mutation is useful in determining both the intra-protein and inter-protein interactions and also aid in finding essential motifs in proteins (Horner et al. 2008; Pazos and Valencia 2002; Saraf et al. 2003; Singer et al. 2002; Tillier et al. 2006; Tress and Valencia 2010).

Coevolutionary rate analysis aims at identifying proteins evolving at same rates which are depicted by their similar phylogenetic trees with similar branch topology and branch lengths. Coevolutionary rate analysis approaches compare the distance matrices rather than the direct comparison of phylogenetic tree to avoid the discrepancies related to phylogenetic tree reconstruction. Distance matrix of a protein family encompasses pairwise information about the divergence of their amino acid sequences. As every individual pairwise entry itself defines the relative evolutionary distance between the two proteins, the distance matrix is a compilation of relative evolutionary rates. Comparison of such matrices forms the basis of many coevolutionary methods for the prediction of PPIs (Clark et al. 2011).

Mirror tree method was first evolved by Pazos and Valencia for the comparison of phylogenetic trees by utilizing their distance matrices (Pazos and Valencia 2001).

This methodology is very commonly applied for the coevolutionary analysis (Craig and Liao 2007; Hakes et al. 2007; Juan et al. 2008a; Sato et al. 2005). It employs Pearson's correlation coefficient for each pair distance entry between two distance matrices of protein to quantify the similarity of trees. The coevolution of two protein families is assessed by first constructing MSA (multiple sequence alignment) using the orthologous sequences from the reference organisms. Mirror tree methodology is then employed to decipher the inter-ortholog distance matrices from the MSA themselves or from the tree derived from the MSA. Further, the similarity between the distance matrices is evaluated using correlation criteria which will serve as a proxy measure of similarity of corresponding trees (De Juan et al. 2013). Higher the correlation coefficient, higher the similarity of distance matrices, which is a sign of interacting partners. Generally, correlation coefficient of more than 0.8 between the two matrices is a decent cut-off value for the prediction of interacting proteins.

One drawback of mirror tree method is that trimming the distance matrices to equal sizes and species composition is required before calculating correlation coefficient (Clark et al. 2011). This method has been successfully used to acquire the *E. coli* protein interaction network (Pazos and Valencia 2001). Other difficulty associated with this method is the background similarity in the phylogenetic trees of protein families which arises due to the global evolutionary relationship among the organisms. This basal similarity among the phylogenetic trees must be considered while calculating protein–protein distances. Some of the approaches have emerged to overcome such discrepancies which include normalizing the evolutionary distances between the protein families by taking into consideration the organism level evolutionary distances (Kann et al. 2007; Pazos et al. 2005; Sato et al. 2005).

The improvement or the extension of mirror tree method is the context mirror method, which aims to remove all the complex evolutionary factors that may result in false positive coevolving protein pairs. Context mirror method has a higher accuracy of prediction of PPIs based on coevolution as it corrects several problems including the tree similarity arises due to speciation. It also has an ability to distinguish between the coevolution trends among the families and coevolution specific to protein pairs (De Juan et al. 2013; Juan et al. 2008a). Other mirror tree-based approaches predict interacting protein pairs within two large protein families. These approaches look for the correspondence between the paralogs of one protein family with the paralogs belonging to the other protein family (Izarzugaza et al. 2008; Jothi et al. 2005; Ramani and Marcotte 2003).

Further, rather than using the whole protein sequences for MSA and phylogenetic tree construction, it has been observed that the trees obtained from the particular domains of proteins can be employed to decipher the interacting domains of interacting protein pairs, as these domains show strong coevolution than the whole protein sequences. It relies on the evidences proving the notion that the interacting domains show strong coevolution patterns than the non-interacting domains with in the same proteins (Jothi et al. 2006).

Matrix match maker (*MMM*) is another approach for coevolution-based prediction of PPIs which is proposed by Tiller and Charlebois (Tillier and Charlebois 2009). This method identifies coevolving protein partners by determining the largest common submatrix in a given pair of distance matrices of two protein families. This method enables the identification of coevolving proteins in the different family sizes and also enables the determination of multiple coevolving partners. It also aids in obtaining lineage-specific coevolution. Tiller and Charlebois employed MMM approach in predicting PPIs in human genome. They showed that physically interacting proteins strongly coevolve than those proteins which only belong to the similar biochemical pathways. Further, they observed the existence of numerous connections in human coevolution network and several other PPIs also persist that are currently known from current experimental techniques. Strong coevolving pairs indicate the interactions are being preserved over long evolutionary periods which imply their essentiality for various cellular activities (Tillier and Charlebois 2009).

MMM approach has shown more accuracy than mirror tree method. The reason behind this accuracy is that MMM has reduced sensitivity to artifacts arose as a result of assembly of protein families which affect the Pearson's correlation coefficient. This coefficient in turn is employed to correlate the distance matrices of two protein families by mirror tree methods. MMM also exhibits low sensitivity to false positive coevolutionary pairs. This is due to the longer internal branches in both the trees containing divergent species. MMM is a better approach than mirror tree method but it requires high computational demand. MMM takes long execution time. To overcome this problem, Radionov et al. have devised a novel approach that uses maximum compatible cliques for detection of protein coevolution (Rodionov et al. 2011). This approach is known as MMMvII, which is 600 times faster than MMM and has comparable accuracy as original MMM. MMMvII allows more extensive and intricate coevolution analysis (Yu et al. 2019).

6.1.2 Gene-Based Methods

Gene-based methodologies predict PPIs utilizing the hidden details contained in genomic sequences. Studying proteins at gene level provides great information about the interaction of proteins. For example, gene-based methods can provide information about how specific protein encoding is governed by a particular gene. Gene-based methodologies are divided into the following three categories, namely: (1) gene fusion/domain fusion methods, (2) gene neighborhood-based methods, and (3) gene co-expression-based methods.

6.1.2.1 Gene Fusion or Domain Fusion Method

Gene fusion/domain fusion methodology depends on the fact that the genes present individually in one organism may fuse and present in a single open reading frame in other organism. This implies that genes fused together are functionally related and are involved in some physical interaction with each other (Enright et al. 1999; Marcotte et al. 1999; Sarkar et al. 2018). Several theories have been proposed behind the gene fusion events. For instance, there is a huge complexity in eukaryotic cells, hence, fusion will aid in maintaining the relative concentration of proteins, without an increase in the production of proteins (Enright et al. 1999; Morilla et al. 2010). It is also suggested that fusions are required to regulate the diffusion rates which are decreased by the hurdles caused due to cytoskeleton (Morilla et al. 2010; Yanai et al. 2001).

Fusions are more frequently observed in prokaryotes where genes are co-expressed in operons (Enright and Ouzounis 2001b; Morilla et al. 2010). All these observations are suggestive of gene fusion events that are dependent on organism phyletic lineages. However, in all the instances, it is recommended that the physical interaction between the proteins is a major reason for the gene fusion events (Morilla et al. 2010). Hence, gene fusion method enables the prediction of PPIs. Several fusion events have been marked for the proteins that are majorly participating in metabolic pathways (Overbeek et al. 1999; Sarkar et al. 2018).

Strategies for the identification of gene fusion events for the PPIs prediction are based on using either the whole gene sequences or the domain information. The methods using the whole sequences are known as gene fusion methods, and those using domain information are known as domain fusion methods.

Enright et al. suggested a sequence comparison-based methodology for the identification of gene fusion events in the whole genomes (Enright et al. 1999). This approach relies on the fact that during the course of evolution, there is a selective pressure on some gene to undergo fusion events for their co-regulation and possible interactions. This information can be used to predict the functional linkage between the proteins. More elaborately, Enright's PPI prediction methodology relies on the notion that in a particular species, some protein families encompass fused domains which usually exist as single complete proteins in other species. They defined individual domains as component proteins and fused proteins as composite proteins. They took full genome sequences of the species into consideration in order to identify gene fusion events so that the orthologous proteins across the species can also be detected. They assumed that in case a composite gene is showing complete identity with the two component genes in other species, then it can be predicted that the two component proteins are most likely involved in some kind of interactions. These interactions can be either direct interactions or indirect interactions.

Direct interaction refers to the physical interaction and indirect refers to the functional interaction between the component proteins. This functional interaction can be either due to their involvement in the same metabolic pathway or their participation in a common biochemical reaction or they may exhibit identical gene regulation. They obtained 215 component proteins that are involved in 88 gene fusion events in three query genomes, namely: *H. influenza*, *E. coli,* and *M. jannaschii*. Out of 88 fusion events, they identified 39 in *E. coli*, 25 in *H. influenza,* and 24 in *M. janaaschii*. Further, 94 composite proteins were identified in four reference genomes which include 36 composite proteins in *S. cerevisiae*, 9 in *M. jannaschii*, 25 in *H. influenza,* and 24 in *E. coli*. Authors found a total of 122 possible pairwise interactions among the proteins and estimated that the methodology possesses ~75% precision (Enright et al. 1999)

Morilla et al. presented a domain fusion-based method to predict PPIs in human interactome (Morilla et al. 2010). This method employs domain-based annotations such as CATH and PFAM. CATH is a manual curation of protein domain structures classification. In contrast, PFAM is a database containing protein domain families each of which is displayed by MSA and hidden Markov models. Eukaryotic genomes contain large expanded families denoting the existence of multiple paralogs, hence the ortholog assignment becomes very noisy. They have developed a method known as CODA (co-occurrence of domain analysis) which utilizes domain fusion data of human genome interactome from both PFAM and CATH and uses a simple score in order to overcome the problems arise due to paralogs in an efficient manner (Morilla et al. 2010; Reid et al. 2010). They employed this method and identified numerous novel PPIs in human.

The performance and accuracy of this method have been tested before applying this method to the real biological query. They benchmarked the CODA performance for prediction of PPIs by taking into consideration various experimental PPI datasets. They observed that domain fusion is highly related to proteins physical interactions as compared to the functional association of proteins such as their involvement in common biological process or pathway. Hence, it is suggested that domain fusion approach is more applicable to the prediction of protein complexes. Further, graph-based analysis of CODA derived PPI networks were performed, it showed that the topological features of the domain fusion derived PPI networks were similar to the topological features of the true biological networks. They applied CODA to study human kinetochore PPI network. They constructed the PPI domain fusion sub-network model for human kinetochore and found that several interactions that are predicted have previously been experimentally observed and several novel PPIs have been predicted that need to be validated experimentally (Morilla et al. 2010). Several undiscovered kinetochore proteins have been revealed while studying the human kinetochore domain fusion sub-network. The outcomes of domain fusion approach suggest that there are several hidden regions in the human interactome containing functional elements playing key roles that need further enlightenment.

Domain or gene fusion methods in several instances are not booming as a resultant of inherent complexity in structure of genomes. Domain fusion approach sometimes results in false positive outcomes in predicting functional association of proteins due to the promiscuous nature of domains. Other demerit of these gene fusion methods is that gene fusion events are not frequent in simple organisms, thus, limiting the usage of these methods in predicting interaction for all the proteins. For instance, Marsh et al. found that only 3.7% of their datasets are related to gene fusion events (Marsh et al. 2013).

6.1.2.2 Gene Cluster/Neighborhood-Based Methods

Gene cluster refers to the set of genes that are clustered together in the genome with their intergenic distance of few thousand base pairs. Gene cluster size can vary from few to hundreds of genes. Genes encoding for the interacting protein partners are mostly transcribed in the form of a single unit known as operon in bacterial species that may show co-regulatory patterns in eukaryotes. Dandekar et al. have found that in eukaryotes, genes indulged in similar functions or participate in same biological reaction or metabolic pathways are located in close proximity of each other (Dandekar et al. 1998).

Gene cluster-based method or gene neighborhood-based method relies on a notion that if the orthologs of genes encoding for the query proteins are found in neighborhood across multiple genomes, then there is a high probability that proteins encoded by such genes are interacting with each other. This approach is also known as domain/gene co-occurrence. One of the disadvantages of this method is that it is not applicable in predicting the interaction between the proteins where the genes encoding for the query proteins are not located close to each other in a genome (Keskin et al. 2016).

Dandekar et al. analyzed three triple genome sets that included proteobacteria (*Escherichia coli*, *Haemophilus influenzae*, and *Helicobacter pylori*), archaea (*Methanobacterium thermoautotrophicum*, *Methanococcus jannaschii*, and *Archaeoglobus fulgidus*) and Gram positive bacteria (*M. pneumoniae*, *M. genitalium*, and *Bacillus subtilis*) (Dandekar et al. 1998). They observed that around 100 genes among each of these three triple set genomes occur as gene clusters or pairs. Direction for the transcription for each of these gene pairs is same for all the genes. They have divided the interaction between the proteins encoded by these gene pairs into three categories, namely: experimentally confirmed, predicted, and putative interactions. They found that the physical interactions between the proteins encoded by 75% of gene pairs have been demonstrated experimentally. The examples include the interaction between the proteins forming the ribosomal particle, RNA polymerase subunits, transporter subunits, cell division proteins, ATP synthetase subunits, etc. Around 20% of the gene pairs are encoding for the proteins that are predicted for physical interaction with each other (Dandekar et al. 1998).

For instance, the gene pairs encoding for proteins conserved in proteobacteria were predicted for their involvement in physical interaction. Of these proteins, one is a homolog of surface exclusion protein and the other is a protein similar to an enzyme CDP ribitol pyrophosphorylase that is involved in synthesis of cell wall surface (Olmsted et al. 1993).

In archaea, genes encoding for proteins, namely mutator protein T (MutT) and transcription factor TFIIS family homolog exist as a conserved gene pair. These proteins are involved in enhancing the fidelity of transcripts. Modeling studies showed that these proteins form complex through their β-sheet interaction and function in a coordinated manner to improve the fidelity of transcript synthesis (Bessman et al. 1996; Jeon and Agarwal 1996).

Further, Dandekar et al. found that there is a chance that the proteins encoded by 5% gene pairs are either not accredited by any function or there is no existence of an interaction between them. For instance, the gene pairs, MG222 and MG221 in *M. genitalium* and their corresponding orthologs in *B. subtilis* and *M. pneumonia* are conserved with genes encoding for the proteins participating in cell division. Encoded proteins of these gene pairs are predicted to form complex and their involvement in cell division has been anticipated. However, the existence of interactions between them is still not proved (Dandekar et al. 1998).

Muley and Ranjan et al. analyzed the interaction among the proteins from 14 biological pathways of *E. coli* reported in KEGG database in order to predict the functional linkage between these proteins. They employed five methods, namely: (1) phylogenetic profiling, (2) gene neighborhood, (3) mirror tree variant, (4) co-occurrence of orthologous genes in gene cluster, (5) co-expression, for such prediction. They observed that gene neighborhood method is the best prediction approach in case of metabolic pathways. They also demonstrated that the efficiency of a specific prediction approach is dependent on the pathway being analyzed (Muley and Ranjan 2013). It has also been reported that gene neighborhood method is simple to adapt, and its reliability increases with increase in the number of genomes considered for the analysis (Keskin et al. 2016).

6.1.2.3 Gene Co-Expression-Based Methods

Genes with similar expression profiles have high probability of interaction, which forms the basis of gene co-expression approach for PPI prediction. Hence, scrutinizing genes which exhibit similar transcription profiles can aid in prediction of PPIs. Algorithms have been designed which clusters genes with identical expression patterns in one cluster. Proteins encoded by the genes residing in same cluster have high chance of interaction than those belonging to other clusters. This consideration is based on yeast transcriptome and proteome datasets (Ge et al. 2001; Jansen et al. 2002). This PPI prediction approach has been developed by Ge et al. which is also well-known as "transcriptome interactome correlation mapping" (Ge et al. 2001).

Grigoriev et al. demonstrated the correlation between the expression of gene pairs and the interaction of proteins encoded by them for the two genomes. They chose considerably two different genomes for their unified analysis of gene expression and protein interactions. These genomes were: (1) genome of bacteriophage T7, a simple genome and (2) genome of yeast *S. cerevisiae*, a complex genome. They observed that the interactions between the pair of proteins encoded by the co-expressed genes are more frequent than for those encoded by the randomly expressed genes. Further, they also found significant similarities for the expression profiles of genes coding for the interacting pair of proteins than for the random protein pairs (Grigoriev 2001).

Bhardwaj and Lu took an initiative for the investigation of global relationship of PPIs with expression of genes (Bhardwaj and Lu 2005). They studied this relationship employing highly divergent species, namely *E. coli*, yeast, mouse, and human, for which the substantial amount of data with respect to gene expression and protein interactions is available. Authors established high correlation for the expression of interacting pairs of proteins in *E. coli*. In contrast to *E. coli*, the other species (yeast, mouse, and human) showed slightly stronger correlation of the expression of interacting pair of protein than the random pairs of proteins. Authors proposed a protocol to integrate interaction and expression data with the ortholog information to further strengthen the correlation. Using this protocol, they found that in yeast the likelihood of predicting the interacting protein pairs has improved from 1.4 to 9.4 and demonstrated that the correlation between the gene expression and protein interaction is strengthened by coupling it via evolutionary information. This study established an evolutionary conservation among the co-expressing and interacting pairs than for the random pairs of proteins (Bhardwaj and Lu 2005).

Several databases have been designed which store gene co-expression data for differential species. These databases include COXPRESdb (Okamura et al. 2015), GEO (Barrett et al. 2013), ATTED (Aoki et al. 2016). ATTED and COXPRESdb are plant and animal species gene co-expression databases, respectively. These databases provide an ample amount of information with respect to gene co-expression and thereby aid in predicting PPIs.

Although there is a strong correlation among PPIs and the gene co-expression, PPIs prediction on the basis of gene co-expression is not devoid of limitations. Two severe pitfalls associated with gene co-expression approach of PPI prediction are: (1) Noise in the expression data affects the prediction efficiency. (2) Lower accuracy as a resultant of lack of absolute correlation between the gene transcription levels and protein levels. Hence, these methods are mostly employed as complementary with experimental outcomes instead of being adopted solely (Ge et al. 2001; Jansen et al. 2002).

6.1.3 Protein-Based Methods

Proteins interact in a highly restricted manner through their three-dimensional structures (Jones and Thornton 1996). The information of specific fold that proteins will acquire is encoded in their amino acid sequence. Further, PPI interface analysis has also shown the existence of shape and amino acid composition complementarity among the interacting proteins (Jones and Thornton 1996; Sheinerman et al. 2000). Recently, it has also been shown that there is a coupling in the evolution of amino acid sequences at the interfacial regions of the interacting partners (Hopf et al. 2014; Ovchinnikov et al. 2014). All these evidences conclude that the composition and order of amino acids in proteins and subsequently the shape the protein attains plays an important role in discerning the nature of interfaces that can interact with each other.

Several protein-based approaches have been established for the prediction and understanding of potential interaction among the proteins. In contrast to phylogeny or gene-based methods, these methods use the information at protein level which makes these methods highly reliable for PPIs prediction. They employ either the protein's amino acid sequences or their structures to figure out the unknown PPIs. Hence protein-based methods for the PPIs prediction have been divided into three classes, namely: (1) sequence-based, (2) structure-based, (3) network-based methods on the basis of the details being used by them.

6.1.3.1 Sequence-Based Methods

The availability of amino acid sequences of proteins as the most complete dataset adds an advantage of developing and employing sequence-based methodologies for both the modeling and the prediction of PPIs (Ben-Hur and Noble 2005; Gomez et al. 2003; Hamp and Rost 2015; Shen et al. 2007). Numerous sequence-based methodologies have been developed that can predict PPIs with high accuracy. Some of them include such as PIPE2 (Pitre et al. 2012), Profppikernel (Hamp and Rost

2015), AutoCorrelation (Guo et al. 2008), SigProd (Martin et al. 2005), and Sun's work (Sun et al. 2017), Yang's work (Yang et al. 2010), Wong's work (Wong et al. 2015), You's work (You et al. 2013), Zhou's work (Zhou et al. 2011).

The major advantage of using sequence-based approach for PPI prediction includes (1) amino acid sequences of proteins can be obtained in a straightforward manner, (2) there is no requirement of additional data such as the protein structures and their phylogenetic profiles, (3) sequence-based methods have major applicability as compared to other methods owing to their more universality. The major disadvantage of methods based only on sequence is that the accuracy of such methods is somewhat lower as compared to other approaches (Keskin et al. 2016). Some of the sequence-based methods developed by various research groups across the globe for the prediction of PPIs are outlined briefly in the following section.

The simplest sequence-based approach is a domain/motif-based method in which the proteins are screened for the possession of motifs or domains or some sequence that are known for interaction with each other in the interacting pairs of proteins. For instance, Becerra et al. have introduced a computational approach to predict motif-mediated PPIs between human and HIV-1 via systematic comparison of linear motif filtering methods (Becerra et al. 2017). The methodology developed by them only uses protein sequences as inputs and there is a scope of extension of this method for prediction of interactions between any other eukaryotic virus and host. This will aid in depicting the virus attack mechanism on the basis of linear motif mimicry (Becerra et al. 2017). Several public databases have been designed which store the information regarding the sequence patterns of PPI sites and functional regions which are known as domains or motifs on the basis of their sequence lengths. These databases include such as InterPro (Finn et al. 2017), ELM (Dinkel et al. 2012), PROSITE (Sigrist et al. 2010), Pfam (Finn et al. 2016), PRINTS (Attwood et al. 2012), and ProDom (Bru et al. 2005; Corpet et al. 1998).

Wojcik and Schachter et al. suggested a new domain-based method known as "interaction domain pair profile method" to predict PPI map of an organism (Wojcik and Schachter 2001). The method utilizes the interaction map of proteins in an organism along with domain interaction details for the prediction of interaction map for another organism. They applied this approach to predict PPI map for *E. coli* from the interaction map available for *Helicobacter pylori* which is a gastric pathogen for humans. Domain-based method showed various advantages over the method that utilizes full length sequences. Domain-based method encompasses high sensitivity by predicting the potential interactions and also eliminates the occurrence of false positive outcomes (Wojcik and Schachter 2001).

Kim et al. presented a statistical scoring system known as PID (potentially interacting domain pairs) matrix score to calculate intractability (interaction probability) between the domains and to yield information about interacting pairs of proteins (Kim et al. 2002). PID matrix showed 98% specificity which shows that there is a 30-fold increment in the information about the interacting protein pairs. Authors demonstrated that the genome-wide interaction network can be obtained via the usage of PID matrix.

Hayashida et al. proposed a novel method using conditional random field for the prediction of PPIs. They focused on utilizing mutual information between the residues and coupled that with conditional random fields (Deng et al. 2003; Hayashida et al. 2011). They performed the computational experiments using PPI datasets for several organisms and calculated the area under the curve. They found that the proposed methodology used with and without mutual information between the domains outperformed the expectation maximization method developed by Deng et al. which is considered to be the best domain–domain interaction-based predictors (Deng et al. 2003; Hayashida et al. 2011).

Kamada et al. utilized protein domain knowledge to predict the strength of PPIs. They employed two machine learning methods, namely: (1) relevance vector machine (RVM) and (2) support vector regression (SVR) to perform computational experiments for the set of data attained from biological experiments (Chen et al. 2006b; Kamada et al. 2014). Authors observed that the results obtained from both SVR and RVM outperformed the best existing method known as association probabilistic method (APM) to measure the strength of PPIs (Chen et al. 2006b; Kamada et al. 2014).

Singhal and Rehat presented a computational approach known as DomainGA which uses domain–domain interaction information for the prediction of PPIs. This multi-parameter optimization method derives quantitative scoring scheme for domain–domain pairs from the available PPI information. Scores obtained are then used to predict the interacting pairs of proteins. DomainGA promises for the wide scale applications across multiple organisms (Singhal and Resat 2007).

Zhang et al. developed a domain-based method which predicts PPIs on the basis of information available on interactions between the domains and proteins. In this approach, an optimization model was created by utilizing the information from various databases including DOMINE, Interdom, Pfam, and 3did databases. Authors predicted 736 probable PPIs for 147 proteins in the integrin adhesome PPI network (Zhang et al. 2016).

Li et al. proposed a computational-based domain–domain method and an SVM (support vector machine) model to predict PPIs. This novel methodology takes into account domain–domain interaction scores and physicochemical property of the domain. A DDI prediction model is created by using physicochemical property of a domain as an SVM feature. Finally, domain–domain interaction score and the DDI prediction model were coupled for the construction of PPI model (Li et al. 2019).

Pitre et al. presented an algorithm that can generate PPI prediction parameters for variety of organisms in an automated manner. This algorithm relies on the notion that there is a co-occurrence of small polypeptide sequences between the interacting pairs of proteins and that remains conserved across different organisms. Further, they have also found that the global PPIs can be predicted from the previously reported PPIs with in the same organism or different organism based on only protein's primary amino acid sequence. They employed their algorithm in order to predict the PPI maps for *C. elegans* and *S. pombe*, which encompasses nearly 37, 500 and 9000 PPIs, respectively (Pitre et al. 2012).

Martin et al. employed pairwise kernel to predict PPIs and represented a sequence by 3-mers (Martin et al. 2005). They converted each 3-mer into a signature in which a middle amino acid is defined as the root and the adjacent amino acids on both the sides as the neighbors that are ordered alphabetically. Protein was then displayed as a signature vector in which each element defines the frequency of one signature. A tensor product of the signature vectors of the proteins was then used to define the feature vector of the protein pair. Finally, the standard Euclidean inner product with in the support vector machine (SVM) classifier was employed to compute the signature product of two protein partners as a kernel function. Authors used this approach for the publically available datasets like for human, mouse, yeast, and *Helicobacter pylori*. 70–80% prediction accuracy rates were obtained for yeast and *H. pylori* datasets. Employed mouse and human datasets showed that their methodology is applicable for the cross species prediction. The yeast dataset was reutilized to further check the applicability of their algorithm for the prediction of domains.

Guo et al. proposed a new methodology which is based on support vector machine (SVM) and auto covariance (AC) (Guo et al. 2008). AC takes into account the interaction between the amino acids which are apart from each other by a fixed number of amino acids in a protein sequence. Thus, their methodology considers the neighboring effect of amino acids thereby making it possible to search for particular patterns along the entire sequence. Amino acids were assigned the numerical values which represents the physicochemical properties of amino acids. These values were then analyzed by AC to decipher the covariance. Ultimately, the construction of SVM method was done by employing the vectors of autocovariance variables as inputs. An optimization experiment showed that a single amino acid along with its 30 vicinal amino acids allows the characterization of PPIs. Authors tested this method on the yeast (*S. cerevisiae*) PPI data and obtained an accuracy of 87.36% in their prediction. Further evaluation of this method using other yeast PPI datasets yielded an accuracy of 88.09% (Guo et al. 2008).

Further, Hamp and Rost established a novel method which is able to predict PPIs from protein sequences alone via usage of their evolutionary profiles through profile kernel support vector machines (SVMs) (Hamp and Rost 2015). The method was optimized for the prediction of those protein pairs that came in close contact to each other at some point of time, and not for predicting any functional association between them. The method showed an improvement in the prediction as compared to other existing methods that solely depend on the protein sequences for the PPI prediction. The substantial improvement was observed in those cases where the proteins were not significantly similar in their sequences. Authors further improvised their method by employing a simple filter that removes the protein pairs not expressed in the same tissue.

Hashemifar et al. presented a new deep learning framework called as DPPI for the prediction and modeling of direct physical PPIs that rely on their primary amino acid sequences (Hashemifar et al. 2018). This methodology skillfully utilizes a deep, Siamese-like convolution neural network in combination with random projection and data augmentation for the PPIs prediction. Authors observed an outperformance of DPPI over the other state-of-the-art methods in terms of area under precisionrecall curve (auPR). They also found higher computational efficiency of DPPI. DPPI is also capable of prediction of homo-dimeric interactions where other methods decline to give accurate results. They have also the found specific applications of DPPI such as in prediction of cytokine-receptor binding affinities. Several advantages of this deep learning framework include: (1) efficient in handling ample amount of data, (2) there is flexibility in training the models for diverse applications without the requirement of major changes in differential parameter, (3) it has the capacity to robustly unravel complex and non-linear relationships among the interacting protein partners.

Yang et al. made an efficient effort in unraveling the essential and challenging human–virus protein–protein interactions (Yang et al. 2020). These efforts provided a mechanistic understanding of viral infections. Authors introduced a computational pipeline known as doc2vec. Doc2vec is an unsupervised protein sequence embedding-based machine learning technique. They trained doc2vec using human-virus PPIs as positive dataset and also compiled negative PPI datasets. They also trained the model using ample number of sequences of proteins to capture the features of proteins for the reliable prediction of human virus PPIs using random forest approach. Authors found higher accuracy for their methodology and facilitate its free accessibility through a web server (http://zzdlab.com/InterSPPI/) as a part of host pathogen PPI prediction platform. This approach not only contributes as an important predictor to accelerate the prediction of host–pathogen interactions but also can provide deep insights of viral infections to humans.

Wang et al. proposed an innovative methodology known as CNN-FSRF for the prediction of PPIs that relies on amino acid sequence of proteins using a combination of deep learning convolution neural network (CNN) with feature selective rotation forest (FSRF) (Wang et al. 2019). This methodology first converts sequence of a protein into PSSM (position specific scoring matrix) that contains evolutionary details of the protein. Further, the usage of CNN extracts the information related to hidden characteristics of the protein. Finally, FSRF filters out the noise by removal of redundant information and yields the prediction results with high accuracy. Method was applied on yeast and *Helicobacter pylori* PPI datasets and CNN-FSRF returned the results with a prediction accuracy of 97.75 and 88.96%, respectively. It has been suggested that CNN-FSRF can complement the experimental outcomes in the identification of interacting partner proteins.

6.1.3.2 Structure-Based Methods

Numerous structure-based methodologies have been developed that aim at predicting PPIs employing the three-dimensional structural details of proteins. These structure-based methods for prediction of PPIs have been classified into two categories, namely: (1) docking-based PPI prediction approaches and (2) template-based approaches to predict assembly of proteins (Keskin et al. 2016). These two structure-based categories are elaborated in the following sections.

Docking-Based Prediction of PPIs

Docking is a computational means to predict the orientation of binding protein partners, calculate their binding energies, and determine their structural assemblies (Vakser 2014). The first step in docking includes scanning all the possible binding orientations between two query proteins. There are two ways to scan for the possible orientations, one is global scan and the other is the local scan. In global scan, one of the proteins kept static and is known as receptor. The other proteins are allowed to rotate around the receptor and that rotating protein is known as ligand. Global scan terminates once all the possible orientations of ligand around the receptor protein in a three-dimensional space have been sampled. Requirement of several transitions and rotations makes the global scan computationally very expensive. Therefore, an approach known as fast Fourier transform has been developed to make global scan strategies cost-effective. In contrast, the local scan involves the search for the surface characteristics including spurs, patch's flatness, solvent excluded regions and pockets in the target proteins. These features from all the proteins are then compared to find the best complementarity. Since there is only the comparison of local surface features, the remaining surfaces of the target proteins can cause steric hindrances. Hence, there is a need to filter out those complexes which exhibit such steric clashes.

An alternate strategy of sampling includes fusion of prior information about the interaction such as biophysical, biochemical, mutation, chemical shift, or computationally predicted details about binding patch at the sampling stage. This method not only enhances the prediction accuracy but it also restricts the sampling space. Sampling stages can be done in two ways either in rigid or flexible way. In rigid sampling approach, modifications in structure associated characteristics such as bond lengths, bond angles, and backbone orientation are not taken into consideration. In contrast, the flexible sampling approach considers the conformational alterations in the structures of proteins.

The task of docking or the prediction of correct orientation of the proteins becomes highly challenging in case of rigid body docking where the target protein faces a substantial change in its conformation upon binding by other protein. Thousands of poses of protein complexes are obtained during the sampling stage and scoring those complexes becomes an essential to obtain the best pose. Scoring and sampling can be performed either simultaneously or they can be done one after the other. Scoring components such as binding free energies obtained by employing different force fields, electrostatic and shape complementarity entities contribute to evaluate the tightness of predicted binding.

Protein–protein docking benchmark serves as valuable source for non-redundant set of the structures of protein complexes and their unbound structural forms. Thus, a docking benchmark aids in systematic evaluation of differential docking methodologies and their scoring methodologies. Docking benchmark should cover all the diverse types of protein interactions. Docking approach should be able to handle all the possible interactions among the proteins and should not be biased toward the particular set of interactions (Hwang et al. 2010). The latest version of docking benchmark is docking benchmark version 5 which includes 230 complexes along with the unbound structure of constituent proteins (Hwang et al. 2010; Vreven et al. 2015). Benchmark includes variety of interacting complexes such as enzyme–substrate complexes, antigen–antibody complexes, etc. Although there is a limitation of number of complexes in the benchmark, this number can be increased via keeping a continuous track of updates in PDB. It has been suggested that docking benchmark can split into three sets, namely: the rigid body docking set, the medium, and the difficult docking sets. This classification is based on the degree of conformational variations observed upon protein binding. Root mean square deviation of Cα atoms of interfacial amino acid residues in a bound and unbound state is an effective means of presentation of difficulty in the docking (Keskin et al. 2016).

Several criteria are being followed to assess the predictions of docking approaches. Calculating the RMSD between the predicted interfaces and native interfaces reveals the accuracy of predicted interface region. Further, superimposition of receptor molecule in the predicted and native complex and determining the Root mean square deviation of ligand molecule predict the degree of accuracy in binding orientation. Calculating the percentage of predicted interface residues to be in the native interface is another means of measuring the prediction accuracy. A simple method known as CAPRI (critical assessment of predicted interactions) has been initiated to evaluate the execution of docking algorithms and scoring methods. CAPRI is not only involved in testing the performance of docking methods, but also contributed significantly to the improvisation of the existing docking methods. CAPRI has also boosted the advancement of novel docking approaches (Janin 2010).

ZDOCK is the most popular rigid body docking program which employs FTT-based global search in order to predict protein complexes. ZDOCK involves the prediction of the possible orientations of ligand and receptor and ranks those orientations using scores calculated by the scoring function. The scoring functions in ZDOCK are based on electrostatics, statistical potential, shape, and complementarity (Pierce et al. 2014).

RosettaDock is an alternate docking tool which is based on Monte Carlo search method (Chaudhury et al. 2011). This tool either begins by random orientation of proteins or by initial user-defined pose. In this step, each side chain is represented as centroid and the step is coarse grained. The method predicts the lowest energy structure after 500-step Monte Carlo search and passes that structure for the high-resolution refinement step. Refinement stage is required to improve the final set of solutions and their corresponding ranks obtained by docking methods. Hence, several refinement techniques have been developed which take into consideration the hits obtained by docking method. Those hits are refined and the final putative complex is optimized via refinement approach which takes into account the flexibility, scoring, and the biological information if available (Mashiach et al. 2010).

HADDOCK is the high ambiguity-driven docking method that utilizes the biophysical and biochemical interaction data including mutagenesis data or chemical shift perturbation data attained from titration experiments conducted using NMR. Details of interacting residues are employed as ambiguous interaction restraints (AIR) to perform docking. Structures obtained are scored or ranked in accordance with the intermolecular energy, which is calculated as the total of electrostatics, van der Waals, and AIR energy values. Dominguez et al. demonstrated the accuracy of their algorithm by solving the structures of three molecular complexes, namely: first is HIV protein gp120 in complex with CD4 protein, second: Enzyme IIA glucose (E2A) in complex with histidine-containing phosphocarrier protein (HPr), and third: N-terminus domain of enzyme 1(E1N) in complex with HPr. Authors observed that HADDOCK generated the best structures of complexes with lowest intermolecular energies (Dominguez et al. 2003). The obtained structures of complexes were close to the previously published structures of those complexes. Structures obtained were within 2.0 Å backbone RMSD with already published structures. Some other docking and refinement approaches are also available that are listed in Table 6.1.

The blind docking is computationally quite cumbersome and generates numerous feasible poses of protein complexes; hence, the usage of classical docking methodology is a daunting task at proteome level. Hence, knowledge-based docking methods have cropped up to assemble the structure of complexes employing the information available about protein binding. Such approaches are also known as template-based docking approaches that are discussed in detail in the following section.

Template-Based Methods for Prediction of PPIs

Template-based docking methods predict structures of protein complexes based on the information available from the experimentally solved structure of complexes (Kundrotas et al. 2012; Tuncbag et al. 2011a). Such approaches have been flourished based on an idea that usually the homologous proteins utilize the same binding interfaces. Further, it has also been notified that protein pairs which are completely different in their structures exhibit structurally conserved binding interface regions (Keskin et al. 2004). Knowledge about both the structural and physiochemical aspects of protein interactions further galvanized that the evolutionary details about the sequence and structural similarity of proteins and the binding interfacial regions can boost the structural modeling process of unknown protein assemblies (Tuncbag et al. 2011b).

Steps involved in template-based prediction approaches are: (1) preparation of library of templates, (2) selection of target set, (3) similarity search for target to template, (4) refinement, and (5) scoring. Of all these steps, template library preparation is the most important step in all these approaches since the prediction performance depends on the template diversity. Performance of template-based docking methods improves with an increase in the coverage of PDB. Annual statistical analysis of PDB showed that there is an exponential increment in the number of experimentally solved protein structures. The second step involves the selection of target set which can include either 2 proteins or all the constituting proteins of a pathway or the whole proteome. Higher computational efficiency of template-based docking approaches makes prediction possible at both gene level and proteome level. Further, in contrast to blind docking, the template-based docking results in limited number of possible conformations due to defined solution space. Template to target level similarity search step has been segregated into two steps, namely: the global and the local similarity-based search. Similarity search can be

Table 6.1 Summary of some of the docking and refinement approaches

Docking tool	URL	Summary	Validation by CAPRI (yes/ no)	References
ClusPro	https://cluspro.bu.edu/login.php	Widely employed protein–protein docking tool requiring only two files in PDB format	Yes	Comeau et al. (2004)
CombDock	http://bioinfo3d.cs.tau.ac.il/CombDock/download/	Employs combinatorial docking program for the assembly of multi-molecular proteins	No	Inbar et al. (2005)
FoXSDock	http://modbase.compbio.ucsf.edu/foxsdock/	Performs docking of two rigid protein structures on the basis of the SAXS (small angle X-ray scattering) profile of the complex of those proteins	Yes	Schneidman-Duhovny et al. (2011)
FireDock	http://bioinfo3d.cs.tau.ac.il/FireDock/	First online server to provide flexible refinement and scoring function for protein–protein docking, facilitates optimization of conformation of side chains	No	Mashiach et al. (2010)
FiberDock	http://bioinfo3d.cs.tau.ac.il/FiberDock/	First online docking refinement server which takes into account the flexibility of both backbone and side chains	No	Mashiach et al. (2010)
F2Dock	https://www.cs.utexas.edu/~bajaj/cvc/software/f2dock.shtml	Fast Fourier protein–protein docking algorithm is an adaptive docking algorithm which allows fast docking in multicore machine	No	Bajaj et al. (2009)
FRODOCK	http://frodock.chaconlab.org/	Novel method to perform fast rotational protein–protein docking	No	Ramírez-Aportela et al. (2016)
GRAMM-X	http://vakser.compbio.ku.edu/resources/gramm/grammx	An extension of original fast GRAMM fast Fourier transformation methodology that includes refinement stage, smoothed potentials and knowledge-based scoring	Yes	Tovchigrechko and Vakser (2006)

(continued)

Table 6.1 (continued)

HADDOCK	https://www.bonvinlab.org/software/haddock2.2/	High ambiguity-driven docking method utilizing the biophysical and biochemical interaction data including mutagenesis data or chemical shift perturbation data attained using NMR	Yes	Dominguez et al. (2003)
HDOCK	http://hdock.phys.hust.edu.cn/	An integrated suite for template-based modeling, homology search, structure prediction, macromolecular docking, and job management for fast protein–protein docking, it completes the docking job in 30 mins	Yes	Yan et al. (2017)
HexServer	http://hexserver.loria.fr/	First protein docking server which is based on Fourier transform which is augmented by graphics processor	No	Macindoe et al. (2010)
MEGADOCK	http://www.bi.cs.titech.ac.jp/megadock/	FT-grid-based protein–protein docking which takes into consideration the benefits of the numerous computation nodes and massively parallel CUDA architecture of NVIDIA GPUs	No	Ohue et al. (2014)
PatchDock	http://bioinfo3d.cs.tau.ac.il/PatchDock/	Docking algorithm based on molecular geometry which aims at determining docking transformations to attain best molecular shape complementarity	Yes	Schneidman-Duhovny et al. (2005)
pyDockWeb	https://life.bsc.es/pid/pydockweb/	Employs FTDock to provide the best rigid-body docking orientations and employs pyDock scoring function that includes desolvation energy, limited van der Waals, and electrostatics contribution for the docking evaluation	Yes	Jiménez-García et al. (2013)

(continued)

Table 6.1 (continued)

PIPER	https://www.schrodinger.com/piper	Based on advanced numerical methods and multi-staged approach which returns accurate structures of protein–protein complexes	Yes	Kozakov et al. (2006)
RosettaDock	https://rosie.graylab.jhu.edu/	Based on Monte Carlo search method for the prediction of lowest energy structures	Yes	Chaudhury et al. (2011)
SymmDock	http://bioinfo3d.cs.tau.ac.il/SymmDock/	Docking algorithm to predict the structures of cyclically symmetric homo-multimer proteins	Yes	Schneidman-Duhovny et al. (2005)
SwarmDock	https://bmm.crick.ac.uk/~svc-bmm-swarmdock/	Combines local docking and particle swarm optimization to obtain low energy positions and orientations of the binding protein partners	Yes	Torchala et al. (2013)
ZDock	http://zdock.umassmed.edu/	Employs FTT-based global search for the prediction of complexes of proteins	No	Pierce et al. (2014)

carried out by either using: sequence alignment, threading or structure alignment. In case of structure alignment, either the complete structures of target and template proteins can be aligned or only the domain or the interface regions of template complexes can be scanned on to the target surfaces.

In sequence-based approach, if the target protein sequences are comparable to the template protein complex sequences, then it is assumed that binding orientation of target proteins will be similar to the orientation of binding of template protein complex. In this approach, matching completely depends on similarity scores calculated for the protein sequences which limit the prediction space of template-based methods. Although it has been observed that proteins with completely different sequences exhibit comparable global folds, it further limits the prediction era of sequence similarity-based approach.

An alternative method of scanning the global structural similarity of proteins on template protein complex has been proposed for the proteins with low sequence identity exhibiting comparable global folds. Domain-based scanning in template proteins is an alternate criterion to predict the structures of target PPIs. Mosca et al. presented a database known as "Interactome 3D", which is not only a repository of structural details of PPIs, but also serves as a platform for modeling of PPIs. Interactome 3D encompasses the structural information of around 12,000 PPIs from eight model organisms. It also allows the users to upload their interaction data and also allows them to choose the best structural templates and construct 3D models in a fully automated mode (Mosca et al. 2013).

Threading is another way which takes into account the known protein structures as template to model the structures of target proteins. In this method, target protein sequence is threaded over the template protein structure in a homology independent manner. Threading-based approaches involve two major steps for modeling the protein assemblies. The first step involves the determination of candidate structures of each target proteins by threading each target protein sequence over the solved structures of template monomers. Then, dimeric threading is carried out in which the target protein sequences are joined and threaded over the structure of target protein complexes. Another step involves superimposition of best ranking templates from monomeric threading and dimeric threading to attain the final list of predicted complexes. Finally, the predicted structural complexes are assessed using the scoring function, which considers the energy calculation and statistical measures for the assessment (Szilagyi and Zhang 2014).

Another way by which completely distinct proteins can interact is through the similarity in their interface motifs, which has unlocked the new avenues to model the structures of protein complexes. Methods have been developed which utilize interface templates for the prediction of PPIs. In these methods, firstly there is a structural comparison of target proteins structure with that of template protein complex. If the surface regions of both target and template are similar, then target proteins are superimposed on template protein interface to obtain a putative protein complex, which is further refined and ranked by scoring functions. The first method that implements protein interface templates to model PPIs is PRISM. PRISM denotes protein interaction by structural matching, which is introduced by Tuncbag et al. (Tuncbag et al. 2011b). This methodology provides user with a protocol for the large-scale prediction of PPIs and for the structural assembly of protein complexes. PRISM works mainly by two steps: (a) comparison of target proteins at rigid body structural level to template protein–protein interfaces for which structures are known and (b) employment of the docking energy functions for the flexible refinement.

In more elaborative way, PRISM first prepares a template library in which there is an extraction of all the protein–protein interfaces contained in the PDB followed by the removal of redundant interfaces. The partner chains from each of the interface are then separated which are then aligned with target protein surfaces. PRISM also takes into consideration the conservation of binding hotspots at structural level in the interfaces while matching structures of target and template proteins. In case, the two target proteins are structurally alike to the protein chains of template interface and if a match of at least one of the binding hotspot between the template and target is observed, then it is considered that the two target chains are interacting. The matching targets are superimposed on template interface to obtain the first putative protein complex. The putative protein complexes obtained are then subjected to refinement procedures in which there is an elimination of complexes exhibiting steric hindrances. Further, there is an optimization of backbone and side chains of the final protein complex via a flexible refinement protocol which increases the accuracy of the prediction. Binding energies of the complexes are calculated to rank and score these complexes.

In the last decade, many new methods employing the structural knowledge of protein interface to predict PPIs were developed which includes: PrePPI (Zhang et al. 2012), IWrap (Hosur et al. 2011), iLoops (Planas-Iglesias et al. 2013), ISEARCH (Günther et al. 2007), EVcomplex (Hopf et al. 2014), KBDock (Ghoorah et al. 2014), PAIRPred (Afsar Minhas et al. 2014).

In summary, structure-based prediction of PPIs has several merits over sequence-based methods for PPIs prediction. Firstly, structure-based methods require information of interfacial residues in comparison with sequence-based methodologies which require the identification of complete amino acid sequence of protein. Secondly, currently, the accuracy of majority of machine learning methods that rely on protein structures is higher than those machine learning algorithms that are based on protein sequences. This is due to the fact that the structure provides the spatial information of interfacial amino acids in protein, which is of utmost importance in macromolecular recognition. Lack of such spatial information of amino acids limits the accuracy of sequence-based machine learning approaches. Thirdly and finally, the prior knowledge and information about geometric complementarity further adds superiority to structure-based methods (Xue et al. 2015).

Structure-based methods are also not devoid of limitations. Some of them include: (1) there are several proteins which are involved only in transient associations, hence they do not have experimentally attained structures. Structures of such proteins are modeled based on their amino acid sequence (Rao et al. 2014). (2) Structure-based methodologies consider the three-dimensional protein structure in their unbound form, they don't take into consideration the occurrence of conformational adjustments in the proteins upon interaction with other protein (Zhou and Qin 2007). (3) Several intrinsically disordered proteins are present in higher organism for which the defined 3D structures are absent. However, such proteins attain a specific structure after associating with their binding partners. In such intrinsically disordered proteins, structure-based methods abort (Dunker et al. 2000).

6.1.3.3 Network-Based Methods

Several network-based methodologies have been suggested for the prediction of PPIs. The methodologies falling under this category employ the existing PPI networks to predict new PPIs. They critically evaluate the topological features of the PPI networks. PPI networks indicate the fascinating pattern of connections between the proteins. In particular, proteins are arranged in small space in which the interacting proteins are close together. In simple terms, PPI networks are graphs which contain nodes and edges (Fig. 6.2). Each node is representing an individual protein and the line/edge between the nodes indicates the association between those proteins/nodes. Many topological properties like direct or indirect interactions between the proteins can be calculated using these graphs.

Proteins with many binding partners are known as hubs in the PPI network. These hubs play an essential role in biological processes. PPI networks in addition to the prediction of functional associations between the proteins also allow the prediction of novel PPIs. It is assumed that proteins with same binding partners in a network participates in a similar biological phenomenon (Chua et al. 2006; Pržulj et al. 2004).

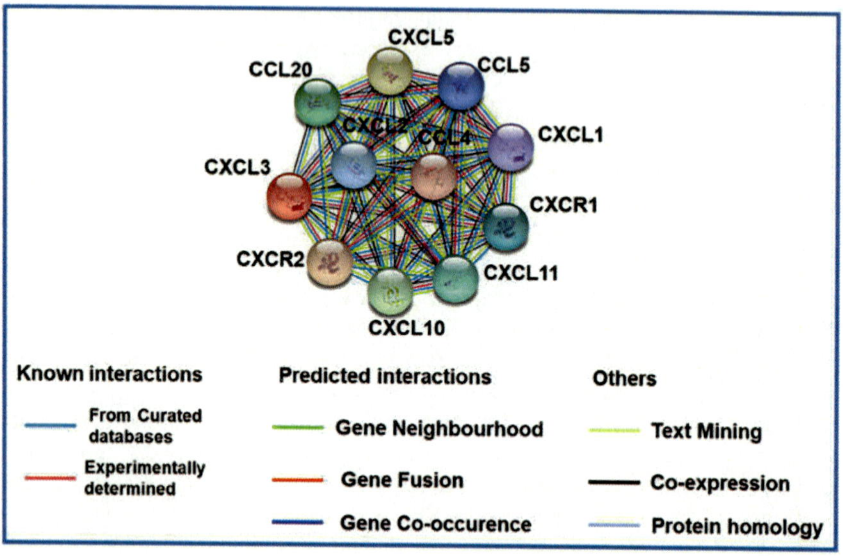

Fig. 6.2 PPI network generated for CXCL3 protein (Gulati et al. 2018) using STRING. Circles indicating nodes represent interacting protein partners for CXCL3. Lines indicate edges which represents association between the protein partners. Different line colors imply differential methods used for the prediction of association between the proteins

Further, it has also been assumed that if two proteins are sharing numerous binding partners, then they may share similar amino acid sequences, structures, folds, and other biochemical features, and it is highly likely that those proteins can also interact with each other. A major advantage of these methodologies is that they are devoid of requirement of sequence and structural properties of proteins. They predict PPIs by considering only the topological properties of existing PPI networks (Goldberg and Roth 2003).

Chen et al. devised a method known as IRAP (Interaction Reliability by Alternative Path), in which a missing interaction is obtained between the protein pairs exhibiting high score. The score here represents the number of common neighboring proteins between the protein pairs in the currently available PPI network (Chen et al. 2006a). Yu et al. described a novel method in which they predict a missing PPI in a defective clique, addition of which will convert the defective clique into a complete clique. Authors formulated an algorithm to apply their methodology to large-scale networks and found that this algorithm is highly efficient and exhibits good predictive ability (Yu et al. 2006).

Liu along with his coworkers proposed a computational protocol to find out spurious as well as missing interactions in the high throughput protein interaction data (Liu et al. 2008). Their method designates a local and a global interacting score to the protein pairs on the basis of their local and global topological information. Local interacting score is computed on the basis of common neighboring proteins for the pair of proteins. Global interacting score is calculated employing globally interacting protein group pairs. A final score known as "LGTweight" is calculated using the combination of both local and global interacting score. Final score denotes the interaction probability of pair of proteins. Authors tested this method on DIP yeast interaction dataset and observed that the interactions ranked top exhibit high functional homogeneity and localization coherence. The method attained five-fold higher sensitivity and precision than other existing methods (Liu et al. 2008).

Kuchaiev et al. presented a novel methodology to appraise the confidence levels of interactions in PPI network attained via experimental efforts (Kuchaiev et al. 2009). The technique has also been used for the prediction of new interactions which were overlooked during experimental studies. Predicted interactions can further aid in designing the needful future biological experiments. The technique is the first of its kind that utilizes the geometric graphs which are known as the best fitting network model for PPI networks. This approach achieved sensitivity and specificity of 90% and 85%, respectively. Authors employed this method for assigning the confidence scores to human PPI network obtained from BioGrid. They attained 251 novel PPIs in human PPI network, and most of which were sharing common GO terms.

Lei and Ruan proposed a unique algorithm based on network topology for the prediction of new PPIs and removal of erroneous interactions in the existing PPI network (Lei and Ruan 2012). They employed a random walk-based approach to find out proteins sharing high order topological features. It is highly likely that such proteins either belong to the same protein complex or they are interacting with each other. The algorithm applied on yeast PPI network data showed that the network reconstructed is much more reliable biologically than the original network. The quality of reconstructed network was assessed using multitude of information such as gene expression, gene ontology, conservation between species and known protein complexes. It was observed that the reconstructed network was of high quality when compared with networks obtained from other existing methods.

Further, continuous efforts are being made to enhance the performance of network-based approaches by merging the sequence and functional details of proteins with network topologies in order to predict PPIs. Phan and Sternberg proposed a method known as PPINLOG (protein–protein interaction network alignment method) (Phan and Sternberg 2012). This method employs the combinatorial power of protein sequence, function, and network topology details to decipher the protein complexes and their functions. Human and yeast genome PPIN were aligned and found that several subnetworks that are engaged in similar biological processes like transcriptional and proteasome related functions were conserved among the species.

Several meta methods have geared up to predict PPIs. Meta methods combine many predictors for the prediction of PPIs. These methodologies are designed to compensate the limitations of one or the other predictors. They blend the outputs of individual predictors to generate a concurrent output with increased accuracy and reliability of outputs than any of the outputs obtained from the individual predictors alone. For instance, MetaPred2CS is a sequence-based meta-server to predict the interaction between the proteins belonging to two-component system of prokaryotes.

This server combines six sequence-based predictors using support vector machine for predicting PPIs. Meta-PPISP is a meta-server to predict the sites of interactions between the proteins (Kara et al. 2016).

This sever was built on Con-PPISP, Promate, and PINUP to increase the robustness and accuracy of PPI prediction. These three methodologies are coupled in a linear regression analysis with the raw scores as input. Reasons to choose these methods are: (1) they are accessible through web servers and (2) they are based on very different attributes as input. Cons-PPISP is a neural network predictor that employs solvent accessibilities and sequence profiles of spatially neighboring residues as inputs. Promate employs a combined probability calculated from factors such as sequence conservation, amino acid pairing, secondary structure, and atom distribution. Empirical energy function consisting of side chain energy term forms the basis of PINUP. Cross validation concluded that Meta-PPISP outperformed all three independent methods (Qin and Zhou 2007).

6.2 Analysis of PPIs

Proteins interact with each other through their interface regions, hence, in order to get detailed understanding of PPIs, it is quintessential to study these protein interface regions in detail. Gaining insights into the characteristics of protein interface also reveals the binding affinities of proteins for their potential partners. The extraction of interface regions is essential to analyze the binding preferences of proteins. Residues at interface regions are conserved than the residues present at other regions (De et al. 2005; Jones et al. 2000; Jones and Thornton 1996; Jones and Thornton 1997; Keskin et al. 2004; Tuncbag et al. 2009b). These residues are either planar or well packed based on the kind of interactions they are involved in. Several approaches such as calculation of atomic distances and calculation of accessible surface area (ASA) of the residues have been developed to decipher the interface regions from three-dimensional protein complex coordinates (Bahadur et al. 2004; Nooren and Thornton 2003a).

Accessible surface area calculation for the residue in a monomeric form and in a complex form reveals that whether a particular residue belongs to an interface region or not. In case the difference between the ASA of a residue in monomeric state and complex state is more than a threshold value of $1 Å^2$, then that particular residue is marked as interface residue (Nooren and Thornton 2003a). Several databases and tools have been created that aid in scanning the interface regions in protein complexes. Some of the prominent are summarized in Table 6.2. Interface databases are enriched with characteristic features of binding surfaces. These databases contain information about both protein–protein interfaces and domain–domain interfaces. Domains are functional substructures of proteins. Intrachain or interchain contacts are being utilized to obtain domain–domain interfaces. Databases listed in Table 6.2 decode interface regions based on the similarity of topology, sequence, and structure.

Table 6.2 Summary of some of the databases and tools employed for analyzing interface regions of proteins

Database/tool	Web link	Summary	References
2P2I inspector	http://2p2idb.cnrs-mrs.fr	Database containing structural properties of PPI surface and also aids in designing of modulators for PPIs	Basse et al. (2012)
3DID	http://2p2idb.cnrs-mrs.fr/	3D domain interaction database provides critical molecular details which are essential for an interaction to occur	Mosca et al. (2014)
ASeDb	http://nic.ucsf.edu/asedb/	Alanine scanning energetics database contains binding energy changes occurred after mutating protein side chains to alanine	Thorn and Bogan (2001)
ANCHOR	http://structure.pitt.edu/anchor/	Tool designed to analyze the protein–protein interface to find out the suitability for the small molecule drug design	Meireles et al. (2010)
BID	http://tsailab.org/BID/	Database containing detailed information about protein interfaces in form of tables, contact maps, and their descriptive functions	Fischer et al. (2003)
Dock ground	http://dockground.compbio.ku.edu/	Comprehensive data resources for designing protein complexes	Kundrotas et al. (2018)
DiMoVo	http://fifi.ibbmc.u-psud.fr/DiMoVo	Machine learning-based method to precisely classify specific and non-specific protein–protein interfaces	Bernauer et al. (2008)
DynaFace	http://safir.prc.boun.edu.tr/dynaface/	Web server on the basis of hypothesis that analyzes the dynamics of proteins can reveal whether the interaction is obligatory/non-obligatory	Haliloglu et al. (2015)
DrugScorePPI	https://cpclab.uni-duesseldorf.de/dsppi/	Knowledge-based scoring function to identify hotspots in protein–protein interface by employing computational alanine scanning	Krüger and Gohlke (2010)

(continued)

Table 6.2 (continued)

FoldX	http://foldxsuite.crg.eu/node/196	An online force field which calculates protein's stability, folding, and their dynamics, also compute free energy for the formation of protein complexes	Schymkowitz et al. (2005)
FTMAP	https://ftmap.bu.edu/login.php	Determine druggable hotspots in proteins by employing Fourier domain correlation technique	Kozakov et al. (2015)
Galinter	http://galinter.bioinf.mpi-inf.mpg.de/	An intuitive method to characterize, analyze, and visualize the modes of binding of proteins and also enables the prediction of binding partners of proteins and also assists in engineering protein interactions and their inhibitors	Zhu et al. (2008)
HotPoint	http://prism.ccbb.ku.edu.tr/hotpoint	Server to predict hotspots for protein interfacial regions	Tunbag et al. (2010)
HotSprint	http://prism.ccbb.ku.edu.tr/hotsprint	Database containing details about computationally predicted hotspots in protein interfacial regions	Guney et al. (2007)
Hot region	http://prism.ccbb.ku.edu.tr/hotregion.	Database that provides information about hot region in the protein interfaces and also describes the structural properties such as pair potential, ASA of interfacial amino acid residues, and their relative ASA in monomeric and complex forms of proteins	Cukuroglu et al. (2012)
IAlign	http://cssb.biology.gatech.edu/iAlign	An efficient and accurate method for the alignment of interface region of proteins	Gao and Skolnick (2010)

(continued)

Table 6.2 (continued)

IBIS	https://www.ncbi.nlm.nih.gov/Structure/ibis/ibis.cgi	Inferred biomolecular interaction server for the prediction of interacting protein partners and their binding sites	Shoemaker et al. (2012)
KDBI	http://xin.cz3.nus.edu.sg/group/kdbi/kdbi.asp	Database that provides kinetic data for the biomolecular interactions including protein–protein interactions	Ji et al. (2003)
KFC, KFC2	http://kfc.mitchell-lab.org	Machine learning-based approach which predicts hotspots for the interaction of proteins	Darnell et al. (2008)
MAPPIS	http://bioinfo3d.cs.tau.ac.il/MAPPIS/	Novel method for the PPI's MSA (multiple structural alignment)	Shulman-Peleg et al. (2008)
NoxClass	http://noxclass.bioinf.mpi-inf.mpg.de/.	Provides a testable hypothesis with respect to the nature of PPIs when the experimental information is not available	Zhu et al. (2006)
PISA	https://www.ebi.ac.uk/pdbe/pisa/	Tool to explore macromolecular interfaces	Krissinel and Henrick (2007)
PQS database	pqs.ebi.ac.uk/	Server for the protein quaternary structures which distinguishes between the biological and crystal contacts features	Henrick and Thornton (1998)
PCAlign	http://brooks.chem.lsa.umich.edu/index.php?page=pcalign&subdir=articles/resources/software	Method for the quantification of physiochemical properties of protein–protein interfaces	Cheng et al. (2015)
iPFAM	http://ipfam.org	Provides interaction details for the domain–domain interactions among the proteins	Finn et al. (2014)
PredHS	http://predhs.denglab.org/	An efficient server to find hotspots on PPI interfaces using structural neighborhood properties	Deng et al. (2014)

(continued)

Table 6.2 (continued)

PCRPi-DB	http://www.bioinsilico.org/PCRPIDB.	Database known as presaging critical residues in protein interfaces database (PCRPi-DB) contains information about the computationally annotated hotspots for those protein complexes for which 3D structures are known	Assi et al. (2010)
ProtCID	http://dunbrack2.fccc.edu/protcid.	Protein common interface database is platform developed for scientist to provide clusters of interfaces of protein chains (full length) in order to identify assembly of macromolecular complexes	Xu et al. (2008)
PISITE	http://pisite.hgc.jp/cgi-bin/top.cgi	Database of protein interaction sites which provides information about interaction sites in proteins	Higurashi et al. (2009)
PINT	http://www.bioinfodatabase.com/pint/	Protein–protein interactions thermodynamic database contains information about the structural and thermodynamic parameters of protein–Protein complexes.	Kumar and Gromiha (2006)
Protein-protein interaction affinity database	https://bmm.crick.ac.uk/~bmmadmin/Affinity/	Contains structural information about 144 protein–protein complexes along with their experimentally attained dissociation constants	Kastritis et al. (2011)
Robetta	http://www.robetta.org/	Facilitates in-silico alanine scanning experiments for the proteins	Kim et al. (2004)
SCORECONS	https://bio.tools/scorecons	Provides information about the conservation of an amino acid residue	Valdar (2002)

(continued)

Table 6.2 (continued)

SKEMPI	https://github.com/guangyu-zhou/MuPIPR	Computes the effect of mutations on properties of PPIs	Moal and Fernández-Recio (2012)
SCOPPI	http://www.scoppi.org/	Structural classification of protein–protein interfaces which provide classification and annotation of domain interactions attained from the protein complexes for which structures are known	Winter et al. (2006)
SCOWLP	http://www.scowlp.org	Tool to predict binding regions in proteins on the basis of structural inferences across the fold space	159

To further understand the nature of interactions between the proteins, it is imperative to study the physicochemical characteristics of protein–protein interfaces which cover their chemical and structural features. For instance, amino acid composition at interface region, their conservation, interaction strengths, distribution of charges, flexibility, pattern of hydrogen bonding and salt bridges at the interface, buried surface area and non-polar fraction between the interacting partners, buried water molecules, residues contributing significantly to free energy of binding (known as hotspots), binding complementarity and shape of interface, secondary structures types are some of the characteristic features of binding site or interface region (Jones and Thornton 1996; Keskin et al. 2008).

Energy distribution at the interface region is one of the most striking features. As mentioned above, amino acid residues with significant energy contribution at the interface regions are labeled as hotspot residues. These hotspots, thus significantly contribute for binding (Bogan and Thorn 1998). Hotspots in the interface regions can be decoded experimentally by a simple technique known as alanine scanning mutagenesis. In this method, a particular residue at the interface regions is mutated to alanine. This simple mutation eliminates the side chain atoms beyond Cβ and does not add any additional conformational freedom (Moreira et al. 2007). This alanine mutation aids in deciphering the role of side chains of original residue in binding. Binding free energies are calculated for the wild-type protein complex and the mutant protein complex. Comparison of these energy values assists in assessing the role of the mutated residues in the interfacial region. If the difference between the binding free energies of a wild-type protein complex and mutant protein complex is more than 2.0 kcal/mol, then the original residue is regarded as hotspot, and is considered to be crucial for binding (Thorn and Bogan 2001).

Hotspots exhibit structural complementarity and fit into each other through salt bridges and hydrophobic interactions (Arkin and Wells 2004). It has also been noticed that such hotspots are mostly buried, non-accessible to solvent and are in close proximity (Keskin et al. 2005; Li et al. 2004). Many research groups across the world aim to target these hotspots for developing drugs to overcome PPIs involved in various diseases. Several computational-based approaches including computational alanine scanning, machine learning-based methods, molecular dynamics-based approaches are also being employed that can rapidly distinguish the hotspots from the non-hotspot residues (Darnell et al. 2008; González-Ruiz and Gohlke 2006; Kortemme and Baker 2002; Ofran and Rost 2007; Tuncbag et al. 2009a; Zhu and Mitchell 2011).

There are several databases like BID (binding interface database) which contains information about hotspots from alanine scanning experiments (Fischer et al. 2003). Tools such as HotPoint (hotspot prediction server for protein complexes), Spoton (hotspots on protein complexes), PredHS (prediction of hotspots) enable the prediction of hotspots (Deng et al. 2014; Moreira et al. 2017; Tuncbag et al. 2010). It has been speculated that mostly aromatic amino acids have preferences for their presence in the binding site (Ma et al. 2003).

Stability and specificity of interactions between the partner proteins rely on the type of interactions between them which can be either: hydrogen bonding, salt bridges, electrostatic or hydrophobic interactions. It has been ascertained that frequency of presence of disulfide bonds in the interface is very low. The contribution of disulfide bonds is noticed in small protein complexes where they confer the stability and rigidity to the complex. The interface of proteins has been divided into two sections, one is the core region and other is rim region. The core region denotes the inner part and rim region is the outer part of interface. The core region of interface resembles with the interior portion of a protein, whereas the rim region that is exposed to solvent resembles the outer surface of protein in terms of composition of amino acid residues (Bahadur et al. 2003). The flexibility of protein binding regions is low in comparison with the surface portion of protein (Keskin et al. 2008).

Further, notable differences have been observed in the types of interactions present in differential interfaces. For instance, hydrophobic interactions are more prevalent in case of interface of permanent complexes than in interface of transient ones (Nooren and Thornton 2003a). Although, the salt bridges and hydrogen bonds are more favorable in transient protein complexes, the amino acid sequence is more conserved in case of interface of obligate protein complexes (Mintseris and Weng 2005). Shape complementarity has not gained much importance in transient protein complexes (Mintseris and Weng 2005). Furthermore, the size of interface region depends on the type of proteins involved in interaction. Large interfaces are present in receptor ligand complexes and globular protein complexes than in transient and oncogenic protein complexes (Kar et al. 2009).

Kar et al. made a significant contribution in studying the properties of interfaces of different proteins (Kar et al. 2009). The authors have investigated the interface properties of both cancer-related and non-cancer-related proteins. They considered 363 cancer-related proteins and 186 non-cancer-related proteins for their analysis. It has been observed that the accessible surface area (ASA) of cancer proteins is much smaller than that of non-cancer proteins. This implies that the complexes formed by the interaction of proteins involved in cancer are less buried, that is the monomeric state of these proteins has less exposed surface area. In general, small interfaces have been observed in transient complexes. Hence, it has been proposed that since cancer proteins use small surface area for their interaction with the partner proteins, and they possess several interacting partners; henceforth, it is highly likely that cancer proteins might have been involved in transient interactions.

Kar et al. have also reported the characteristics on the complementarity and closeness of interfaces in protein complexes (Kar et al. 2009). The measurement of empty space between the interacting proteins indicates the closeness and complementarity of interfaces of proteins. This measuring parameter is termed as "gap volume." Another parameter is "gap volume index" (GVI), which is the ratio of gap volume versus interface area size. In simple words, GVI is the gap volume normalized by interface area size. Smaller GVI indicates better complementarity (Bahadur et al. 2004). Authors discerned that cancer-related proteins exhibit smaller gap volume than non-cancer proteins. Cancer proteins observed with higher GVI than non-cancer proteins imply that the complementarity is less optimized in case of cancer proteins than in non-cancer proteins. This also confirms that the complementarity and closeness of the interface are quite different in cancer and non-cancer protein.

It is also imperative to study the shape of interface. Planarity index is a measure of shape of the interface. Interfaces of the interacting partners either remain flat or twist to interact. The assessment of how flat or twisted the protein interfaces is a measure of how much the interface atoms deviate from the plane. This measurement is known as planarity. The planarity index (PI) is calculated as root mean square deviation of atoms of interface from least square plane fitted through interface atoms. Larger PI indicates less planar interface. In contrast, smaller PI implies more planar interface (Jones and Thornton 1996). For cancer-related protein complexes, smaller PI was observed as compared to that for non-cancer protein complexes which implies more planarity of interface in cancer protein complexes (Kar et al. 2009). Planarity of interface is also known to exhibit higher correlation with their ASA. Increase in ASA of interface increases the planarity index of interfaces. More planar interfaces have been ascertained for transient complexes (Nooren and Thornton 2003b). This implies that the cancer protein complexes involve more planar interfaces for their interaction (Kar et al. 2009).

It is also vital to study the type of interaction between the interfaces of proteins. Kar et also took an initiative in describing the type of interactions involved in cancer-related protein complexes. Authors reported lesser hydrophobic and more charged interactions in cancer-related protein complexes which indicates more hydrophilic interfaces in cancer-related protein complexes than in non-cancer protein complexes (Kar et al. 2009). Indeed, previous studies reported that small interfaces involve reduced hydrophobic interactions (Tsai et al. 1997), although it has been well established that hydrophobic interactions are predominant in PPIs (Hu et al. 2000).

Studies reporting the significance of hydrophilic interactions in the interface regions have been conducted. Tormo et al. mentioned the highly hydrophilic nature of interface of C-type-lectin-like receptor family. Charged interactions are more prevalent in the interfaces of these receptors. They encompass 43% charged, 22% neutral polar, and 35% non-polar atoms at their interface (Tormo et al. 1999). In contrast, an average PPI is composed of 15% charged, 29% neutral-polar and 56% non-polar (Lo Conte et al. 1999). Kar et al. also found that intrinsically disordered cancer associated proteins exhibit less hydrophobic interactions than other proteins (Kar et al. 2009).

6.2.1 Role of Post-translational Modifications in PPIs

PPIs are also highly influenced by the post-translational modifications (PTMs). PTMs are the chemical modifications of proteins that are made after their translation from RNA. These modifications involve the attachment of some chemical moieties at particular amino acid. Several different PTM types including phosphorylation, acetylation, methylation, ubiquitination, amidation, glycosylation, etc. have been recognized that affect the various cellular processes such as signal transduction, metabolism, protein stability, etc. (Deribe et al. 2010; Zhao et al. 2010). PTMs like phosphorylation, ubiquitination, acetylation, methylation at the interface regions exert huge effects on interacting partner proteins (Keskin et al. 2016).

There do exist a functional connectivity between the PTM and PPI, as PTM specifically modulates the interaction of proteins by varying the structural or electrostatic properties of sites involved in interactions (Seet et al. 2006). Analysis of PTM data from dbPTM (database of PTM) showed that 60% of PTM sites are placed in functional domains of proteins that are directly engaged in PPIs (Lu et al. 2013). Different PTMs modify the protein in distinct manner, consequently, their effects on PPIs will also be disparate. It is also observed that proteins with a particular PTM type have a high propensity to interact with partner proteins that underwent same type of PTM (Duan and Walther 2015).

Phosphorylation is one of the most intensively investigated PTM types (Beltrao et al. 2013). It has been suggested that in both yeast and *Arabidopsis thaliana*, phosphorylated proteins are more involved in PPIs than non-phosphorylated proteins (Duan et al. 2013; Yachie et al. 2011). Several studies showed that phosphorylation plays a significant role in the formation of protein complexes and thereby also regulate their functioning. It is also noted that sites of phosphorylation at the interface regions are evolutionarily more conserved than other interfacial sites (Nishi et al. 2011). The sites of phosphorylation are quite distinctive in different types of protein complexes. For instance, sites for phosphorylation are mostly located in the binding interface of hetero-oligomeric and weak transient homo-oligomeric protein complexes. The binding hotspots have a high tendency to get phosphorylated in case of hetero-oligomers. This phosphorylation-based modification at the interface attunes the nature and strength of interactions (Nishi et al. 2011). The binding energy of a complex gets directly affected by the occurrence of a phosphorylation at or in vicinity of the binding interface. In contrast, the presence of phosphorylation site outside the binding interface exerts long range effects on conformation of proteins, which amends the binding for its partner proteins through allosteric mechanism (Nishi et al. 2011).

Duan and Walther took an initiative to comprehensively investigate the role of PTMs in protein interaction networks. They analyzed 12 different types of PTMs and their subtypes across nine different species. They scrutinized the placement of proteins with particular PTM type in the network of PPIs. They proposed that the proteins with PTM occupy central positions in the network than non-PTM proteins. Further, among all other 12 PTMs, phosphorylated proteins are more populated at the central locations in the network. Glycosylated proteins were found at the peripheral regions of the network. These proteins with fewer interacting partners are mostly localized in plasma membrane and are involved in receptor or transporter activities. Strong association between the proteins with particular PTM type involved in disease processes and protein interaction network properties has also been elucidated. This amalgamated study of PTMs and protein interaction networks provides a platform to unravel the role of PTMs in PPIs and thereby in differential cellular processes (Duan and Walther 2015).

In a nutshell, the interaction properties of proteins modulated by PTM are quite different from the non-PTM proteins. Consequently, the location of PTM proteins in the network of PPIs is specific and imperative in deciphering their actions in coordinating the molecular processes carried out by the interacting proteins partners.

6.3 PPI Storage Databases

Enormous amount of PPI data has been generated as an outcome of high-throughput experiments and computational methods for prediction of PPIs. This prompted the scientist community for a need to store, access, and analyze PPI data for the future research endeavors. Researchers took a serious initiative to develop variety of databases with an aim to unify the available PPI data (Table 6.3).

The differential databases have been divided broadly into four categories: (1) databases of experimentally verified proteins like BioGRID (Stark et al. 2011), IntAct (Kerrien et al. 2012), (2) databases like GENEMANIA (Warde-Farley et al. 2010) and STRING (Szklarczyk et al. 2011), that include computationally predicted PPIs and text mining which is devoid of manual assistance, (3) pathway databases including KEGG, Reactome, in which there is a collection of consensus knowledge by curators (Croft et al. 2011; Kanehisa et al. 2010), (4) databases with distinct goals include HPRD (human protein reference database), Innatedb database (Fig. 6.3). Some of the databases from each of the above-mentioned categories have been described in detail in the following sections.

Several other protein interaction databases are also available that focus either on a particular disease (HDBase for Huntington's disease (Goodman et al. 2003), CIDeR for multifunction interaction networks in human disease (Lechner et al. 2012)), organism (human protein reference database and human protein pedia for humans (Goel et al. 2012), Saccharomyces genome database for *Saccharomyces cerevisiae* (Cherry et al. 2012), DroIID for *Drosophila melanogaster* (Murali et al. 2011)), or

Table 6.3 Summary of some of the databases used for the storage of PPIs

Database	Website	Summary	References
AraPPISite	http://systbio.cau.edu.cn/arappisite/	Provides fine-grained interactions of *A. thaliana*	Li et al. (2016)
ANCHOR	http://structure.pitt.edu/anchor/	Analysis of PPI interaction hotspots for drug discovery	Meireles et al. (2010)
Adan	http://adan.crg.es/.	Provides an opportunity to obtain scoring matrices calculated using FoldX and attain putative binding partners	Encinar et al. (2009)
BIND	http://bind.ca	Facilitates to determine the protein functions via providing details on interactions of protein with its binding partners at molecular level	Bader et al. (2003)
BioGRID	https://thebiogrid.org/	Dedicated to store genetic, chemical, and protein interactions for human and all other major organisms	Oughtred et al. (2019)
ChiTaRS	http://chitars.md.biu.ac.il/	Insights into the biological characterization of chimeric proteins in normal and cancer associated cells	Fenkel-Morgenstern et al. (2013)
CORUM	http://mips.helmholtz-muenchen.de/corum/.	Manually curated database containing experimentally determined protein complexes from mammals including mouse, rat, and human	Giurgiu et al. (2019)
CCSB-Interactome database	http://interactome.dfci.harvard.edu/	Encompasses PPIs related to human, viruses, plants along with PPIs involving only independent modular protein domains	Ruepp et al. (2010)
Consensus PathDB	http://cpdb.molgen.mpg.de/	Meta-database which combines interaction details from a variety of interaction data resources	Kamburov et al. (2011)
DIP	https://dip.doe-mbi.ucla.edu/dip/Main.cgi	Comprehensive tool to efficiently extract information about PPIs and PPINs in biological processes	Xenarios et al. (2000)
Death domain DB	http://www.deathdomain.org	Database containing PPIs related to death domain superfamily of proteins	(Kwon et al. 2012)
FPPI	http://csb.shu.edu.cn/fppi	Contains information about PPIs attained for F. graminearum on the basis of interlogs information from PPI databases of seven species and also include domain–domain interactions that are attained experimentally on the basis of structures of proteins.	Zhao et al. (2010)
HAPPI	http://bio.informatics.iupui.edu/HAPPI/	Comprehensive compilation of human PPIs and also provides quality scoring to the data required for further emerging of PPI network studies	Chen et al. (2009)

(continued)

Table 6.3 (continued)

Database	Website	Summary	References
HPID	http://www.hpid.org	Facilitates the prediction of proteins queried by users, also provides information about human PPIs	Han et al. (2004)
HitPredict	http://hintdb.hgc.jp/htp	Source of experimentally attained physical PPIs along with scoring system as measure of reliability of data	Lopez et al. (2015)
HPRD	http://www.hprd.org/	Repository of information about experimentally verified human PPIs including post-translational modification, their association with diseases, expression in tissues, and enzyme substrate complex relationships.	Goel et al. (2012)
HIPPIE	http://cbdm.mdc-berlin.de/tools/hippie	Integrates human PPI datasets from different sources and calculates confidence scores for these PPIs to reflect the quality of PPI data	Schaefer et al. (2012)
I2D	http://ophid.utoronto.ca/ophidv2.204/index.jsp	Integrates experimentally validated and computationally predicted PPIs for humans and five model organisms including *C. elegans, S. cerevisiae, D. melanogaster, M. musculus*, and *R. norvegicus*	Brown and Jurisica (2007)
IID	http://ophid.utoronto.ca/iid	First database to provide tissue-specific view of PPIs for humans and 6 other species including *C. elegans, S. cerevisiae, D. melonogaster, M. musculus*, and *R. norvegicus*.	Kotlyar et al. (2016)
IntAct	http://www.ebi.ac.uk/intact	An open source for PPI data and also provides a tool kit to store, visualize, and analyze the PPI data	Hermjakob et al. (2004b)
InterDom	http://InterDom.lit.org.sg	Focuses on providing evidences to support the protein interactions detected on the basis of interacting domains	Ng et al. (2003)
Interpare	http://interpare.net	An open source containing information about the interface regions of proteins for which the three-dimensional structures are available	Gong et al. (2005)
IMID	www.integrativebiology.org.	Gathers interaction information about proteins/genes from the literature by automatic extraction and also from the databases with manual annotations	Balaji et al. (2012)
IsoBase	http://isobase.csail.mit.edu/.	Identifies functionally related proteins based on sequence data and PPI network data	Park et al. (2011)

(continued)

Table 6.3 (continued)

Database	Website	Summary	References
iRefIndex	http://irefindex.uio.no	Aims at generating keys for protein interaction and for the participating protein	Razick et al. (2008)
iRefWeb	http://wodaklab.org/iRefWeb/	Provides an interface to 10 protein interaction databases including BioGRID, CORUM, BIND, IntACt, HPRD, OPHID, MINT, MPPI, MPact. Allows user to analyze aggregated interaction data for protein of interest	Turner et al. (2010)
JET2Viewer	http://www.jet2viewer.upmc.fr/	Presents putative binding interfaces/ sites for all the 3D-protein structures available in PDB, employs JET^2 computational strategy	Ripoche et al. (2017)
KEGG	http://www.kegg.jp	Encyclopedia of genes and genomes, contains interaction details in the form of KEGG pathway maps, KEGG modules, and BRITE hierarchies	Kanehisa et al. (2017)
MINT	http://cbm.bio.uniroma2.it/mint/index.html	Stores large amount of protein interaction details in a concise way to easily retrieve and analyze the PPI data	Zanzoni et al. (2002)
MiMI	http://mimi.ncibi.org	Merges protein interaction data from different sources on a single platform	Jayapandian et al. (2007)
NCI-PID	http://pid.nci.nih.gov	Allows browsing predefined pathway data and enables creation of interaction network maps	Schaefer et al. (2009)
OPHID	http://ophid.utoronto.ca	Contains predicted interactions for human proteins	Brown and Jurisica (2005)
PIPS	http://www.compbio.dundee.ac.uk/www-pips	Encompasses protein interactions in human, out of 37,000 total predicted protein interactions contained in the database, 34,000 interactions are not reported in databases including HPRD, OPHID, DIP, and BIND	McDowall et al. (2009)
PCPPI	http://bdg.hfut.edu.cn/pcppi/index.html.	Source of Penicillium crop PPIs	Yue et al. (2017)
PPIM	http://comp-sysbio.org/ppim	Includes not only predicted PPIs for maize but also includes maize PPIs reported in literature	Zhu et al. (2016)
PARPS	http://sourceforge.net/projects/parpdb/	Web-based tool which includes protein sequence management, protein database searching, experiment annotation, and data mining of identified proteins and peptides	Droit et al. (2007)

(continued)

Table 6.3 (continued)

Database	Website	Summary	References
PCRPi-DB	http://www.bioinsilico.org/PCRPIDB	Designed to find hotspots in the protein interfacial regions	Segura and Fernandez-Fuentes (2011)
PDZBase	http://icb.med.cornell.edu/services/pdz/start	Developed to contain all the PPIs that are mediated by PDZ domain	Beuming et al. (2005)
PRIMOS	http://primos.fh-hagenberg.at.	Unifies six major PPI databases including HPRD,MINT, BIND, IntAct, MIPS, DIP	Rid et al. (2013)
PIBASE	http://salilab.org/pibase/suppinfo.html	Contains information about structurally defined interfaces between pair of protein domains	Davis and Sali (2005)
ProtCom	http://www.ces.clemson.edu/compbio/protcom	Compiles known 3D structures of protein–protein complexes and also contains domain–domain structures created artificially using available PDB entries	Kundrotas and Alexov (2007)
PDBSum	http://www.ebi.ac.uk/pdbsum	Provides structural information about the entries contained in PDB	Laskowski et al. (2018)
Reactome	http://www.reactome.org	An open source database which allows the analysis of molecular interaction data and construction of molecular pathways	Croft et al. (2011)
SpirPro	https://pubmed.ncbi.nlm.nih.gov/26220682/	Repository for storage, visualization and analysis of PPI data for *Spirulina* (Arthrospira) *platensis*	Senachak et al. (2015)
SPIDer	https://pubmed.ncbi.nlm.nih.gov/17254300/	Database contains PPIs predicted for Saccharomyces cerevisiae and also provides links to other databases like MIPs, DIP, BIND and allows the construction of PPINs	Wu et al. (2006)
STRING	http://www.bork.embl-heidelberg.de/STRING/	Database containing functional and genomic association of genes/proteins	von Mering et al. (2003)
SNAPPI-DB	http://www.compbio.dundee.ac.uk/SNAPPI/downloads.js	High performance database for the structure, interface, and alignment of PPIs	Jefferson et al. (2007)
SCOPPI	http://www.scoppi.org	Comprehensive database for the classification and annotation of interacting domains that belongs to the proteins with known structures	Winter et al. (2006)
SPIKE	http://www.cs.tau.ac.il/&~spike/	Database containing information about interactions involved in human signaling pathways	Paz et al. (2011)
TRIP database 2.0	http://www.trpchannel.org	Repository containing comprehensive details about PPIs for TRP (transient receptor potential) channels	Shin et al. (2011)

(continued)

Table 6.3 (continued)

Database	Website	Summary	References
xComb	http://phenyx.proteomics.washington.edu/CXDB/index.cgi	Database of cross-linked peptides for the analysis of PPIs	Panchaud et al. (2010)
2P2Idb v2	http://2p2idb.cnrs-mrs.fr.	Manually curated database for the structures of PPIs along with their known inhibitors	Bourgeas et al. (2010)
3DID	http://3did.irbbarcelona.org	Provides atomic details of PPIs for which the structures are known	Stein et al. (2011)

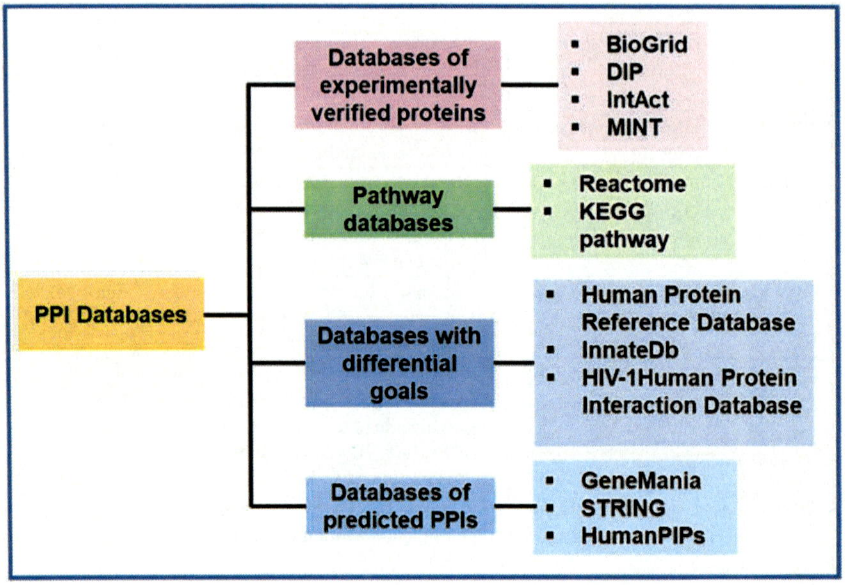

Fig. 6.3 Broad classification of some of the common databases containing information about PPIs

interaction type (PhosphoELM (database of phosphorylation sites (Dinkel et al. 2011)). Besides these specific databases, there are various databases which in addition to PPI data also comprise the interaction data of proteins with small drug molecules, metabolites, and other macromolecules like DNA, RNA (Caspi et al. 2008; Croft et al. 2011; Kanehisa et al. 2010).

An interesting feature of PPI databases is the graphical visualization of interaction network of proteins in spite of just displaying the list of interacting protein partners. Such a graphical representation gives user a sketch of first interacting neighbors, which further aids in visualizing other higher-order interactions. Although most databases provide a means of visualizing these interaction networks, in few databases the interaction network file is downloadable and can be imported in other visualization tools like NAVIGaTOR (Brown et al. 2009) or Cytoscape (Smoot et al. 2011). These tools allow user to perform variety of analysis of networks like term enrichment analysis, clustering, etc. (Gene Ontology 2010). PPI resources are also being widely utilized for the hypothesis testing and interpretation of data.

6.3.1 Databases of Experimentally Verified Proteins

It is quite hard to retrieve the amount of information available for a particular interaction as the data with respect to the molecular interactions is available in the form of text in research articles, tables, or supplementary files. Although manual curation of data contained in literature is an accurate way of extraction of primary interaction data, it is a cumbersome task to integrate and interpret the morass of data in a codified manner. There was an exacerbation of this problem owing to the lack of tools to analyze, manage, and integrate the data. This shortfall was overcome by the development of BioGRID (The Biological General Repository for Interaction Datasets) by Stark et al. in 2011 (Stark et al. 2011).

6.3.1.1 BioGrid Database

BioGRID is an open source database which encompasses the compilation of protein as well as genetic interactions comprehensively curated in a systematic way from the available biomedical literature (Stark et al. 2011). It provides a user-friendly interface to construe the desired interaction. It also enables the user to federate the new datasets with embedded datasets. These datasets aid in construction of networks to interpret different data types and to adumbrate novel biological functions (Breitkreutz et al. 2010; Huttenhower et al. 2009). The latest version of BioGRID is 4.0.189 that searched 73,056 publications in order to decipher 1,891,411 protein and genetic interactions, 874,796 PTMs, and 28,093 chemical interactions between proteins/genes and biologically active small molecules from human and other leading model organisms including *E. coli, S. cerevisiae, Schizosaccharomyces pombe, Drosophila melanogaster, Mus musculus, Danio rerio, C. elegans,* and many others. All the data contained in BioGRID are freely accessible through the search index and are also downloadable in specific formats.

BioGRID is devoid of a native network visualization system but it allows the user to export data in Cytoscape compatible format (Oughtred et al. 2019). BioGRID is also making edgy efforts to release new datasets relating to SARS-CoV, SARS-CoV-2, and MERS-CoV to assist in development of new therapeutics. The current version BioGRID 4.0.189 has been updated with corona virus related interactions which includes 3502 interactions from 175 publications and preliminary reports (https://thebiogrid.org/news/245).

BioGRID works in conjunction with MODs (model organism databases like FlyBase: fruit fly *D. melanogaster* (Gramates et al. 2017), PomBase: *S. pombe* database (McDowall et al. 2015), Bio-Analytic Resource for Plant Biology (BAR) (Waese and Provart 2017)), genetic databases (Wormbase) (Howe et al. 2012), chemical databases (DrugBank (Wishart et al. 2018), CTD (Davis et al. 2017), BindingDB (Gilson et al. 2016), ChEMBL (Nowotka et al. 2017), PharmGKB (Barbarino et al. 2018)), and other meta-databases (Pathway Commons (Cerami et al. 2011), FlyAtlas (Leader et al. 2018), InnateDB (Breuer et al. 2013) Gene Mania (Warde-Farley et al. 2010), and STRING (Szklarczyk et al. 2017)), in order to facilitate the propagation of data incorporated in it and to foster the curation efforts (Oughtred et al. 2019).

6.3.1.2 DIP Database

Like BioGRID, DIP (database of interacting proteins) catalogs variety of experimentally proven PPIs (Xenarios et al. 2000). The interaction information from different sources is combined to create single freely accessible platform for PPIs. Keeping in mind the differential reliabilities of experimentally obtained PPIs, quality assessment tools are being followed to determine the most impeccable interactions. The data contained in the database is curated manually as well as automatically by utilizing computational methodologies employing the most reliable set of protein interaction networks data stored in DIP. Initially DIP contained data for binary protein–protein interactions obtained from research articles. The advent of technology resulted in high loads of interaction data between proteins (Ito et al. 2001; Uetz et al. 2000) and recognition of multi-protein complexes (Gavin et al. 2002; Ho et al. 2002). This has not only elicited the development of DIP but also boosted the advancements in quality assessment methods of PPIs (Deane et al. 2002).

DIP is not focused on the data from particular organism, although 30% of the interaction data contained in it is from *Saccharomyces cerevisiae*. DIP provides a provision to the user for the visualization and analysis of the stored interactions in Cytoscape's MiSink plug-in (Salwinski and Eisenberg 2007). It also enables the users to evaluate the external interaction data including the comparative analysis of (1) paralogous proteins, (2) expression profiles of proteins, (3) specific domain content in the proteins, etc. using the tool services provided by DIP (Deane et al. 2002; Deng et al. 2002). DIP data is accessible online as well as downloadable in variety of formats that can be directly opened in spreadsheet applications. DIP data is also available in MIF (molecular interaction format) which is a standard format developed for the exchange of PPI data (Hermjakob et al. 2004a).

6.3.1.3 IntAct Database

IntAct is a central repository that gathers and integrates the PPI data. It includes both manually curated data from the literature and also allows the producers of the data to submit their PPI data as a piece of publication (Kerrien et al. 2012). All the data contained in IntAct are freely available and the software developed by the IntAct developers is free, and can be employed, modified, redistributed under the terms and conditions of Apache Software license. In addition to PPI data, IntAct also made an initiative to enclose the other interaction data that includes protein-nucleic acid, protein-small molecule, and protein-gene loci (Kerrien et al. 2012). The major reference resources for these datasets are ChEBI (de Matos et al. 2010), Ensembl/ Ensembl Genomes (Flicek et al. 2011; Kersey et al. 2010), INSDC (Cochrane et al. 2011).

IntAct works in close association with Gene Ontology Annotation (GOA) project and UniProt consortium (9). IntAct exports the binary pairs to Annotation section of both GOA and UniProt (Kerrien et al. 2012). A confidence score is given to each experimentally identified binary interactions which is cumulative and weighted on the basis of evidences for the interaction. Users are allowed to search for an interaction on the basis of number of categories including Reactome pathway association, chromosome location, gene ontology terms (Gene Ontology 2010), ChEBI ontology (Hastings et al. 2013), taxonomy, protein domains, and chromosomal location. IntAct also assists users in visualizing interactions in CytoscapeWeb (Lopes et al. 2010), which is customized to furnish users with additional functions like uncluttering large networks to merge edges or to obtain different graph layouts to simplify the complex analyses.

6.3.1.4 MINT Database

MINT (molecular interactions database) is one of the major repositories of PPIs. It encompasses the experimentally validated interactions contained in peer-reviewed journals. All the interactions are curated by the expert curators (Ceol et al. 2010). In addition to PPIs, MINT also captures the interactions between the proteins and nucleic acid. MINT provides a user-friendly interface with the advancements in query and graphic tools. MINT is a member of the international molecular exchange consortium (IMEx), a body developed for the integration and exchange of molecular interactions data. The goal of IMEx is to disseminate the curation workload among the participating databases in order to avert the data redundancy and enhance the literature survey. Apart from MINT, many others including IntAct, DIP, I2D, MBinfo, BioGrid, InnateDB, PrimesDB, Molecular Connections (MC), MBinfo, SIB, UCL-BHF, HPIDB, and MatrixDb are also the part of IMEx. All these databases are following a common curation manual explaining the amount and form of information required for the database. Further, the interactions stored in MINT are scored which ranges between 0 and 1, depending on the reliability of interaction. The interactions which are more reliable and are supported by strong evidences exhibit score closer to 1. Thus, a scoring system allows the user to assess the quality of interaction. The results of user query can be viewed in MINT Viewer as a list or as a network which is amenable for modification. MINT Viewer allows one for the arrangement of nodes, expansion of network as well as filters or shows links between the nodes in the network. It also allows the access to the sister databases (Ceol et al. 2010). MINT automatically imports the curated and validated data into its sister databases.

MINT has three sister databases, namely: (1) HomoMINT (Persico et al. 2005), (2) DOMINO (Ceol et al. 2007), (3) VirusMINT (Chatr-aryamontri et al. 2009). HomoMINT is a database specific for human interactions. Interactions related to human proteins either verified experimentally or deciphered using a model organism deposited in MINT are imported in HomoMINT (Persico et al. 2005). DOMINO is a database of domain peptide interactions (Ceol et al. 2007). It aims at storage of interactions involving protein domains. Whenever MINT receives an entry in which the interaction is carried out by the domains of proteins, the entry is automatically imported into DOMINO. DOMINO additionally comprises interactions between peptides and domains which are not the part of the protein (Ceol et al. 2007). Finally, all the virus–host and virus–virus interaction entries in MINT are transferred in VirusMINT. VirusMINT provides user a special interface for visualizing virus interactomes along with host interactions (Chatr-aryamontri et al. 2009).

6.3.2 Pathway Databases

Pathway databases are those bioinformatic databases that contain information about the biochemical reactions and also encompass details with respect to the components including enzymes and substrates involved in those reactions. The major goal of these databases includes systematic organization of proteins and their functions and linking them to display the network of interactions between them. Some of these databases including Reactome, KEGG Pathway are described in detail in the following sections.

6.3.2.1 Reactome

Reactome is an open access pathway database developed by expert curators. In contrast to other interaction databases, the major goal of Reactome is veracity in the interactome data. Reactome containing primarily the human pathway data also allows the access to the cross references of pathway annotations to genes (Ensembl, NCBI, HapMap, UCSC), proteins (UniProt), literature (PubMed), small molecules (CheBI, and KEGG), and GO (Flicek et al. 2010; Gene Ontology 2010; International HapMap et al. 2007; Jain et al. 2009; Kanehisa et al. 2010; Rhead et al. 2010; Sayers et al. 2010).

Reactome also provides a provision for the modeling of pathways, analysis of genomes, fundamental research, and education. Reactome pathway data has also extended to include annotations from Gallus and Drosophila. Gallus Reactome initially contained annotations for 127 reactions involving 133 proteins belonging to DNA repair and metabolism domains. Drosophila Pathway database which includes variety of signaling pathway data is also maintained using the hardware and software modules of Reactome. Although Reactome being a highly devoted podium for the pathway data and its analysis, it is lagging behind in covering the human proteins. To overcome this, Reactome made an effort to blend the molecular interaction (MI) data and network details into the pathway data. Interacting protein partners will be visualized using pathway visualization tool in the workspace. "Analyze, annotate, and upload" feature of pathway workspace allows the overlay of all the interactors for all the pathway proteins. PSICQUIC interface is being employed which imports the binary data of MI into Reactome pathway figures. PSICQUIC is already being used by many interaction databases including MINT, BioGRID, STRING, IntAct, ChEMBL, and iRefindex (Aranda et al. 2010;

Breitkreutz et al. 2008; Ceol et al. 2010; Jensen et al. 2009; Overington 2009; Razick et al. 2008). Edges and nodes in the network that are displayed provide direct links to the interaction databases and physical entities as required.

Reactome manages two sets of interaction data: one is from "Reactome" itself and the other is from "Rectome-FIs". Datasets in original "Reactome" are derived from the molecular interaction data directly from reaction and complexes contained in Reactome. In contrast, datasets in Reactome-FIs (functional interactions) includes interactions from other pathway databases including Panther, BioCyc, KEGG, PID, and The Cancer Cell Map (Caspi et al. 2010; Kanehisa et al. 2010; Mi et al. 2010; Schaefer et al. 2009).

The Interaction data contained in Reactome is annotated deeply which includes details like reaction type, stoichiometry, substrates, localization, any disease association, etc. Reactome allows the user to navigate through the pathway, clicking on the particular part of pathway will return the genes associated with it. Reactome provides an in-depth description of the pathway including the corresponding references. Reactome also allows user for the pathway enrichment analysis for the uploaded gene sets. This database also lets the users upload the gene expression data and map it on the pathway with genes color coded as per user instructions (Croft et al. 2011).

6.3.2.2 KEGG Pathway Database

KEGG Pathway (Kyoto Encyclopedia of Genes and Genomes) is one of the most widely employed pathway resources. KEGG database project began in 1995 as part of Japanese human genome program (Kanehisa 1997). Knowledge about cellular functions and organism behaviors is organized as hierarchical lists in BRITE functional hierarchies and in the form of molecular networks in KEGG pathway maps. This organization of data in the form of network maps and lists is easy to biologically interpret the large datasets. KEGG is an integrated database resource, comprising 16 main databases including KEGG genes, KEGG genome, KEGG ORTHOLOGY, KEGG SSDB, KEGG DGENES and EGENES, KEGG GLYCAN, KEGG COMPOUND, KEGG RPAIR, KEGG REACTION, KEGG ENZYME, categorized on the basis of information contained in them (Kanehisa et al. 2010).

KEGG PATHWAY contains the system information in the form of pathway map. Here, system information indicates the molecular systems that carry out cellular processes and also govern the behavior of organisms. The knowledge about such entities is collected from the literature and stored in the form of "simple list" (membership), "pathway map" (graph), or "hierarchical list" (ontology). Pathway map in KEGG PATHWAY database denotes our current understanding about variety of molecular networks which include molecular interaction networks related to metabolic processes, gene-related, environmental, disease, drug development, or other cell-related processes.

The network contains both edges and nodes. Nodes represent either: a protein, gene, orthologs, or small molecules, etc. Edges reflect interactions, reactions, or other relationship between the nodes. Simple list denotes the set of nodes displaying the list of member molecules. This type of presentation of information is employed in those cases where there is a lack of details about the molecular system. Hierarchical list notifies the hierarchically organized information of members and is also known as ontology. This format is used to display the knowledge about the functional hierarchies inherent in variety of systems including molecular, cellular, and organism levels.

KEGG PATHWAY is one of the most comprehensive databases in the context of species, as each pathway specifies the complete knowledge about all the reactions contained in it. Each node in a unified manually curated map (reference pathway) represents not a single gene or protein but specifies the group of genes from distinct species. For building species-specific pathways, reference pathway is used to map the genes from the specific organism. Color coding is used in the pathway to differentiate the genes from a particular species to other. KEGG PATHWAY coordinates with other KEGG databases for the retrieval of information (Kanehisa et al. 2010).

6.3.3 Databases with Differential Goals

Bio-informaticians have designed several databases that have specific aims. For instance, HPRD contains information about the human proteins. InnateDb is created to encompass details about the innate immune-related proteins. Human Immunodeficiency Virus Type 1 (HIV-1), Human Protein Interaction Database focuses on the interactions between the HIV-1 proteins and Human proteins. These databases are described in detail in the following sections.

6.3.3.1 Human Protein Reference Database (HPRD)

HPRD is a resource for experimentally validated interactions of human proteins. HPRD also contains details like post-translational modifications, localization and expression sites, motifs/domain structures, disease association, and enzyme/substrate relations (Goel et al. 2012). HPRD is also a comprehensive source of phosphoproteins documentation. Each data associated with protein reported in HPRD is linked to the literature where it is originally described. HPRD contains curated details of more than 39,194 PPIs and 109,513 PTMs. HPRD provides a user-friendly interface in which the protein sequences are displayed. Presence of domains and motifs, also modifications, corresponding to genomic regions are also depicted. HPRD provides extensive browsing means using protein name, weight, length, location, motif, domain, PTM, and expression site.

HPRD employs gene ontology for including details like molecular functions, subcellular localization, and biological phenomenon. HPRD has also contributed in development of proteomic standard initiative-molecular interaction (PSI-MI) vocabularies (Orchard and Hermjakob 2008). In HPRD, the interactions are classified into two categories: one is direct which corresponds to binary interaction and the other is complex type in which the topology of interaction is not known. All the interactions are linked to their source literature. HPRD also supplements with isoform-specific details if available for the above-mentioned annotations. HPRD also provides cross references to Netpath pathway database and Human Proteinpedia for all the proteins (Kandasamy et al. 2010). HPRD also enables the search for motifs in the submitted proteins using Phosphomotif finder. Along with motifs found in protein sequence, HPRD provides the details about the matched sequence and matched motif and publication link from which the motif details are obtained (Goel et al. 2012).

6.3.3.2 InnateDb Database

InnateDb is an open source database that primarily focuses on innate immunity-related manually curated interactions, pathways, and their involved molecules. InnateDb initially contained around 3500 interactions when it was first released publicly in 2008 (Breuer et al. 2013). Currently, InnateDb contains 3,27,527 interactions validated experimentally and 4,62,421 predicted interactions (https://www.innatedb.com/statistics.do?s). InnateDb covers interactions mainly from three species including *M. musculus*, *H. sapiens,* and *Bos taurus*. InnateDb also took an initiative in annotating interactions and pathways related to asthma and allergic reactions. All the interactions are supplemented with contextual details including the type of interaction, its detection method, tissue or cell type in which the interaction occurred, etc. The information contained in the database follows MIMIx (molecular interaction experiment) standards.

Newly annotated interaction data is added in the database on a weekly basis. Data contained in InnateDb can be used to unravel the new relationships between the innate immunity and other immune processes, for the identification of novel members regulating those processes. Further, users can also analyze their own data from the network viewpoint. Apart from interaction data, InnateDb also contains annotated data for genes related to innate immune system. InnateDb is also the member of IMEx consortium, which as discussed previously defines rules for annotating the interaction data to make them available in public domain (Orchard et al. 2012).

InntaDb also works in close association with other databases including genome, interaction, and pathway databases. InnateDb to date integrated 3,40,355 interactions from databases including MINT, DIP, BioGRID, IntAct, and BIND. InnateDb encompasses cross references to 5485 mouse, human and bovine pathways form PID (the NCI-Nature Pathway Interaction Database), KEGG, INOH (Integrating Network Objects with Hierarchies), Reactome, and NetPath databases. User can visit the InnateDb website for the current statistics about the interactions.

InnateDb provides a user-friendly interface to search for a variety of genes or proteins using different search fields. User can also upload the list of genes or proteins to perform more complex search and analysis. User can quickly obtain transcription factor binding sites, gene ontologies, and the pathways over-represented in their data. User can also construct, visualize, and analyze the molecular interaction networks form the list of genes/proteins uploaded (Breuer et al. 2013).

All the interactions contained in InnateDb can be downloaded in a variety of formats including sif (simple interaction format), tab, csv, xls (text-based formats), MITAB, PSI-MI XML 2.5 formats. Interaction data in InnateDb can be visualized in

Cytoscape plugins: CYOog (Shannon et al. 2003) and cerebral (Barsky et al. 2007). InnateDb has also integrated with BioLayout Express 3D 2.2 for the visualization and analysis of large networks (Theocharidis et al. 2009).

6.3.3.3 Human Immunodeficiency Virus Type 1 (HIV-1), Human Protein Interaction Database

HIV-1 human protein interaction database was developed to focus on interactions between HIV-1 proteins and human proteins (Fu et al. 2009). Cataloging these interactions in the database serves scientific community in designing novel drugs and therapeutics against HIV infections. The database provides details such as brief description about the interaction, PUBMED ID numbers (identification) for the original articles, Entrez gene ID numbers, reference sequence protein accession numbers for each HIV human protein interactions stored in the database.

The major goal of this database is to provide the above-mentioned information about the interaction in a concise manner. The database is also integrated with other databases like Entrez gene to provide ample information about the interaction. This database also allows the visualization of interaction networks between the HIV-1 proteins and human proteins (Fu et al. 2009). Information contained in the database is publically available through NCBI web browser. In order to retrieve the related data to these interactions, the database provides links to other interaction databases including MINT (Chatr-aryamontri et al. 2007), DIP (Salwinski et al. 2004), LANL (Los Alamos National Laboratories (LANL) HIV Databases) (Kuiken et al. 2003), and binding database (Liu et al. 2007).

6.3.4 Databases of Predicted PPIs

Databases of predicted PPIs contain information about the interactions that are predicted for the proteins. This interaction data includes both the genetic and physical interactions, pathway data, co-expression data. Some of the databases in this category also provide comprehensive information about both computationally predicted and experimentally validated interactions among the proteins. Databases falling under this category are described in the following sections.

6.3.4.1 GeneMANIA

GeneMANIA provides a user-friendly interface to access the predicted interactions between the proteins, gene co-expression data, physical and genetic interactions, other molecular interaction data, and pathway data (Warde-Farley et al. 2010). The database covers majorly six organisms including *M. musculus*, *H. sapiens*, *S. cerevisiae*, *A. thaliana*, *D. melanogaster*, *C. elegans*. Datasets contained in GeneMANIA are obtained from public databases including BioGRID (Breitkreutz et al. 2008), GEO (gene expression omnibus) (Barrett et al. 2009), I2D (Brown and Jurisica 2005), MINT (Ceol et al. 2010), Reactome (Vastrik et al. 2007), HumanCyc (Romero et al. 2005), Human protein reference database (Keshava Prasad et al.

2009), IntAct (Aranda et al. 2010), NCI (mature pathway interaction database) (Schaefer et al. 2009).

Networks are constructed either in a direct manner (as in genetic or protein interactions) or using an in-built pipeline in which the profiles are converted into functionally associated networks (Mostafavi et al. 2008). Web interface provides an advanced option panel to the users where the users can make a choice for the default set of networks chosen for each of the organism for their analysis. One can select or deselect all the networks from a given source using the check boxes system.

Database also allows the user to upload their datasets which are accessible to the user only during user's session and can be used along with the already loaded datasets in GeneMANIA. The composite network constructed in GeneMANIA is weighted in accordance with the independent data sources. These weights are non-negative, their sum is 100%, and displays the confidence levels of each data source for the predicted data. Further, the user is also free to select/deselect particular dataset or prediction methods for their contribution in the network construction. GeneMANIA can also be queried with multiple proteins at one time. Automated Gene Ontology annotations are also generated for the proteins in the network (Warde-Farley et al. 2010).

6.3.4.2 STRING (Search Tool for the Retrieval of Interacting Genes)

STRING is an online database which provides users the comprehensive details for both experimentally validated and predicted protein interactions (Szklarczyk et al. 2011). The major goal of STRING is to act as one-stop shop for the retrieval of all the functional interactions of proteins. The distinctive features of STRING include its comprehensiveness, interactive user-friendly interface, and its confidence scoring system. STRING comprises information for more than 5090 organisms ranging from bacteria to Homo sapiens. STRING encompasses their complete sequenced genomes, which are utilized by the variety of algorithms for the prediction of interactions.

User can easily query for the protein of interest in the STRING website. User will be presented with the network of protein partners for the queried protein as output. Each node in the network is provided with pop-up windows which contains detailed information about the proteins along with the evidences for the interaction with the input protein. These networks can be adjusted by the addition or the removal of particular proteins. Confidence levels can also be altered as per the user requirements.

STRING also provides an interactive network viewer, which is based on Adobe's flash player that allows user for the repositioning of nodes in the network (Szklarczyk et al. 2011). Advanced options of the viewer also enable the usage of clustering algorithms in the network (de Hoon et al. 2004; Enright et al. 2002; Pavlopoulos et al. 2009). Protein information pop up window also provides 3D structure content of proteins. User can export the interaction data in the form of high resolution images for the publication. Data can also be exported in the form of PSI-MI format (Hermjakob et al. 2004a).

STRING provides cross links to the widely used websites like SMART (Letunic et al. 2009), UniProt (Apweiler et al. 2010), Swiss Model repository (Kiefer et al. 2009), and GeneCards (Safran et al. 2010). The data from STRING can be visualized and analyzed in Cytoscape. Users are enabled to access STRING networks from Cytoscape software through a "StringWSClient" plugin. Variety of analysis including clustering of networks, PPI enrichment, protein domain enrichment, gene ontology term enrichment, and pathway enrichment can be carried out using the network analysis toolbox provided by STRING (Szklarczyk et al. 2011; Szklarczyk et al. 2019). STRING is also associated with other databases including (1) STITCH which facilitates the study of interactions of proteins with small molecules (Kuhn et al. 2012) and (2) eggNOG, which involves in prediction of orthologous relations among the genes/proteins (Powell et al. 2012).

6.3.4.3 Human PIPs (Protein–Protein Interaction Prediction Database)

Human PIPs database has been developed especially for describing PPIs predicted for humans. PIPs employ Bayesian prediction method that calculates PPIs on the basis of information from variety of sources including post-translational modification, cellular localization, domain occurrence, expression, and orthology. This method also considers the topological features of predicted interaction networks. Involvement of "Transitive" module reflects uniqueness of Bayesian prediction method. Transitive module collects the details about an interaction by analyzing the common predictors for the pair of proteins, which results in a set of interactions that are orthogonal to those in other PPI databases (Scott and Barton 2007).

The web interface of PIPs allows the user to visualize the complete path for each predicted interaction. Thus, PIPs not only serve as a resource of predicted PPIs, but also act as a tool to figure out the hidden features of predicted interaction networks for the proteins being studied. PIPs database can be queried using the UniProt, IPI, or RefSeq Ids for the proteins. PIPs also allow the advanced search options in which the protein sequence can be used as a query. Magic Match is employed to compare query protein sequence with those proteins stored in the database. The database returns those sequences which exactly matches the query protein sequence. PIPs also enable the user to search for multiple proteins simultaneously using IPI IDs. A protein summary page will be displayed for each query protein analyzed. This page summarizes the number of predicted interactions above a threshold value and also displays the links to access other protein databases including UniProt (Apweiler et al. 2010), HPRD (Mishra et al. 2006; Peri et al. 2003), Entrez (Wheeler et al. 2008), and RefSeq (Pruitt et al. 2007).

6.4 PPI Network Construction and Visualization

As discussed above, considerable amount of PPI data obtained via both experimental and computational methodologies is stored in different types of databases. This PPI data is used to probe the network of interactions depending on their functional or physical links (Barabasi and Oltvai 2004; Grindrod and Kibble 2004). This network of protein–protein interactions is known as protein–protein interaction networks (PPINs). PPIN is the graphical representation of interactions of proteins in which a protein is linked to all its interacting protein partners. As discussed in above Sect. 6.1.3.3, PPINs integrate the PPI data in the mathematical form comprising nodes and edges. It is essential to analyze these networks in order to get better insights into the protein interaction data for deciphering the novel functions of proteins and for identification of conserved molecular interaction patterns (Jaeger et al. 2010; Sharan et al. 2005). Additionally, it enables the understanding of biological pathways or processes that involve multitude of proteins, which further aids in improvement of other biomedical and biological applications.

6.4.1 Concept of Network Graph Theory

The notion of network graph theory was first given by Leonard Euler in the eighteenth century. However, the network theory was not applied in the biological networks until the development in computing system (Bollobás and Béla 2001). Huge advancements have been made in the network theory. The major breakthrough in graph theory includes (1) random graph theory, (2) small world networks, (3) scale free networks.

According to the random graph theory model, probability of putting an edge between the two nodes is not concerned with the probability of locating other edges in the graph. This model does not fit with the real-world applications, thus other two models, namely small world networks and scale free networks were proposed. Small world networks exhibit two features including high clustering coefficient and shorter than expected path length. This model has been applied in numerous real-world applications like in social networks and power grids. However, this model is not pertinent for most of the complex protein networks owing to its transient and socio-temporal dynamics. Hence, the third model, scale free networks, in which the major characteristic is that the degree distribution follows the power law rule. According to this rule, most of the nodes have lesser connections with other nodes; in contrast, the number of nodes with high degrees of connectivity (large number of connections with other nodes) is lower than expected. Such nodes with ample number of connections are also known as "hubs." It has been implied that the PPINs follow this power law distribution rule of scale free networks (Albert et al. 2000).

Analysis of protein interaction networks has captivated the great attention of scientists across the globe since 1990 when it was just initiated. Analysis of PPINs deciphered the complex relationships among the proteins involved in replication and transcription (Bartel et al. 1996; Fromont-Racine et al. 1997). Further, the connectivity between different pathways has unraveled that the signal transduction is not a resultant of series of isolated linear pathways but is an outcome of interconnectivities between differential pathways (Letamendia et al. 2001; Shuai 2000). Hence, efforts are being made in the direction of unraveling large amount of PPIs and thereby constructing PPINs.

6.4.2 Structure of Protein Interaction Networks

Studies have shown that the PPINs are *scale free* regardless of species while analyzing the structure of protein interaction networks in variety of species. This implies that there are some proteins (known as hubs) in the network which are involved in large number of connections, in contrast other proteins (non-hubs) are involved in only small proportion of connections (Albert 2005). Attempts have also been made in the direction of integrating the analysis of PPINs with dynamics of gene expression to decipher the changes in the networks in different biological states. A model known as "just in time" was proposed for describing the dynamic nature of protein complexes. According to this model, the proteins involved in a dynamic process are co-expressed regardless of the stage while they are inactive. The key elements of a protein complex are expressed at a particular period to activate the complex for a specific task. Thus, it has been proposed that protein networks exhibit dynamic modular structures that are not species specific (Bertin et al. 2007; Taylor et al. 2009).

6.4.3 Network Topology

The structure of networks is characterized by the network topology. The topological indices that are used to describe the network structure include: (1) Average degree (K), which represents an average value of all the degrees for the nodes in the network, where degree is nothing but the number of links for a node. (2) Clustering coefficient (C), which denotes the fraction of links between the nodes within i-neighborhood divided by the number of links that can be possible between them. (3) Average path length that represents the number of steps required to link each pair of node through the shortest path, and (4) Diameter that specifies the minimum number of links that are used to connect the two most distant nodes in the network (Li et al. 2006; Yook et al. 2004).

The four distributions that are used to comprehensively describe the topological features of PPINs include (1) degree distribution *P*(k), (2) degree distribution of cluster coefficients *C*(k), (3) shortest path distribution SP(*i*), (4) topological coefficient distribution TC(*k*). Additionally, these criteria also aid in gaining deep insights into the network stability, molecular functions, dynamic responses, and cellular evolution (Bergmann et al. 2004; Boccaletti et al. 2006; Wuchty et al. 2006; Zhu and Qin 2005).

6.4.4 Tools for Analysis and Visualization of PPINs

It is quintessential to accurately investigate the protein interaction networks with an aim to predict the functional relevance of proteins which is based on the hypothesis that proteins which are physically related have high probabilities of exhibiting functional relativity (Franzot and Carugo 2003; Gerber et al. 2009). Several

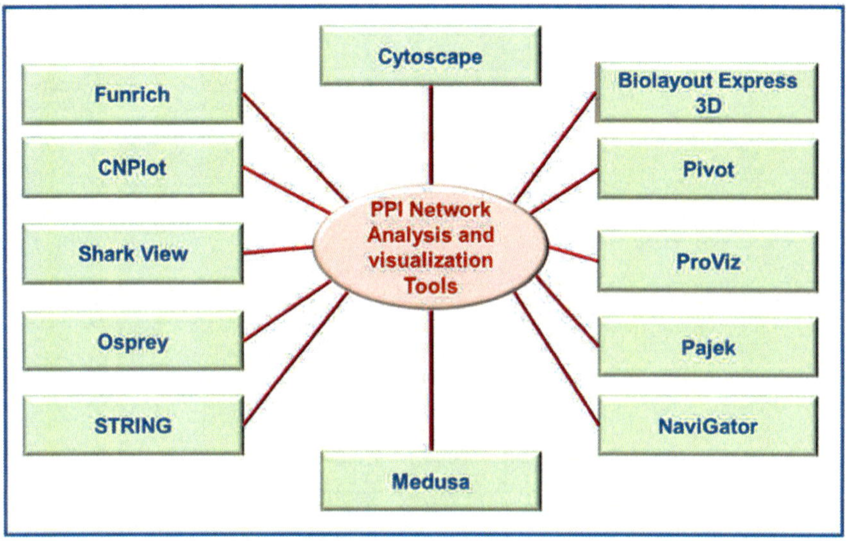

Fig. 6.4 Schematic illustrating the differential tools for the analysis and visualization of PPI networks

computational tools have been introduced to visualize and thereby analyze the complex protein networks (Fig. 6.4, Table 6.4).

The major goal of these visualization tools is to design an intuitive and interactive visualization of complex protein networks which should not be lost within the wired complexity of networks. Several protein interaction databases such as STRING, PINV (Protein Interaction Network Visualizer), UniHI (Unified Human Interactome) are available which have in-built web-based visualization tools (Kalathur et al. 2014; Salazar et al. 2014; Szklarczyk et al. 2015).

In addition, there are variety of independent and dedicated network visualization tools including Cytoscape (Shannon et al. 2003), BioLayout Express3D (Enright and Ouzounis 2001a), NAViGaTOR (Brown et al. 2009) which are freely accessible online. Every tool has its own advantages and disadvantages, thus it is imperative for the users to understand the merits and demerits of these visualization tools to make an appropriate choice of the tool for the desired analysis of data. The most frequently employed visualization tools which ease the analysis and interpretation of PPI data are summarized below.

6.4.4.1 Cytoscape

Cytoscape is an open source toolbox for data integration, analysis, and visualization of the protein interaction networks, high throughput expression data, and metabolomics data (Shannon et al. 2003). It is widely applicable owing to its association with variety of databases for PPIs, genetic interactions, and protein–DNA interactions. Cytoscape is easily downloadable from its home page (http://www.cytoscape.org/) and provides a very user-friendly interface for visualizing the

Table 6.4 Tools used for analysis and visualization of PPI networks

Tool	URL	Summary	Reference
Cytoscape	https://cytoscape.org/	Offers to integrate data in form of graphs and allows to filter and analyze the data and also provides an access to outdoor methods in form of plugins	Shannon et al. (2003)
Medusa	https://sites.google.com/site/medusa3visualization	Simple and attractive tool for analyzing any type of interaction, however, it cannot handle visualization of large datasets.	Hooper and Bork (2005)
NAViGATOR	http://ophid.utoronto.ca/navigator/	Interactive tool for both two-dimensional and three-dimensional visualization of molecular networks, also enables user to visualize and annotate large networks	Brown et al. (2009)
FunRich	http://www.funrich.org/	Freely available standalone tool both for the functional enrichment and interaction network analysis, facilitates the analysis of data against three databases (UniProt, custom, and FunRich) in the background	Pathan et al. (2015)
Osprey	https://osprey.thebiogrid.org/	Perform functional comparisons between the datasets in addition to display interactions in expandable and flexible graphical layouts	Breitkreutz et al. (2003)
ProViz	http://cbi.labri.fr/eng/proviz.htm	Fast and open tool to explore large complex biological networks by providing access to various applications through different plugins	Iragne et al. (2005)
Pivot	http://acgt.cs.tau.ac.il/pivot/	Aids in visualization of protein interaction data collected from differential sources and also allows to decipher relationship between the proteins using graph theory algorithms	Orlev et al. (2004)
Pajek	http://mrvar.fdv.uni-lj.si/pajek/	Networks with up to one billion vertices can be visualized and analyzed, less popular for analyzing PPI networks	Mrvar and Batagelj (2016)
Biolayout express 3D	http://biolayout.org/	Enables the creation of network using tabulated data, employs stochastic petri net algorithm for the investigation of predesigned networks	Freeman et al. (2007)

complex wired networks that contain numerous nodes and edges (Merico et al. 2009).

The core software of Cytoscape aids in providing the basic functions for integrating the data on graph, visualization of those graphs, filtering, and selection options for analyzing the data and also supplements the user with links to outdoor methods as plugins. Cytoscape provides the visualization of nodes and edges of molecular interaction data in the form of 2D networks (Battista et al. 1998).

Cytoscape furnishes the user with three differential algorithms, namely circular layout, hierarchical layout, and spring embedded layout to visualize their molecular interaction data in three distinct formats. Spring embedder is one of the most commonly applied algorithms for the arrangement of 2D graphs (Eades 1984). In this model, the edges of the graph are displayed as springs which create a repulsive force between the nearby nodes and an attractive force between those nodes which are distantly related.

Cytoscape also allows the user to change the appearance of nodes and edges in a desired way. It allows the user to define the shape, color, size, border color, and border thickness of nodes along with edge length, color and its thickness. Further, Cytoscape also serves the user to curtail the intricacies associated with large protein interaction networks. The subtypes of nodes and edges can be selected for display.

Cytoscape provides a toolbox containing variety of filtering ways for displaying only subset of nodes and edges that include: (1) Minimum neighbor filter: which displays nodes with minimum number of neighbors within specified network distance, (2) Local distance filter: that displays the subgroups of nodes that lie within the specified distance in network of preselected nodes, (3) Differential expression filter allowing the user to display nodes in accordance with their expression data, (4) Combination filter let the users to display nodes by combining other filters (Shannon et al. 2003).

Additionally, Cytoscape interface provides links to various applications for the network analysis which includes *Analyzer* for carrying out network, nodes, and edge statistics, *CHAT* for recognizing hubs in the biological networks, *Dynet* for the comparison of networks to identify the most rewire nodes. *DynNetwork* for visualizing the dynamic networks and network comparisons, *Network Randomizer* for creating random networks and to compare them with real networks, *PEWCC* for identification of protein/gene complexes (dense subnetworks) in the PPI/gene networks. All the applications provided by Cytoscape for network analysis can be accessed through the given link (http://apps.cytoscape.org/apps/with_tag/networkanalysis). Besides all these advantages of visualizing and analyzing the complex biological networks on large scale in Cytoscape, one limitation associated with Cytoscape is that it does not provide three-dimensional representations of molecular networks.

6.4.4.2 Medusa

Medusa is employed for the visualization of PPI networks in two dimensions (2D). It runs as a standalone program and does not require any additional packages, however, it works only in systems in which Java is installed (Hooper and Bork 2005). It also runs as an applet for differential applications in web interfaces. The applet version of Medusa is also available for its direct use in web pages. The online version of Medusa has limited functions as it is designed just for the easy visualization of graphs from the web-based databases such as STRING. It is also being employed by

EMBL as an in-house tool. Medusa can export the graphs in differential formats such as in Pajek format for the further analysis of data (Hooper and Bork 2005).

It provides several benefits over other graph analysis tools (Cytoscape (Shannon et al. 2003), Pajek (Batagelj and Mrvar 1998), Osprey (Breitkreutz et al. 2003), CNPlot (Batada 2004), SHARKview (Pinney et al. 2005), ProViz (Iragne et al. 2005). Some of the advantages of Medusa include: (1) Its ability of presenting ten multiple edges simultaneously between the nodes by employing Bezier curves, (2) Quality of figures can be enhanced by insertion of background images, (3) Edges can be shown or hidden based on the type of interactions they are involved in, and (4) Properties of nodes such as annotation, color, shape, position can be defined by expanding or collapsing the nodes.

The major aim of designing Medusa is to provide a simple and attractive tool for the user for the analysis of any type of interactions. The major disadvantage of Medusa is that it cannot handle the visualization of large datasets. Henceforth, the analysis of interactions containing large number of nodes is not possible in Medusa (Hooper and Bork 2005).

6.4.4.3 NAViGATOR (Network Analysis, Visualization and Graphing Toronto)

NAViGATOR is an appealing graphical application for the interactive visualization of molecular networks in both two and three dimensions (Brown et al. 2009). It also enables the visualization and annotations of large graphs owing to its OpenGL hardware acceleration, scalable layout algorithms, and other visual mark-up tools. It also supports various formats including PSI-XML, GML, BioPax, tab delimited text files which makes this tool to operate with other existing tools. Data can also be loaded from other online databases including cPATH and I2D.

Users are free to customize the graphs according to their choice. Users are also allowed to employ automated layouts such as linear, circle on the subset of nodes or on the entire network. There is an option of spreadsheet view in which one can select or deselect the nodes and edges based on any attributes. Grouping of nodes and edges is also possible which enables the group of nodes or edges to get expanded and collapsed or to get manipulated by some set of operations altogether.

Further, NAViGaTOR also provides an analysis toolbox which allows the identifications of hubs, articulation points, and cliques, shortest paths between the nodes and edges, other detailed connectivity information between nodes and edges and also a brief summary regarding network statistics. Random networks for enrichment analyses can also be generated efficiently using multi-threaded implementation in NAViGaTOR. Its user interface tools aid in simplifying hair ball structure of networks. A technique known as alpha blending is used to fade the non-essential parts and focus on essential parts of the networks. Pan/zoom navigation system further allows the user to quickly simplify or explore the complex molecular networks (Brown et al. 2009).

6.4.4.4 Pajek

Pajek is a free non-commercial program designed for the visualization and analysis of giant networks (Mrvar and Batagelj 2016). Networks with up to one billion vertices can be visualized and analyzed in Pajek. The documentation, program, and supporting material for Pajek is accessible freely from its web page (http://mrvar.fdv.uni-lj.si/pajek/). The development of Pajek was started in 1996 by Andrej Mrvar. Pajek is being utilized broadly for analyzing and visualizing general networks including social networks, however, this software is less popular for studying PPI networks.

The analysis and visualization of networks in Pajek is carried out by employing six data types, namely: vector (numerical properties of vertices), permutation (reordering of vertices, ordinal properties), cluster (subset of vertices), partition (ordinal or nominal properties of vertices), hierarchy (tree structure on vertices), and network (graph).

There does exist a requirement to execute some basic operations on the network for obtaining results in Pajek, hence, it is also called network calculator by its users. Some basic operations contained in Pajek are: (1) extraction of subnetworks, (2) shrink selected portions of networks, (3) compute shortest paths, (4) finding connected components either strong, weak, or bi-connected, (5) search centralities of vertices and networks, and (6) searching fragments and clusters in the network. Pajek is the only program that includes generalized block modeling methodology which is used to partition smaller networks (Batagelj et al. 2004).

Pajek is not only utilized for analyzing directed or undirected networks, but it is also useful for other networks including signed networks (networks containing positive and negative lines), temporal networks, two-mode networks (networks amenable for change with time), acyclic networks and multi-relational networks (containing multiple relations among the same set of vertices) (Mrvar and Batagelj 2016).

A special version of Pajek has been developed known as Pajek XXI, which consumes less memory. Pajek XXI is used for analyzing huge networks. After obtaining some interesting parts from the large networks, the standard version can be used for the further analysis. Pajek contains different algorithms including VOS mapping, Fruchterman Reingold optimization, Pivot MDS, Kamada-Kawai optimization, FishEye transformation, for obtaining distinct network layouts. The major disadvantage of Pajek is that it is not connected to any database, which remains the main factor for its limited use (Mrvar and Batagelj 2016).

6.4.4.5 Biolayout Express 3D

Biolayout is an impressive platform developed for the visualization and exploration of complex biological datasets (Enright and Ouzounis 2001a). Biolayout also provides an option to create networks from any tabulated data or discrete data values. It is also used for investigating the predesigned networks by employing stochastic Petri net algorithm. It follows other algorithms like Fruchterman-Reingold and FMMM for the advanced analysis of data. The interaction networks can be visualized in both 2D and 3D formats depending on the user's desire. Data in

Biolayout can be loaded in various formats including: .layout, .csv, .expression, . graphml, and .gml. Thus, Biolayout allows for the integration and analysis of data from different platforms. Outcomes of analysis in Biolayout can also be exported in different formats like PNG, TIF, and JPG.

Biolayout also couples with several other algorithms to provide deeper insights and understanding of data to the users. It also facilitates the user to construct additional nodes and edges by simply providing their names in text format. It aids in designing weighted graphs. Rotation, Zoom in/out, and movement options are also available to visualize networks in different forms. The latest version of Bioexpress 3D has included Markov clustering algorithm to perform clustering analysis. The major drawbacks of Biolayout include: (1) it does not support any plugin, thus cannot integrate with other databases, (2) color of nodes can be defined by users but cannot be saved, (3) it uses graphics renderer which limits the size of data that can be processed in Biolayout (Lu et al. 2013; Lua et al. 2014).

6.4.4.6 FunRich (Functional Enrichment and Interaction Analysis Tool)

Funrich is a windows-based freely available standalone tool both for the functional enrichment and interaction network analysis (Pathan et al. 2015). Funrich facilitates the user to analyze their data against three databases (UniProt, Custom, and FunRich) in the background. Funrich exclusively contains human-specific genes/ proteins data, hence allows user to perform enrichment analysis against human genes/proteins. Further, users can identify the enriched and depleted terms in their dataset against 20 distinct taxonomical levels using UniProt which is also accessible through FunRich. Any desired organism's data can be downloaded from UniProt and can be loaded in FunRich analysis tool to carry out advanced analysis.

An important highlight of FunRich is that it allows the access to custom database using which the enrichment analysis can be carried out irrespective of type of organism. It also assists the user to customize their graphical data to generate publication quality images. Graphs can be presented in many different forms including bar graphs, heatmap, pie charts, column graphs, Venn diagrams, etc. Other significant feature of FunRich is that it allows the user to sort, mask, and display the desirable enrichment terms in the output in addition to customize the size, color, axis labels, and legends of the graphs.

The PPI networks can be visualized in several differential formats including circular, planetary, square, and packed. The nodes which are enriched in specific biological pathways can be highlighted and can be used to construct subnetworks. The subnetworks can also be extended by adding more interacting partners by choosing those nodes that are in direct connection with highlighted nodes. The networks can also be decorated using different sizes, colors, and labels of nodes in a user-defined manner (Pathan et al. 2015).

6.4.4.7 Osprey

Breitkreutz et al. have developed network visualization platform known as Osprey for the manipulation of complex biological networks (Breitkreutz et al. 2003). Osprey is designed to perform functional comparisons between the datasets in addition to display interactions in expandable and flexible graphical layouts. Osprey is accessible as a standalone program and as an add-on viewer in online databases. Fully personalized networks can be constructed with addition of differential datasets in different formats from existing interaction databases. Osprey utilizes automatically the datasets contained in GRID (General Repository for Interaction Datasets as a database) for the construction of networks. User can select nodes and edges to be included or excluded in the network. Resulting networks can be saved as tab delimited text file to perform further manipulations or can be exported in different formats such as PNG, SVG, and JPEG to prepare high quality images.

The major advantages of Osprey include: (1) It allows to search for a particular gene in large networks using gene names in text search queries, (2) Functional information about the nodes in the network can be retrieved, and (3) Various filters such as source, experimental, function, and connectivity can be applied to the interaction networks. For instance, interactions obtained from a particular source can be filtered and displayed. The complex networks can be simplified using node relaxation options employed by the users. Further, different network layouts like circle, spoke, dual ring, and concentric circles can be used to present the interaction network data. The position of chosen nodes and edges can also be locked to anchor the network. Osprey also facilitates the comparison of several complex datasets by an overlaying option (Breitkreutz et al. 2003).

6.4.4.8 ProViz

Iragne et al. have presented a software known as ProViz (protein interaction visualization and exploration tool) for interactive visualization and exploration of large networks (Iragne et al. 2005). ProViz is designed to furnish fast, open tool to explore large complex biological networks by providing access to various applications through different plugins. Proteins and interactions of interest can be found in the large networks using keyword search option. Graphs from different species can be compared. Sub-graphs can be generated for further analysis. Related proteins and their interactions can be clustered together. Proviz is fast which facilitates the manipulation of networks with a million of elements and allows screen updates within 50 ms on a standard workstation.

ProViz works in integration with IntAct database (Hermjakob et al. 2004b). Links for each node and edge are provided which connects back to IntAct web services. Interface of ProViz is user-friendly and uncluttered intentionally. Different network layout options are available to view the network in desired fashion. Plugins are available for dozens of layout algorithms. Of all these, three can be used directly owing to their capability of focusing on biologically relevant information. These algorithms include: (1) Hierarchical layout is used to decipher the nodes which shares common ancestors, hence it is used to study metabolic pathway networks or cascade-type interactions. (2) GEM is a force-based graph drawing algorithm, which clusters the related nodes and allows a quick access to the proteins performing desired function. (3) Circular layout is a neutral choice which does not provide any details about edge relations. ProViz is developed on Tulip platform, hence it can also provide a three-dimensional view of networks (Iragne et al. 2005).

6.4.4.9 Pivot (Protein Interaction and Visualization Tool)

Pivot is a novel software developed by Orlev et al. to visualize protein interaction data collected from differential sources and also allows to decipher relationship between the proteins using graph theory algorithms (Orlev et al. 2004). The visualization interface of pivot represents proteins as nodes and edges as interactions as displayed by other software as discussed above. One additional feature of Pivot interface is that some nodes are marked with dot, which marks that the particular node/protein exhibits other interactions also which are not yet displayed in the current network view. In order to show those interactions in the current network, those nodes/proteins are required to be clicked twice.

Functional information will be displayed for the desired protein in networks when user points on that particular protein using mouse cursor. Users can also create their own networks by combining their own interaction results with the interaction information collected from the databases. Pivot provides an easy way of following interactions in dense areas of network. If an option known as "neighbor's marker" is on, then all the proteins directly interacting with a desired protein will be displayed when user will point on that specifically desired protein. Pivot also aids in deducing homology maps in which the known interactions between the proteins in one organism are employed to obtain the interactions between their homologs in other species (Orlev et al. 2004).

6.5 Conclusions

PPIs play an imperative role in all the major biological processes. Both experimental and computational efforts are being made by the researcher in solving the compelling puzzle of PPIs. Emergence of fast computational techniques have also geared up the prediction of novel and missing PPIs. As a consequence, ample amount of PPI data has been generated for which special databases have been designed to store that data. Further, attempts have also been made to visualize and analyze large network of proteins in the form of PPI networks to decipher their involvement in real biological phenomena and to observe how differential proteins synchronize in different pathways. Different methods along with their specific features have their own merits and pitfalls, hence methodologies for prediction of PPIs are selected on the basis of distinct requirements. In order to achieve more reliable outcomes in this rapidly blooming era of PPIs, strategies amalgamating different methodologies and meta-servers are also evolving. It is highly expected that in future well-defined proteome scale maps will be procured by the integration of various approaches.

References

Afsar Minhas FuA, Geiss BJ, Ben-Hur A. (2014). PAIRpred: Partner-specific prediction of interacting residues from sequence and structure. *Proteins: Structure, Function, and Bioinformatics*, *82*, 1142-1155.

Albert R. (2005). Scale-free networks in cell biology. *Journal of cell science, 118*, 4947-4957.
Albert R, Jeong H, Barabasi AL. (2000). Error and attack tolerance of complex networks. *Nature, 406*, 378-382.
Aoki Y, Okamura Y, Tadaka S, Kinoshita K, Obayashi T. (2016). ATTED-II in 2016: A Plant Coexpression Database Towards Lineage-Specific Coexpression. *Plant & cell physiology, 57*, e5.
Apweiler R et al. (2010). The universal protein resource (UniProt) in 2010. *Nucleic acids research, 38*, D142-D148.
Aranda B et al. (2010). The IntAct molecular interaction database in 2010. *Nucleic acids research, 38*, D525-531.
Argos P. (1988). An investigation of protein subunit and domain interfaces. *Protein Engineering, Design and Selection, 2*, 101-113.
Arkin MR, Wells JA. (2004). Small-molecule inhibitors of protein–protein interactions: progressing towards the dream. *Nature reviews Drug discovery, 3*, 301-317.
Assi SA, Tanaka T, Rabbitts TH, Fernandez-Fuentes N. (2010). PCRPi: Presaging Critical Residues in Protein interfaces, a new computational tool to chart hot spots in protein interfaces. *Nucleic acids research, 38*, e86-e86.
Attwood TK et al. (2012). The PRINTS database: a fine-grained protein sequence annotation and analysis resource—its status in 2012. *Database, 2012*.
Aytuna AS, Gursoy A, Keskin O. (2005). Prediction of protein–protein interactions by combining structure and sequence conservation in protein interfaces. *Bioinformatics, 21*, 2850-2855.
Bader GD, Betel D, Hogue CW. (2003). BIND: the Biomolecular Interaction Network Database. *Nucleic acids research, 31*, 248-250.
Bahadur RP, Chakrabarti P, Rodier F, Janin J. (2003). Dissecting subunit interfaces in homodimeric proteins. *Proteins, 53*, 708-719.
Bahadur RP, Chakrabarti P, Rodier F, Janin J. (2004). A dissection of specific and non-specific protein–protein interfaces. *Journal of molecular biology, 336*, 943-955.
Bajaj CL, Chowdhury R, Siddahanavalli V. (2009). $ F^ 2$ Dock: Fast Fourier Protein-Protein Docking. *IEEE/ACM Transactions on Computational Biology and Bioinformatics, 8*, 45-58.
Balaji S, McClendon C, Chowdhary R, Liu JS, Zhang J. (2012). IMID: integrated molecular interaction database. *Bioinformatics, 28*, 747-749.
Barabasi AL, Oltvai ZN. (2004). Network biology: understanding the cell's functional organization. *Nature reviews. Genetics, 5*, 101-113.
Barbarino JM, Whirl-Carrillo M, Altman RB, Klein TE. (2018). PharmGKB: A worldwide resource for pharmacogenomic information. *Wiley interdisciplinary reviews. Systems biology and medicine, 10*, e1417.
Barrett T et al. (2009). NCBI GEO: archive for high-throughput functional genomic data. *Nucleic acids research, 37*, D885-890.
Barrett T et al. (2013). NCBI GEO: archive for functional genomics data sets--update. *Nucleic acids research, 41*, D991-995.
Barsky A, Gardy JL, Hancock RE, Munzner T. (2007). Cerebral: a Cytoscape plugin for layout of and interaction with biological networks using subcellular localization annotation. *Bioinformatics, 23*, 1040-1042.
Bartel PL, Roecklein JA, SenGupta D, Fields S. (1996). A protein linkage map of Escherichia coli bacteriophage T7. *Nature genetics, 12*, 72-77.
Basse MJ et al. (2012). 2P2Idb: a structural database dedicated to orthosteric modulation of protein–protein interactions. *Nucleic acids research, 41*, D824-D827.
Batada NN. (2004). CNplot: visualizing pre-clustered networks. *Bioinformatics, 20*, 1455-1456.
Batagelj V, Mrvar A. (1998). Pajek-program for large network analysis. *Connections, 21*, 47-57.
Batagelj V, Mrvar A, Ferligoj A, Doreian P. (2004). Generalized blockmodeling with Pajek. *Metodoloski zvezki, 1*, 455.
Battista GD, Eades P, Tamassia R, Tollis IG (1998) Graph drawing: algorithms for the visualization of graphs. Prentice Hall PTR,

Becerra A, Bucheli VA, Moreno PA. (2017). Prediction of virus-host protein-protein interactions mediated by short linear motifs. *BMC bioinformatics, 18*, 163.

Beltrao P, Bork P, Krogan NJ, van Noort V. (2013). Evolution and functional cross-talk of protein post-translational modifications. *Molecular systems biology, 9*, 714.

Ben-Hur A, Noble WS. (2005). Kernel methods for predicting protein–protein interactions. *Bioinformatics, 21*, i38-i46.

Bergmann S, Ihmels J, Barkai N. (2004). Similarities and differences in genome-wide expression data of six organisms. *PLoS biology, 2*, E9.

Bernauer J, Bahadur RP, Rodier F, Janin J, Poupon A. (2008). DiMoVo: a Voronoi tessellation-based method for discriminating crystallographic and biological protein–protein interactions. *Bioinformatics, 24*, 652-658.

Bertin N et al. (2007). Confirmation of organized modularity in the yeast interactome. *PLoS biology, 5*, e153.

Bessman MJ, Frick DN, O'Handley SF. (1996). The MutT proteins or "Nudix" hydrolases, a family of versatile, widely distributed,"housecleaning" enzymes. *Journal of Biological Chemistry, 271*, 25059-25062.

Beuming T, Skrabanek L, Niv MY, Mukherjee P, Weinstein H. (2005). PDZBase: a protein-protein interaction database for PDZ-domains. *Bioinformatics, 21*, 827-828.

Bhardwaj N, Lu H. (2005). Correlation between gene expression profiles and protein-protein interactions within and across genomes. *Bioinformatics, 21*, 2730-2738.

Bloom JD, Lu Z, Chen D, Raval A, Venturelli OS, Arnold FH. (2007). Evolution favors protein mutational robustness in sufficiently large populations. *BMC biology, 5*, 29.

Boccaletti S, Latora V, Moreno Y, Chavez M, Hwang D-U. (2006). Complex networks: Structure and dynamics. *Physics reports, 424*, 175-308.

Bogan AA, Thorn KS. (1998). Anatomy of hot spots in protein interfaces. *Journal of molecular biology, 280*, 1-9.

Bollobás B, Béla B (2001) Random graphs. vol 73. Cambridge university press,

Bourgeas R, Basse MJ, Morelli X, Roche P. (2010). Atomic analysis of protein-protein interfaces with known inhibitors: the 2P2I database. *PLoS One, 5*, e9598.

Breitkreutz A et al. (2010). A global protein kinase and phosphatase interaction network in yeast. *Science, 328*, 1043-1046.

Breitkreutz BJ et al. (2008). The BioGRID Interaction Database: 2008 update. *Nucleic acids research, 36*, D637-640.

Breitkreutz BJ, Stark C, Tyers M. (2003). Osprey: a network visualization system. *Genome Biol, 4*, R22.

Breuer K et al. (2013). InnateDB: systems biology of innate immunity and beyond--recent updates and continuing curation. *Nucleic acids research, 41*, D1228-1233.

Brown KR, Jurisica I. (2005). Online predicted human interaction database. *Bioinformatics, 21*, 2076-2082.

Brown KR, Jurisica I. (2007). Unequal evolutionary conservation of human protein interactions in interologous networks. *Genome biology, 8*, R95.

Brown KR et al. (2009). NAViGaTOR: Network Analysis, Visualization and Graphing Toronto. *Bioinformatics, 25*, 3327-3329.

Bru C, Courcelle E, Carrere S, Beausse Y, Dalmar S, Kahn D. (2005). The ProDom database of protein domain families: more emphasis on 3D. *Nucleic acids research, 33*, D212-215.

Caspi R et al. (2010). The MetaCyc database of metabolic pathways and enzymes and the BioCyc collection of pathway/genome databases. *Nucleic acids research, 38*, D473-479.

Caspi R et al. (2008). The MetaCyc Database of metabolic pathways and enzymes and the BioCyc collection of Pathway/Genome Databases. *Nucleic acids research, 36*, D623-631.

Ceol A, Chatr-aryamontri A, Santonico E, Sacco R, Castagnoli L, Cesareni G. (2007). DOMINO: a database of domain-peptide interactions. *Nucleic acids research, 35*, D557-560.

Ceol A et al. (2010). MINT, the molecular interaction database: 2009 update. *Nucleic acids research, 38*, D532-539.

Cerami EG et al. (2011). Pathway Commons, a web resource for biological pathway data. *Nucleic acids research*, *39*, D685-690.

Chatr-aryamontri A, Ceol A, Palazzi LM, Nardelli G, Schneider MV, Castagnoli L, Cesareni G. (2007). MINT: the Molecular INTeraction database. *Nucleic acids research*, *35*, D572-574.

Chatr-aryamontri A et al. (2009). VirusMINT: a viral protein interaction database. *Nucleic acids research*, *37*, D669-673.

Chaudhury S, Berrondo M, Weitzner BD, Muthu P, Bergman H, Gray JJ. (2011). Benchmarking and analysis of protein docking performance in Rosetta v3. 2. *PloS one*, *6*, e22477.

Chen J, Hsu W, Lee ML, Ng SK. (2006a). Increasing confidence of protein interactomes using network topological metrics. *Bioinformatics*, *22*, 1998-2004.

Chen JY, Mamidipalli S, Huan T. (2009). HAPPI: an online database of comprehensive human annotated and predicted protein interactions. *BMC genomics*, *10 Suppl 1*, S16.

Chen L, Wu LY, Wang Y, Zhang XS. (2006b). Inferring protein interactions from experimental data by association probabilistic method. *Proteins*, *62*, 833-837.

Cheng S, Zhang Y, Brooks CL. (2015). PCalign: a method to quantify physicochemical similarity of protein-protein interfaces. *BMC bioinformatics*, *16*, 33.

Cherry JM et al. (2012). Saccharomyces Genome Database: the genomics resource of budding yeast. *Nucleic acids research*, *40*, D700-D705.

Chothia C, Janin J. (1975). Principles of protein–protein recognition. *Nature*, *256*, 705-708.

Chua HN, Sung W-K, Wong L. (2006). Exploiting indirect neighbours and topological weight to predict protein function from protein–protein interactions. *Bioinformatics*, *22*, 1623-1630.

Clark GW, Bezginov A, Yang JM, Charlebois RL, Tillier ER (2011) Using coevolution to predict protein–protein interactions. In: Network Biology. Springer, pp 237-256

Cochrane G, Karsch-Mizrachi I, Nakamura Y, International Nucleotide Sequence Database C. (2011). The International Nucleotide Sequence Database Collaboration. *Nucleic acids research*, *39*, D15-18.

Comeau SR, Gatchell DW, Vajda S, Camacho CJ. (2004). ClusPro: a fully automated algorithm for protein–protein docking. *Nucleic acids research*, *32*, W96-W99.

Corpet F, Gouzy J, Kahn D. (1998). The ProDom database of protein domain families. *Nucleic acids research*, *26*, 323-326.

Craig RA, Liao L. (2007). Phylogenetic tree information aids supervised learning for predicting protein-protein interaction based on distance matrices. *BMC bioinformatics*, *8*, 1-12.

Croft D et al. (2011). Reactome: a database of reactions, pathways and biological processes. *Nucleic acids research*, *39*, D691-697.

Cukuroglu E, Gursoy A, Keskin O. (2012). HotRegion: a database of predicted hot spot clusters. *Nucleic acids research*, *40*, D829-D833.

Dandekar T, Snel B, Huynen M, Bork P. (1998). Conservation of gene order: a fingerprint of proteins that physically interact. *Trends in biochemical sciences*, *23*, 324-328.

Darnell SJ, LeGault L, Mitchell JC. (2008). KFC Server: interactive forecasting of protein interaction hot spots. *Nucleic acids research*, *36*, W265-W269.

Davis AP et al. (2017). The Comparative Toxicogenomics Database: update 2017. *Nucleic acids research*, *45*, D972-D978.

Davis FP, Sali A. (2005). PIBASE: a comprehensive database of structurally defined protein interfaces. *Bioinformatics*, *21*, 1901-1907.

de Hoon MJ, Imoto S, Nolan J, Miyano S. (2004). Open source clustering software. *Bioinformatics*, *20*, 1453-1454.

De Juan D, Pazos F, Valencia A. (2013). Emerging methods in protein coevolution. *Nature Reviews Genetics*, *14*, 249-261.

de Matos P et al. (2010). Chemical Entities of Biological Interest: an update. *Nucleic acids research*, *38*, D249-254.

De S, Krishnadev O, Srinivasan N, Rekha N. (2005). Interaction preferences across protein-protein interfaces of obligatory and non-obligatory components are different. *BMC structural biology*, *5*, 15.

Deane CM, Salwinski L, Xenarios I, Eisenberg D. (2002). Protein interactions: two methods for assessment of the reliability of high throughput observations. *Molecular & cellular proteomics : MCP, 1*, 349-356.

Deng L, Zhang QC, Chen Z, Meng Y, Guan J, Zhou S. (2014). PredHS: a web server for predicting protein–protein interaction hot spots by using structural neighborhood properties. *Nucleic acids research, 42*, W290-W295.

Deng M, Mehta S, Sun F, Chen T. (2002). Inferring domain-domain interactions from protein-protein interactions. *Genome research, 12*, 1540-1548.

Deng M, Zhang K, Mehta S, Chen T, Sun F. (2003). Prediction of protein function using protein-protein interaction data. *Journal of computational biology : a journal of computational molecular cell biology, 10*, 947-960.

Deribe YL, Pawson T, Dikic I. (2010). Post-translational modifications in signal integration. *Nature structural & molecular biology, 17*, 666-672.

Ding Z, Kihara D. (2018). Computational Methods for Predicting Protein-Protein Interactions Using Various Protein Features. *Current protocols in protein science, 93*, e62.

Dinkel H, Chica C, Via A, Gould CM, Jensen LJ, Gibson TJ, Diella F. (2011). Phospho.ELM: a database of phosphorylation sites--update 2011. *Nucleic acids research, 39*, D261-267.

Dinkel H et al. (2012). ELM--the database of eukaryotic linear motifs. *Nucleic acids research, 40*, D242-251.

Dominguez C, Boelens R, Bonvin AM. (2003). HADDOCK: a protein-protein docking approach based on biochemical or biophysical information. *Journal of the American Chemical Society, 125*, 1731-1737.

Droit A, Hunter JM, Rouleau M, Ethier C, Picard-Cloutier A, Bourgais D, Poirier GG. (2007). PARPs database: a LIMS systems for protein-protein interaction data mining or laboratory information management system. *BMC bioinformatics, 8*, 483.

Duan G, Walther D. (2015). The roles of post-translational modifications in the context of protein interaction networks. *PLoS computational biology, 11*, e1004049.

Duan G, Walther D, Schulze WX. (2013). Reconstruction and analysis of nutrient-induced phosphorylation networks in Arabidopsis thaliana. *Frontiers in plant science, 4*, 540.

Dudkina NV, Kouril R, Bultema JB, Boekema EJ. (2010). Imaging of organelles by electron microscopy reveals protein-protein interactions in mitochondria and chloroplasts. *FEBS letters, 584*, 2510-2515.

Dunker AK, Romero P, Obradovic Z, Garner EC, Brown CJ. (2000). Intrinsic protein disorder in complete genomes. *Genome informatics, 11*, 161-171.

Dutkowski J, Tiuryn J. (2007). Identification of functional modules from conserved ancestral protein-protein interactions. *Bioinformatics, 23*, i149-158.

Eades P. (1984). A heuristic for graph drawing. *Congressus numerantium, 42*, 149-160.

Encinar JA, Fernandez-Ballester G, Sánchez IE, Hurtado-Gomez E, Stricher F, Beltrao P, Serrano L. (2009). ADAN: a database for prediction of protein–protein interaction of modular domains mediated by linear motifs. *Bioinformatics, 25*, 2418-2424.

Enright AJ, Iliopoulos I, Kyrpides NC, Ouzounis CA. (1999). Protein interaction maps for complete genomes based on gene fusion events. *Nature, 402*, 86-90.

Enright AJ, Ouzounis CA. (2001a). BioLayout--an automatic graph layout algorithm for similarity visualization. *Bioinformatics, 17*, 853-854.

Enright AJ, Ouzounis CA. (2001b). Functional associations of proteins in entire genomes by means of exhaustive detection of gene fusions. *Genome biology, 2*, 1-7.

Enright AJ, Van Dongen S, Ouzounis CA. (2002). An efficient algorithm for large-scale detection of protein families. *Nucleic acids research, 30*, 1575-1584.

Finn RD et al. (2017). InterPro in 2017-beyond protein family and domain annotations. *Nucleic acids research, 45*, D190-D199.

Finn RD et al. (2016). The Pfam protein families database: towards a more sustainable future. *Nucleic acids research, 44*, D279-285.

Finn RD, Miller BL, Clements J, Bateman A. (2014). iPfam: a database of protein family and domain interactions found in the Protein Data Bank. *Nucleic acids research, 42*, D364-D373.
Fischer TB et al. (2003). The binding interface database (BID): a compilation of amino acid hot spots in protein interfaces. *Bioinformatics, 19*, 1453-1454.
Flicek P et al. (2010). Ensembl's 10th year. *Nucleic acids research, 38*, D557-562.
Flicek P et al. (2011). Ensembl 2011. *Nucleic acids research, 39*, D800-806.
Folador EL, Hassan SS, Lemke N, Barh D, Silva A, Ferreira RS, Azevedo V. (2014). An improved interolog mapping-based computational prediction of protein–protein interactions with increased network coverage. *Integrative Biology, 6*, 1080-1087.
Franzot G, Carugo O. (2003). Computational approaches to protein-protein interaction. *Journal of structural and functional genomics, 4*, 245-255.
Freeman TC et al. (2007). Construction, visualisation, and clustering of transcription networks from microarray expression data. *PLoS computational biology, 3*, 2032-2042.
Frenkel-Morgenstern M et al. (2013). ChiTaRS: a database of human, mouse and fruit fly chimeric transcripts and RNA-sequencing data. *Nucleic acids research, 41*, D142-151.
Fromont-Racine M, Rain JC, Legrain P. (1997). Toward a functional analysis of the yeast genome through exhaustive two-hybrid screens. *Nature genetics, 16*, 277-282.
Fu W, Sanders-Beer BE, Katz KS, Maglott DR, Pruitt KD, Ptak RG. (2009). Human immunodeficiency virus type 1, human protein interaction database at NCBI. *Nucleic acids research, 37*, D417-422.
Gao M, Skolnick J. (2010). iAlign: a method for the structural comparison of protein–protein interfaces. *Bioinformatics, 26*, 2259-2265.
Garcia-Garcia J, Schleker S, Klein-Seetharaman J, Oliva B. (2012). BIPS: BIANA Interolog Prediction Server. A tool for protein-protein interaction inference. *Nucleic acids research, 40*, W147-151.
Gavin AC et al. (2002). Functional organization of the yeast proteome by systematic analysis of protein complexes. *Nature, 415*, 141-147.
Ge H, Liu Z, Church GM, Vidal M. (2001). Correlation between transcriptome and interactome mapping data from Saccharomyces cerevisiae. *Nature genetics, 29*, 482-486.
Geisler-Lee J, O'Toole N, Ammar R, Provart NJ, Millar AH, Geisler M. (2007). A predicted interactome for Arabidopsis. *Plant physiology, 145*, 317-329.
Gene Ontology C. (2010). The Gene Ontology in 2010: extensions and refinements. *Nucleic acids research, 38*, D331-335.
Gerber D, Maerkl SJ, Quake SR. (2009). An in vitro microfluidic approach to generating protein-interaction networks. *Nature methods, 6*, 71-74.
Ghoorah AW, Devignes M-D, Smaïl-Tabbone M, Ritchie DW. (2014). KBDOCK 2013: a spatial classification of 3D protein domain family interactions. *Nucleic acids research, 42*, D389-D395.
Gilson MK, Liu T, Baitaluk M, Nicola G, Hwang L, Chong J. (2016). BindingDB in 2015: A public database for medicinal chemistry, computational chemistry and systems pharmacology. *Nucleic acids research, 44*, D1045-1053.
Giurgiu M et al. (2019). CORUM: the comprehensive resource of mammalian protein complexes-2019. *Nucleic acids research, 47*, D559-D563.
Goel R, Harsha HC, Pandey A, Prasad TS. (2012). Human Protein Reference Database and Human Proteinpedia as resources for phosphoproteome analysis. *Molecular bioSystems, 8*, 453-463.
Goh C-S, Bogan AA, Joachimiak M, Walther D, Cohen FE. (2000). Coevolution of proteins with their interaction partners. *Journal of molecular biology, 299*, 283-293.
Goh C-S, Cohen FE. (2002). Coevolutionary analysis reveals insights into protein–protein interactions. *Journal of molecular biology, 324*, 177-192.
Goldberg DS, Roth FP. (2003). Assessing experimentally derived interactions in a small world. *Proceedings of the National Academy of Sciences, 100*, 4372-4376.
Gomez SM, Noble WS, Rzhetsky A. (2003). Learning to predict protein–protein interactions from protein sequences. *Bioinformatics, 19*, 1875-1881.

Gong S et al. (2005). A protein domain interaction interface database: InterPare. *BMC bioinformatics, 6*, 207.

González-Ruiz D, Gohlke H. (2006). Targeting protein-protein interactions with small molecules: challenges and perspectives for omputational binding epitope detection and ligand finding. *Current medicinal chemistry, 13*, 2607-2625.

Goodman N et al. (2003). Plans for HDBase—a research community website for Huntington's Disease. *Clinical Neuroscience Research, 3*, 197-217.

Gramates LS et al. (2017). FlyBase at 25: looking to the future. *Nucleic acids research, 45*, D663-D671.

Grigoriev A. (2001). A relationship between gene expression and protein interactions on the proteome scale: analysis of the bacteriophage T7 and the yeast Saccharomyces cerevisiae. *Nucleic acids research, 29*, 3513-3519.

Grindrod P, Kibble M. (2004). Review of uses of network and graph theory concepts within proteomics. *Expert review of proteomics, 1*, 229-238.

Gu H, Zhu P, Jiao Y, Meng Y, Chen M. (2011). PRIN: a predicted rice interactome network. *BMC bioinformatics, 12*, 161.

Gulati K, Gangele K, Agarwal N, Jamsandekar M, Kumar D, Poluri KM. (2018). Molecular cloning and biophysical characterization of CXCL3 chemokine. *International journal of biological macromolecules, 107*, 575-584.

Guney E, Tuncbag N, Keskin O, Gursoy A. (2007). HotSprint: database of computational hot spots in protein interfaces. *Nucleic acids research, 36*, D662-D666.

Günther S, May P, Hoppe A, Frömmel C, Preissner R. (2007). Docking without docking: ISEARCH—prediction of interactions using known interfaces. *Proteins: Structure, Function, and Bioinformatics, 69*, 839-844.

Guo Y, Yu L, Wen Z, Li M. (2008). Using support vector machine combined with auto covariance to predict protein–protein interactions from protein sequences. *Nucleic acids research, 36*, 3025-3030.

Hakes L, Lovell SC, Oliver SG, Robertson DL. (2007). Specificity in protein interactions and its relationship with sequence diversity and coevolution. *Proceedings of the National Academy of Sciences, 104*, 7999-8004.

Haliloglu T, Ben-Tal N, Garzon J, Ozbek P, Soner S. (2015). DynaFace: Discrimination between Obligatory and Non-obligatory Protein-Protein Interactions Based on the Complex's Dynamics.

Hamp T, Rost B. (2015). Evolutionary profiles improve protein-protein interaction prediction from sequence. *Bioinformatics, 31*, 1945-1950.

Han K, Park B, Kim H, Hong J, Park J. (2004). HPID: the human protein interaction database. *Bioinformatics, 20*, 2466-2470.

Hashemifar S, Neyshabur B, Khan AA, Xu J. (2018). Predicting protein-protein interactions through sequence-based deep learning. *Bioinformatics, 34*, i802-i810.

Hastings J et al. (2013). The ChEBI reference database and ontology for biologically relevant chemistry: enhancements for 2013. *Nucleic acids research, 41*, D456-463.

Hayashida M, Kamada M, Song J, Akutsu T. (2011). Conditional random field approach to prediction of protein-protein interactions using domain information. *BMC systems biology, 5 Suppl 1*, S8.

Henrick K, Thornton JM. (1998). PQS: a protein quaternary structure file server. *Trends in biochemical sciences, 23*, 358-361.

Herman D, Ochoa D, Juan D, Lopez D, Valencia A, Pazos F. (2011). Selection of organisms for the coevolution-based study of protein interactions. *BMC bioinformatics, 12*, 363.

Hermjakob H et al. (2004a). The HUPO PSI's molecular interaction format--a community standard for the representation of protein interaction data. *Nature biotechnology, 22*, 177-183.

Hermjakob H et al. (2004b). IntAct: an open source molecular interaction database. *Nucleic acids research, 32*, D452-455.

Higurashi M, Ishida T, Kinoshita K. (2009). PiSite: a database of protein interaction sites using multiple binding states in the PDB. *Nucleic acids research, 37*, D360-D364.

Ho Y et al. (2002). Systematic identification of protein complexes in Saccharomyces cerevisiae by mass spectrometry. *Nature*, *415*, 180-183.
Hooper SD, Bork P. (2005). Medusa: a simple tool for interaction graph analysis. *Bioinformatics*, *21*, 4432-4433.
Hopf TA et al. (2014). Sequence coevolution gives 3D contacts and structures of protein complexes. *Elife*, *3*, e03430.
Horner DS, Pirovano W, Pesole G. (2008). Correlated substitution analysis and the prediction of amino acid structural contacts. *Briefings in bioinformatics*, *9*, 46-56.
Hoskins J, Lovell S, Blundell TL. (2006). An algorithm for predicting protein–protein interaction sites: abnormally exposed amino acid residues and secondary structure elements. *Protein Science*, *15*, 1017-1029.
Hosur R, Xu J, Bienkowska J, Berger B. (2011). iWRAP: an interface threading approach with application to prediction of cancer-related protein–protein interactions. *Journal of molecular biology*, *405*, 1295-1310.
Howe K et al. WormBase: annotating many nematode genomes. In: Worm, 2012. vol 1. Taylor & Francis, pp 15-21
Hu Z, Ma B, Wolfson H, Nussinov R. (2000). Conservation of polar residues as hot spots at protein interfaces. *Proteins*, *39*, 331-342.
Huang TW et al. (2004). POINT: a database for the prediction of protein-protein interactions based on the orthologous interactome. *Bioinformatics*, *20*, 3273-3276.
Huttenhower C et al. (2009). Detailing regulatory networks through large scale data integration. *Bioinformatics*, *25*, 3267-3274.
Hwang H, Vreven T, Janin J, Weng Z. (2010). Protein-protein docking benchmark version 4.0. *Proteins*, *78*, 3111-3114.
Inbar Y, Benyamini H, Nussinov R, Wolfson HJ. (2005). Combinatorial docking approach for structure prediction of large proteins and multi-molecular assemblies. *Physical biology*, *2*, S156.
International HapMap C et al. (2007). A second generation human haplotype map of over 3.1 million SNPs. *Nature*, *449*, 851-861.
Iragne F, Nikolski M, Mathieu B, Auber D, Sherman D. (2005). ProViz: protein interaction visualization and exploration. *Bioinformatics*, *21*, 272-274.
Ito T, Chiba T, Ozawa R, Yoshida M, Hattori M, Sakaki Y. (2001). A comprehensive two-hybrid analysis to explore the yeast protein interactome. *Proceedings of the National Academy of Sciences of the United States of America*, *98*, 4569-4574.
Izarzugaza JM, Juan D, Pons C, Pazos F, Valencia A. (2008). Enhancing the prediction of protein pairings between interacting families using orthology information. *BMC bioinformatics*, *9*, 35.
Jaeger S, Sers CT, Leser U. (2010). Combining modularity, conservation, and interactions of proteins significantly increases precision and coverage of protein function prediction. *BMC genomics*, *11*, 717.
Jain E et al. (2009). Infrastructure for the life sciences: design and implementation of the UniProt website. *BMC bioinformatics*, *10*, 136.
Janin J. (2010). Protein-protein docking tested in blind predictions: the CAPRI experiment. *Molecular bioSystems*, *6*, 2351-2362.
Janin J, Chothia C. (1990). The structure of protein-protein recognition sites. *Journal of Biological Chemistry*, *265*, 16027-16030.
Janin J, Miller S, Chothia C. (1988). Surface, subunit interfaces and interior of oligomeric proteins. *Journal of molecular biology*, *204*, 155-164.
Jansen R, Greenbaum D, Gerstein M. (2002). Relating whole-genome expression data with protein-protein interactions. *Genome research*, *12*, 37-46.
Jayapandian M et al. (2007). Michigan Molecular Interactions (MiMI): putting the jigsaw puzzle together. *Nucleic acids research*, *35*, D566-571.
Jefferson ER, Walsh TP, Roberts TJ, Barton GJ. (2007). SNAPPI-DB: a database and API of Structures, iNterfaces and Alignments for Protein-Protein Interactions. *Nucleic acids research*, *35*, D580-589.

Jensen LJ et al. (2009). STRING 8--a global view on proteins and their functional interactions in 630 organisms. *Nucleic acids research, 37*, D412-416.

Jeon C, Agarwal K. (1996). Fidelity of RNA polymerase II transcription controlled by elongation factor TFIIS. *Proceedings of the National Academy of Sciences, 93*, 13677-13682.

Ji ZL et al. (2003). KDBI: kinetic data of bio-molecular interactions database. *Nucleic acids research, 31*, 255-257.

Jiménez-García B, Pons C, Fernández-Recio J. (2013). pyDockWEB: a web server for rigid-body protein–protein docking using electrostatics and desolvation scoring. *Bioinformatics, 29*, 1698-1699.

Jones S, Marin A, M. Thornton J. (2000). Protein domain interfaces: characterization and comparison with oligomeric protein interfaces. *Protein Engineering, 13*, 77-82.

Jones S, Thornton JM. (1995). Protein-protein interactions: a review of protein dimer structures. *Progress in biophysics and molecular biology, 63*, 31-65.

Jones S, Thornton JM. (1996). Principles of protein-protein interactions. *Proceedings of the National Academy of Sciences, 93*, 13-20.

Jones S, Thornton JM. (1997). Analysis of protein-protein interaction sites using surface patches. *Journal of molecular biology, 272*, 121-132.

Jothi R, Cherukuri PF, Tasneem A, Przytycka TM. (2006). Coevolutionary analysis of domains in interacting proteins reveals insights into domain–domain interactions mediating protein–protein interactions. *Journal of molecular biology, 362*, 861-875.

Jothi R, Kann MG, Przytycka TM. (2005). Predicting protein–protein interaction by searching evolutionary tree automorphism space. *Bioinformatics, 21*, i241-i250.

Juan D, Pazos F, Valencia A. (2008a). Coevolution and co-adaptation in protein networks. *FEBS letters, 582*, 1225-1230.

Juan D, Pazos F, Valencia A. (2008b). High-confidence prediction of global interactomes based on genome-wide coevolutionary networks. *Proceedings of the National Academy of Sciences, 105*, 934-939.

Kalathur RKR, Pinto JP, Hernandez-Prieto MA, Machado RS, Almeida D, Chaurasia G, Futschik ME. (2014). UniHI 7: an enhanced database for retrieval and interactive analysis of human molecular interaction networks. *Nucleic acids research, 42*, D408-D414.

Kamada M, Sakuma Y, Hayashida M, Akutsu T. (2014). Prediction of protein-protein interaction strength using domain features with supervised regression. *TheScientificWorldJournal, 2014*, 240673.

Kamburov A, Pentchev K, Galicka H, Wierling C, Lehrach H, Herwig R. (2011). ConsensusPathDB: toward a more complete picture of cell biology. *Nucleic acids research, 39*, D712-717.

Kandasamy K et al. (2010). NetPath: a public resource of curated signal transduction pathways. *Genome Biol, 11*, R3.

Kanehisa M. (1997). A database for post-genome analysis. *Trends in genetics : TIG, 13*, 375-376.

Kanehisa M, Furumichi M, Tanabe M, Sato Y, Morishima K. (2017). KEGG: new perspectives on genomes, pathways, diseases and drugs. *Nucleic acids research, 45*, D353-D361.

Kanehisa M, Goto S, Furumichi M, Tanabe M, Hirakawa M. (2010). KEGG for representation and analysis of molecular networks involving diseases and drugs. *Nucleic acids research, 38*, D355-360.

Kann MG, Jothi R, Cherukuri PF, Przytycka TM. (2007). Predicting protein domain interactions from coevolution of conserved regions. *Proteins: Structure, Function, and Bioinformatics, 67*, 811-820.

Kar G, Gursoy A, Keskin O. (2009). Human cancer protein-protein interaction network: a structural perspective. *PLoS computational biology, 5*, e1000601.

Kara A, Vickers M, Swain M, Whitworth DE, Fernandez-Fuentes N. (2016). MetaPred2CS: a sequence-based meta-predictor for protein-protein interactions of prokaryotic two-component system proteins. *Bioinformatics, 32*, 3339-3341.

Kastritis PL, Moal IH, Hwang H, Weng Z, Bates PA, Bonvin AM, Janin J. (2011). A structure-based benchmark for protein–protein binding affinity. *Protein Science*, *20*, 482-491.

Kenworthy AK. (2001). Imaging protein-protein interactions using fluorescence resonance energy transfer microscopy. *Methods*, *24*, 289-296.

Kerrien S et al. (2012). The IntAct molecular interaction database in 2012. *Nucleic acids research*, *40*, D841-846.

Kersey PJ et al. (2010). Ensembl Genomes: extending Ensembl across the taxonomic space. *Nucleic acids research*, *38*, D563-569.

Keshava Prasad TS et al. (2009). Human Protein Reference Database--2009 update. *Nucleic acids research*, *37*, D767-772.

Keskin O, Gursoy A, Ma B, Nussinov RJCr. (2008). Principles of protein– protein interactions: What are the preferred ways for proteins to interact? *Chem Rev*, *108*, 1225-1244.

Keskin O, Ma B, Nussinov R. (2005). Hot regions in protein–protein interactions: the organization and contribution of structurally conserved hot spot residues. *Journal of molecular biology*, *345*, 1281-1294.

Keskin O, Tsai CJ, Wolfson H, Nussinov R. (2004). A new, structurally nonredundant, diverse data set of protein–protein interfaces and its implications. *Protein Science*, *13*, 1043-1055.

Keskin O, Tuncbag N, Gursoy A. (2016). Predicting protein–protein interactions from the molecular to the proteome level. *Chemical reviews*, *116*, 4884-4909.

Kiefer F, Arnold K, Kunzli M, Bordoli L, Schwede T. (2009). The SWISS-MODEL Repository and associated resources. *Nucleic acids research*, *37*, D387-392.

Kim DE, Chivian D, Baker D. (2004). Protein structure prediction and analysis using the Robetta server. *Nucleic acids research*, *32*, W526-W531.

Kim WK, Park J, Suh JK. (2002). Large scale statistical prediction of protein-protein interaction by potentially interacting domain (PID) pair. *Genome informatics. International Conference on Genome Informatics*, *13*, 42-50.

Kobe B et al. (2008). Crystallography and protein–protein interactions: biological interfaces and crystal contacts. *Biochemical Society Transactions*, *36*, 1438-1441.

Kortemme T, Baker D. (2002). A simple physical model for binding energy hot spots in protein–protein complexes. *Proceedings of the National Academy of Sciences*, *99*, 14116-14121.

Kotlyar M, Pastrello C, Sheahan N, Jurisica I. (2016). Integrated interactions database: tissue-specific view of the human and model organism interactomes. *Nucleic acids research*, *44*, D536-541.

Kozakov D, Brenke R, Comeau SR, Vajda S. (2006). PIPER: an FFT-based protein docking program with pairwise potentials. *Proteins: Structure, Function, and Bioinformatics*, *65*, 392-406.

Kozakov D et al. (2015). The FTMap family of web servers for determining and characterizing ligand-binding hot spots of proteins. *Nature protocols*, *10*, 733-755.

Krüger DM, Gohlke H. (2010). DrugScorePPI webserver: fast and accurate in silico alanine scanning for scoring protein–protein interactions. *Nucleic acids research*, *38*, W480-W486.

Krissinel E, Henrick K. (2007). Inference of macromolecular assemblies from crystalline state. *Journal of molecular biology*, *372*, 774-797.

Kuchaiev O, Rasajski M, Higham DJ, Przulj N. (2009). Geometric de-noising of protein-protein interaction networks. *PLoS computational biology*, *5*, e1000454.

Kuhn M, Szklarczyk D, Franceschini A, von Mering C, Jensen LJ, Bork P. (2012). STITCH 3: zooming in on protein-chemical interactions. *Nucleic acids research*, *40*, D876-880.

Kuiken C, Korber B, Shafer RW. (2003). HIV sequence databases. *AIDS reviews*, *5*, 52-61.

Kumar MS, Gromiha MM. (2006). PINT: protein–protein interactions thermodynamic database. *Nucleic acids research*, *34*, D195-D198.

Kundrotas PJ, Alexov E. (2007). PROTCOM: searchable database of protein complexes enhanced with domain-domain structures. *Nucleic acids research*, *35*, D575-579.

Kundrotas PJ, Anishchenko I, Dauzhenka T, Kotthoff I, Mnevets D, Copeland MM, Vakser IA. (2018). Dockground: a comprehensive data resource for modeling of protein complexes. *Protein Science*, *27*, 172-181.

Kundrotas PJ, Zhu Z, Janin J, Vakser IA. (2012). Templates are available to model nearly all complexes of structurally characterized proteins. *Proceedings of the National Academy of Sciences*, *109*, 9438-9441.

Kwon D et al. (2012). A comprehensive manually curated protein–protein interaction database for the Death Domain superfamily. *Nucleic acids research*, *40*, D331-D336.

Laskowski RA, Jablonska J, Pravda L, Varekova RS, Thornton JM. (2018). PDBsum: Structural summaries of PDB entries. *Protein science : a publication of the Protein Society*, *27*, 129-134.

Leader DP, Krause SA, Pandit A, Davies SA, Dow JAT. (2018). FlyAtlas 2: a new version of the Drosophila melanogaster expression atlas with RNA-Seq, miRNA-Seq and sex-specific data. *Nucleic acids research*, *46*, D809-D815.

Lechner M et al. (2012). CIDeR: multifactorial interaction networks in human diseases. *Genome biology*, *13*, R62.

Lee S-A, Chan C-h, Tsai C-H, Lai J-M, Wang F-S, Kao C-Y, Huang C-YF. (2008). Ortholog-based protein-protein interaction prediction and its application to inter-species interactions. *BMC bioinformatics*, *9*, S11.

Lei C, Ruan J A random walk based approach for improving protein-protein interaction network and protein complex prediction. In: 2012 IEEE International Conference on Bioinformatics and Biomedicine, 2012. IEEE, pp 1-6

Letamendia A, Labbe E, Attisano L. (2001). Transcriptional regulation by Smads: crosstalk between the TGF-beta and Wnt pathways. *The Journal of bone and joint surgery. American volume*, *83-A Suppl 1*, S31-39.

Letunic I, Doerks T, Bork P. (2009). SMART 6: recent updates and new developments. *Nucleic acids research*, *37*, D229-232.

Li D et al. (2006). Protein interaction networks of Saccharomyces cerevisiae, Caenorhabditis elegans and Drosophila melanogaster: large-scale organization and robustness. *Proteomics*, *6*, 456-461.

Li H, Yang S, Wang C, Zhou Y, Zhang Z. (2016). AraPPISite: a database of fine-grained protein-protein interaction site annotations for Arabidopsis thaliana. *Plant molecular biology*, *92*, 105-116.

Li X, Keskin O, Ma B, Nussinov R, Liang J. (2004). Protein–protein interactions: hot spots and structurally conserved residues often locate in complemented pockets that pre-organized in the unbound states: implications for docking. *Journal of molecular biology*, *344*, 781-795.

Li X, Yang L, Zhang X, Jiao X. (2019). Prediction of Protein-Protein Interactions Based on Domain. *Computational and mathematical methods in medicine*, *2019*, 5238406.

Lin J-S, Lai E-M (2017) Protein–protein interactions: co-immunoprecipitation. In: Bacterial Protein Secretion Systems. Springer, pp 211-219

Lin T-W, Wu J-W, Chang DT-H. (2013). Combining phylogenetic profiling-based and machine learning-based techniques to predict functional related proteins. *PloS one*, *8*, e75940.

Liu G, Li J, Wong L. (2008). Assessing and predicting protein interactions using both local and global network topological metrics. *Genome informatics. International Conference on Genome Informatics*, *21*, 138-149.

Liu T, Lin Y, Wen X, Jorissen RN, Gilson MK. (2007). BindingDB: a web-accessible database of experimentally determined protein-ligand binding affinities. *Nucleic acids research*, *35*, D198-201.

Lo Conte L, Chothia C, Janin J. (1999). The atomic structure of protein-protein recognition sites. *J Mol Biol*, *285*, 2177-2198.

Lopes CT, Franz M, Kazi F, Donaldson SL, Morris Q, Bader GD. (2010). Cytoscape Web: an interactive web-based network browser. *Bioinformatics*, *26*, 2347-2348.

Lopez Y, Nakai K, Patil A. (2015). HitPredict version 4: comprehensive reliability scoring of physical protein-protein interactions from more than 100 species. *Database : the journal of biological databases and curation, 2015.*

Lu CT et al. (2013). DbPTM 3.0: an informative resource for investigating substrate site specificity and functional association of protein post-translational modifications. *Nucleic acids research, 41*, D295-305.

Lua RC, Marciano DC, Katsonis P, Adikesavan AK, Wilkins AD, Lichtarge O. (2014). Prediction and redesign of protein–protein interactions. *Progress in biophysics and molecular biology, 116*, 194-202.

Ma B, Elkayam T, Wolfson H, Nussinov R. (2003). Protein-protein interactions: structurally conserved residues distinguish between binding sites and exposed protein surfaces. *Proceedings of the National Academy of Sciences of the United States of America, 100*, 5772-5777.

Macindoe G, Mavridis L, Venkatraman V, Devignes M-D, Ritchie DW. (2010). HexServer: an FFT-based protein docking server powered by graphics processors. *Nucleic acids research, 38*, W445-W449.

Marcotte EM, Pellegrini M, Ng H-L, Rice DW, Yeates TO, Eisenberg D. (1999). Detecting protein function and protein-protein interactions from genome sequences. *Science, 285*, 751-753.

Marsh JA, Hernandez H, Hall Z, Ahnert SE, Perica T, Robinson CV, Teichmann SA. (2013). Protein complexes are under evolutionary selection to assemble via ordered pathways. *Cell, 153*, 461-470.

Martin S, Roe D, Faulon J-L. (2005). Predicting protein–protein interactions using signature products. *Bioinformatics, 21*, 218-226.

Mashiach E, Nussinov R, Wolfson HJ. (2010). FiberDock: flexible induced-fit backbone refinement in molecular docking. *Proteins: Structure, Function, and Bioinformatics, 78*, 1503-1519.

Matthews LR et al. (2001). Identification of potential interaction networks using sequence-based searches for conserved protein-protein interactions or "interologs". *Genome research, 11*, 2120-2126.

McDowall MD et al. (2015). PomBase 2015: updates to the fission yeast database. *Nucleic acids research, 43*, D656-661.

McDowall MD, Scott MS, Barton GJ. (2009). PIPs: human protein-protein interaction prediction database. *Nucleic acids research, 37*, D651-656.

Meireles LM, Dömling AS, Camacho CJ. (2010). ANCHOR: a web server and database for analysis of protein–protein interaction binding pockets for drug discovery. *Nucleic acids research, 38*, W407-W411.

Merico D, Gfeller D, Bader GD. (2009). How to visually interpret biological data using networks. *Nature biotechnology, 27*, 921-924.

Mi H, Dong Q, Muruganujan A, Gaudet P, Lewis S, Thomas PD. (2010). PANTHER version 7: improved phylogenetic trees, orthologs and collaboration with the Gene Ontology Consortium. *Nucleic acids research, 38*, D204-210.

Mintseris J, Weng Z. (2005). Structure, function, and evolution of transient and obligate protein–protein interactions. *Proceedings of the National Academy of Sciences, 102*, 10930-10935.

Mishra GR et al. (2006). Human protein reference database--2006 update. *Nucleic acids research, 34*, D411-414.

Moal IH, Fernández-Recio J. (2012). SKEMPI: a Structural Kinetic and Energetic database of Mutant Protein Interactions and its use in empirical models. *Bioinformatics, 28*, 2600-2607.

Moreira IS, Fernandes PA, Ramos MJ. (2007). Hot spots—A review of the protein–protein interface determinant amino-acid residues. *Proteins: Structure, Function, and Bioinformatics, 68*, 803-812.

Moreira IS et al. (2017). SpotOn: high accuracy identification of protein-protein interface hot-spots. *Scientific reports, 7*, 1-11.

Morilla I, Lees JG, Reid AJ, Orengo C, Ranea JA. (2010). Assessment of protein domain fusions in human protein interaction networks prediction: application to the human kinetochore model. *New biotechnology, 27*, 755-765.

Mosca R, Ceol A, Aloy P. (2013). Interactome3D: adding structural details to protein networks. *Nature methods*, *10*, 47-53.

Mosca R, Ceol A, Stein A, Olivella R, Aloy P. (2014). 3did: a catalog of domain-based interactions of known three-dimensional structure. *Nucleic acids research*, *42*, D374-D379.

Mostafavi S, Ray D, Warde-Farley D, Grouios C, Morris Q. (2008). GeneMANIA: a real-time multiple association network integration algorithm for predicting gene function. *Genome Biol*, *9 Suppl 1*, S4.

Mrvar A, Batagelj V. (2016). Analysis and visualization of large networks with program package Pajek. *Complex Adaptive Systems Modeling*, *4*, 6.

Muley VY, Ranjan A. (2013). Evaluation of physical and functional protein-protein interaction prediction methods for detecting biological pathways. *PLoS One*, *8*, e54325.

Murali T, Pacifico S, Yu J, Guest S, Roberts GG, 3rd, Finley RL, Jr. (2011). DroID 2011: a comprehensive, integrated resource for protein, transcription factor, RNA and gene interactions for Drosophila. *Nucleic acids research*, *39*, D736-743.

Ng SK, Zhang Z, Tan SH, Lin K. (2003). InterDom: a database of putative interacting protein domains for validating predicted protein interactions and complexes. *Nucleic acids research*, *31*, 251-254.

Ngounou Wetie AG, Sokolowska I, Woods AG, Roy U, Loo JA, Darie CC. (2013). Investigation of stable and transient protein-protein interactions: Past, present, and future. *Proteomics*, *13*, 538-557.

Nikolovska-Coleska Z (2015) Studying protein-protein interactions using surface plasmon resonance. In: Protein-Protein Interactions. Springer, pp 109-138

Nishi H, Hashimoto K, Panchenko AR. (2011). Phosphorylation in protein-protein binding: effect on stability and function. *Structure*, *19*, 1807-1815.

Nooren IM, Thornton JM. (2003a). Diversity of protein–protein interactions. *The EMBO journal*, *22*, 3486-3492.

Nooren IM, Thornton JM. (2003b). Structural characterisation and functional significance of transient protein-protein interactions. *J Mol Biol*, *325*, 991-1018.

Nowotka MM, Gaulton A, Mendez D, Bento AP, Hersey A, Leach A. (2017). Using ChEMBL web services for building applications and data processing workflows relevant to drug discovery. *Expert opinion on drug discovery*, *12*, 757-767.

Ofran Y, Rost B. (2007). Protein–protein interaction hotspots carved into sequences. *PLoS computational biology*, *3*, e119.

Ohue M, Shimoda T, Suzuki S, Matsuzaki Y, Ishida T, Akiyama Y. (2014). MEGADOCK 4.0: an ultra–high-performance protein–protein docking software for heterogeneous supercomputers. *Bioinformatics*, *30*, 3281-3283.

Okamura Y, Aoki Y, Obayashi T, Tadaka S, Ito S, Narise T, Kinoshita K. (2015). COXPRESdb in 2015: coexpression database for animal species by DNA-microarray and RNAseq-based expression data with multiple quality assessment systems. *Nucleic acids research*, *43*, D82-86.

Olmsted S, Erlandsen S, Dunny GM, Wells CL. (1993). High-resolution visualization by field emission scanning electron microscopy of Enterococcus faecalis surface proteins encoded by the pheromone-inducible conjugative plasmid pCF10. *Journal of bacteriology*, *175*, 6229-6237.

Orchard S, Hermjakob H. (2008). The HUPO proteomics standards initiative--easing communication and minimizing data loss in a changing world. *Brief Bioinform*, *9*, 166-173.

Orchard S et al. (2012). Protein interaction data curation: the International Molecular Exchange (IMEx) consortium. *Nature methods*, *9*, 345-350.

Orlev N, Shamir R, Shiloh Y. (2004). PIVOT: protein interacions visualizatiOn tool. *Bioinformatics*, *20*, 424-425.

Oughtred R et al. (2019). The BioGRID interaction database: 2019 update. *Nucleic acids research*, *47*, D529-D541.

Ovchinnikov S, Kamisetty H, Baker D. (2014). Robust and accurate prediction of residue–residue interactions across protein interfaces using evolutionary information. *Elife*, *3*, e02030.

Overbeek R, Fonstein M, D'souza M, Pusch GD, Maltsev N. (1999). The use of gene clusters to infer functional coupling. *Proceedings of the National Academy of Sciences, 96*, 2896-2901.

Overington J. (2009). ChEMBL. An interview with John Overington, team leader, chemogenomics at the European Bioinformatics Institute Outstation of the European Molecular Biology Laboratory (EMBL-EBI). Interview by Wendy A. Warr. *Journal of computer-aided molecular design, 23*, 195-198.

Panchaud A, Singh P, Shaffer SA, Goodlett DR. (2010). xComb: a cross-linked peptide database approach to protein-protein interaction analysis. *Journal of proteome research, 9*, 2508-2515.

Park D, Singh R, Baym M, Liao CS, Berger B. (2011). IsoBase: a database of functionally related proteins across PPI networks. *Nucleic acids research, 39*, D295-300.

Pathan M et al. (2015). FunRich: An open access standalone functional enrichment and interaction network analysis tool. *Proteomics, 15*, 2597-2601.

Pavlopoulos GA, Moschopoulos CN, Hooper SD, Schneider R, Kossida S. (2009). jClust: a clustering and visualization toolbox. *Bioinformatics, 25*, 1994-1996.

Paz A et al. (2011). SPIKE: a database of highly curated human signaling pathways. *Nucleic acids research, 39*, D793-799.

Pazos F, Ranea JA, Juan D, Sternberg MJ. (2005). Assessing protein coevolution in the context of the tree of life assists in the prediction of the interactome. *Journal of molecular biology, 352*, 1002-1015.

Pazos F, Valencia A. (2001). Similarity of phylogenetic trees as indicator of protein–protein interaction. *Protein engineering, 14*, 609-614.

Pazos F, Valencia A. (2002). In silico two-hybrid system for the selection of physically interacting protein pairs. *Proteins: Structure, Function, and Bioinformatics, 47*, 219-227.

Pellegrini M, Marcotte EM, Thompson MJ, Eisenberg D, Yeates TO. (1999). Assigning protein functions by comparative genome analysis: protein phylogenetic profiles. *Proceedings of the National Academy of Sciences, 96*, 4285-4288.

Peri S et al. (2003). Development of human protein reference database as an initial platform for approaching systems biology in humans. *Genome research, 13*, 2363-2371.

Persico M, Ceol A, Gavrila C, Hoffmann R, Florio A, Cesareni G. (2005). HomoMINT: an inferred human network based on orthology mapping of protein interactions discovered in model organisms. *BMC bioinformatics, 6 Suppl 4*, S21.

Phan HT, Sternberg MJ. (2012). PINALOG: a novel approach to align protein interaction networks—implications for complex detection and function prediction. *Bioinformatics, 28*, 1239-1245.

Piehler J. (2005). New methodologies for measuring protein interactions in vivo and in vitro. *Current opinion in structural biology, 15*, 4-14.

Pierce BG, Wiehe K, Hwang H, Kim B-H, Vreven T, Weng Z. (2014). ZDOCK server: interactive docking prediction of protein–protein complexes and symmetric multimers. *Bioinformatics, 30*, 1771-1773.

Pinney JW, Shirley MW, McConkey GA, Westhead DR. (2005). metaSHARK: software for automated metabolic network prediction from DNA sequence and its application to the genomes of Plasmodium falciparum and Eimeria tenella. *Nucleic acids research, 33*, 1399-1409.

Pitre S et al. (2012). Short co-occurring polypeptide regions can predict global protein interaction maps. *Scientific reports, 2*, 239.

Planas-Iglesias J, Marin-Lopez MA, Bonet J, Garcia-Garcia J, Oliva B. (2013). iLoops: a protein–protein interaction prediction server based on structural features. *Bioinformatics, 29*, 2360-2362.

Powell S et al. (2012). eggNOG v3.0: orthologous groups covering 1133 organisms at 41 different taxonomic ranges. *Nucleic acids research, 40*, D284-289.

Pruitt KD, Tatusova T, Maglott DR. (2007). NCBI reference sequences (RefSeq): a curated non-redundant sequence database of genomes, transcripts and proteins. *Nucleic acids research, 35*, D61-65.

Pržulj N, Wigle DA, Jurisica I. (2004). Functional topology in a network of protein interactions. *Bioinformatics, 20*, 340-348.

Qin S, Zhou HX. (2007). meta-PPISP: a meta web server for protein-protein interaction site prediction. *Bioinformatics, 23*, 3386-3387.

Ramani AK, Marcotte EM. (2003). Exploiting the coevolution of interacting proteins to discover interaction specificity. *Journal of molecular biology, 327*, 273-284.

Ramírez-Aportela E, López-Blanco JR, Chacón P. (2016). FRODOCK 2.0: fast protein–protein docking server. *Bioinformatics, 32*, 2386-2388.

Rao VS, Srinivas K, Sujini G, Kumar G. (2014). Protein-protein interaction detection: methods and analysis. *International journal of proteomics, 2014*.

Razick S, Magklaras G, Donaldson IM. (2008). iRefIndex: a consolidated protein interaction database with provenance. *BMC bioinformatics, 9*, 405.

Reid AJ, Ranea JA, Clegg AB, Orengo CA. (2010). CODA: accurate detection of functional associations between proteins in eukaryotic genomes using domain fusion. *PloS one, 5*, e10908.

Rhead B et al. (2010). The UCSC Genome Browser database: update 2010. *Nucleic acids research, 38*, D613-619.

Rid R et al. (2013). PRIMOS: an integrated database of reassessed protein-protein interactions providing web-based access to in silico validation of experimentally derived data. *Assay and drug development technologies, 11*, 333-346.

Ripoche H, Laine E, Ceres N, Carbone A. (2017). JET2 Viewer: a database of predicted multiple, possibly overlapping, protein-protein interaction sites for PDB structures. *Nucleic acids research, 45*, 4278.

Rodionov A, Bezginov A, Rose J, Tillier ER. (2011). A new, fast algorithm for detecting protein coevolution using maximum compatible cliques. *Algorithms for molecular biology, 6*, 1-9.

Romero P, Wagg J, Green ML, Kaiser D, Krummenacker M, Karp PD. (2005). Computational prediction of human metabolic pathways from the complete human genome. *Genome Biol, 6*, R2.

Ruepp A et al. (2010). CORUM: the comprehensive resource of mammalian protein complexes—2009. *Nucleic acids research, 38*, D497-D501.

Safran M et al. (2010). GeneCards Version 3: the human gene integrator. *Database : the journal of biological databases and curation, 2010*, baq020.

Sahu SS, Weirick T, Kaundal R. (2014). Predicting genome-scale Arabidopsis-Pseudomonas syringae interactome using domain and interolog-based approaches. *BMC bioinformatics, 15 Suppl 11*, S13.

Salazar GA, Meintjes A, Mazandu GK, Rapanoël HA, Akinola RO, Mulder NJ. (2014). A web-based protein interaction network visualizer. *BMC bioinformatics, 15*, 1-8.

Salwinski L, Eisenberg D. (2007). The MiSink Plugin: Cytoscape as a graphical interface to the Database of Interacting Proteins. *Bioinformatics, 23*, 2193-2195.

Salwinski L, Miller CS, Smith AJ, Pettit FK, Bowie JU, Eisenberg D. (2004). The Database of Interacting Proteins: 2004 update. *Nucleic acids research, 32*, D449-451.

Saraf MC, Moore GL, Maranas CD. (2003). Using multiple sequence correlation analysis to characterize functionally important protein regions. *Protein Engineering, 16*, 397-406.

Sarkar S, Gulati K, Kairamkonda M, Mishra A, Poluri KM. (2018). Elucidating protein-protein interactions through computational approaches and designing small molecule inhibitors against them for various diseases. *Current topics in medicinal chemistry, 18*, 1719-1736.

Sato T, Yamanishi Y, Horimoto K, Kanehisa M, Toh H. (2006). Partial correlation coefficient between distance matrices as a new indicator of protein–protein interactions. *Bioinformatics, 22*, 2488-2492.

Sato T, Yamanishi Y, Kanehisa M, Toh H. (2005). The inference of protein–protein interactions by coevolutionary analysis is improved by excluding the information about the phylogenetic relationships. *Bioinformatics, 21*, 3482-3489.

Sayers EW et al. (2010). Database resources of the National Center for Biotechnology Information. *Nucleic acids research, 38*, D5-16.

Schaefer CF, Anthony K, Krupa S, Buchoff J, Day M, Hannay T, Buetow KH. (2009). PID: the Pathway Interaction Database. *Nucleic acids research, 37*, D674-679.

Schaefer MH, Fontaine JF, Vinayagam A, Porras P, Wanker EE, Andrade-Navarro MA. (2012). HIPPIE: Integrating protein interaction networks with experiment based quality scores. *PLoS One*, *7*, e31826.

Schneidman-Duhovny D, Hammel M, Sali A. (2011). Macromolecular docking restrained by a small angle X-ray scattering profile. *Journal of structural biology*, *173*, 461-471.

Schneidman-Duhovny D, Inbar Y, Nussinov R, Wolfson HJ. (2005). PatchDock and SymmDock: servers for rigid and symmetric docking. *Nucleic acids research*, *33*, W363-W367.

Schymkowitz J, Borg J, Stricher F, Nys R, Rousseau F, Serrano L. (2005). The FoldX web server: an online force field. *Nucleic acids research*, *33*, W382-W388.

Scott MS, Barton GJ. (2007). Probabilistic prediction and ranking of human protein-protein interactions. *BMC bioinformatics*, *8*, 239.

Seet BT, Dikic I, Zhou MM, Pawson T. (2006). Reading protein modifications with interaction domains. *Nature reviews. Molecular cell biology*, *7*, 473-483.

Segura J, Fernandez-Fuentes N. (2011). PCRPi-DB: a database of computationally annotated hot spots in protein interfaces. *Nucleic acids research*, *39*, D755-760.

Senachak J, Cheevadhanarak S, Hongsthong A. (2015). SpirPro: A Spirulina proteome database and web-based tools for the analysis of protein-protein interactions at the metabolic level in Spirulina (Arthrospira) platensis C1. *BMC bioinformatics*, *16*, 233.

Shannon P et al. (2003). Cytoscape: a software environment for integrated models of biomolecular interaction networks. *Genome research*, *13*, 2498-2504.

Sharan R et al. (2005). Conserved patterns of protein interaction in multiple species. *Proceedings of the National Academy of Sciences of the United States of America*, *102*, 1974-1979.

Sheinerman FB, Norel R, Honig B. (2000). Electrostatic aspects of protein–protein interactions. *Current opinion in structural biology*, *10*, 153-159.

Shen J et al. (2007). Predicting protein–protein interactions based only on sequences information. *Proceedings of the National Academy of Sciences*, *104*, 4337-4341.

Shin YC, Shin SY, So I, Kwon D, Jeon JH. (2011). TRIP Database: a manually curated database of protein-protein interactions for mammalian TRP channels. *Nucleic acids research*, *39*, D356-361.

Shoemaker BA et al. (2012). IBIS (Inferred Biomolecular Interaction Server) reports, predicts and integrates multiple types of conserved interactions for proteins. *Nucleic acids research*, *40*, D834-D840.

Shuai K. (2000). Modulation of STAT signaling by STAT-interacting proteins. *Oncogene*, *19*, 2638-2644.

Shulman-Peleg A, Shatsky M, Nussinov R, Wolfson HJ. (2008). MultiBind and MAPPIS: webservers for multiple alignment of protein 3D-binding sites and their interactions. *Nucleic acids research*, *36*, W260-W264.

Sigrist CJ, Cerutti L, de Castro E, Langendijk-Genevaux PS, Bulliard V, Bairoch A, Hulo N. (2010). PROSITE, a protein domain database for functional characterization and annotation. *Nucleic acids research*, *38*, D161-166.

Singer MS, Vriend G, Bywater RP. (2002). Prediction of protein residue contacts with a PDB-derived likelihood matrix. *Protein Engineering*, *15*, 721-725.

Singh A. (2019). PPI discovery using proteome coevolution. *Nature methods*, *16*, 804.

Singhal M, Resat H. (2007). A domain-based approach to predict protein-protein interactions. *BMC bioinformatics*, *8*, 199.

Smoot ME, Ono K, Ruscheinski J, Wang PL, Ideker T. (2011). Cytoscape 2.8: new features for data integration and network visualization. *Bioinformatics*, *27*, 431-432.

Stark C et al. (2011). The BioGRID Interaction Database: 2011 update. *Nucleic acids research*, *39*, D698-704.

Stein A, Ceol A, Aloy P. (2011). 3did: identification and classification of domain-based interactions of known three-dimensional structure. *Nucleic acids research*, *39*, D718-723.

Sun J, Li Y, Zhao Z. (2007). Phylogenetic profiles for the prediction of protein–protein interactions: how to select reference organisms? *Biochemical and Biophysical Research Communications, 353*, 985-991.

Sun J, Xu J, Liu Z, Liu Q, Zhao A, Shi T, Li Y. (2005). Refined phylogenetic profiles method for predicting protein–protein interactions. *Bioinformatics, 21*, 3409-3415.

Sun T, Zhou B, Lai L, Pei J. (2017). Sequence-based prediction of protein protein interaction using a deep-learning algorithm. *BMC bioinformatics, 18*, 1-8.

Szilagyi A, Zhang Y. (2014). Template-based structure modeling of protein-protein interactions. *Current opinion in structural biology, 24*, 10-23.

Szklarczyk D et al. (2011). The STRING database in 2011: functional interaction networks of proteins, globally integrated and scored. *Nucleic acids research, 39*, D561-568.

Szklarczyk D et al. (2015). STRING v10: protein–protein interaction networks, integrated over the tree of life. *Nucleic acids research, 43*, D447-D452.

Szklarczyk D et al. (2019). STRING v11: protein-protein association networks with increased coverage, supporting functional discovery in genome-wide experimental datasets. *Nucleic acids research, 47*, D607-D613.

Szklarczyk D et al. (2017). The STRING database in 2017: quality-controlled protein-protein association networks, made broadly accessible. *Nucleic acids research, 45*, D362-D368.

Taylor IW et al. (2009). Dynamic modularity in protein interaction networks predicts breast cancer outcome. *Nature biotechnology, 27*, 199-204.

Theocharidis A, van Dongen S, Enright AJ, Freeman TC. (2009). Network visualization and analysis of gene expression data using BioLayout Express(3D). *Nat Protoc, 4*, 1535-1550.

Thorn KS, Bogan AA. (2001). ASEdb: a database of alanine mutations and their effects on the free energy of binding in protein interactions. *Bioinformatics, 17*, 284-285.

Tillier ER, Biro L, Li G, Tillo D. (2006). Codep: maximizing co-evolutionary interdependencies to discover interacting proteins. *Proteins: Structure, Function, and Bioinformatics, 63*, 822-831.

Tillier ER, Charlebois RL. (2009). The human protein coevolution network. *Genome research, 19*, 1861-1871.

Torchala M, Moal IH, Chaleil RA, Fernandez-Recio J, Bates PA. (2013). SwarmDock: a server for flexible protein–protein docking. *Bioinformatics, 29*, 807-809.

Tormo J, Natarajan K, Margulies DH, Mariuzza RA. (1999). Crystal structure of a lectin-like natural killer cell receptor bound to its MHC class I ligand. *Nature, 402*, 623-631.

Tovchigrechko A, Vakser IA. (2006). GRAMM-X public web server for protein–protein docking. *Nucleic acids research, 34*, W310-W314.

Tress ML, Valencia A. (2010). Predicted residue–residue contacts can help the scoring of 3D models. *Proteins: Structure, Function, and Bioinformatics, 78*, 1980-1991.

Tsai CJ, Lin SL, Wolfson HJ, Nussinov R. (1997). Studies of protein-protein interfaces: a statistical analysis of the hydrophobic effect. *Protein science : a publication of the Protein Society, 6*, 53-64.

Tuncbag N, Gursoy A, Keskin O. (2009a). Identification of computational hot spots in protein interfaces: combining solvent accessibility and inter-residue potentials improves the accuracy. *Bioinformatics, 25*, 1513-1520.

Tuncbag N, Gursoy A, Keskin O. (2011a). Prediction of protein–protein interactions: unifying evolution and structure at protein interfaces. *Physical biology, 8*, 035006.

Tuncbag N, Gursoy A, Nussinov R, Keskin O. (2011b). Predicting protein-protein interactions on a proteome scale by matching evolutionary and structural similarities at interfaces using PRISM. *Nature protocols, 6*, 1341.

Tuncbag N, Kar G, Keskin O, Gursoy A, Nussinov R. (2009b). A survey of available tools and web servers for analysis of protein–protein interactions and interfaces. *Briefings in bioinformatics, 10*, 217-232.

Tuncbag N, Keskin O, Gursoy A. (2010). HotPoint: hot spot prediction server for protein interfaces. *Nucleic acids research, 38*, W402-W406.

Turner B et al. (2010). iRefWeb: interactive analysis of consolidated protein interaction data and their supporting evidence. *Database : the journal of biological databases and curation, 2010*, baq023.

Uetz P et al. (2000). A comprehensive analysis of protein-protein interactions in Saccharomyces cerevisiae. *Nature, 403*, 623-627.

Vakser IA. (2014). Protein-protein docking: from interaction to interactome. *Biophysical journal, 107*, 1785-1793.

Valdar WS. (2002). Scoring residue conservation. *Proteins: structure, function, and bioinformatics, 48*, 227-241.

Vastrik I et al. (2007). Reactome: a knowledge base of biologic pathways and processes. *Genome Biol, 8*, R39.

Vinogradova O, Qin J (2011) NMR as a unique tool in assessment and complex determination of weak protein–protein interactions. In: NMR of Proteins and Small Biomolecules. Springer, pp 35-45

von Mering C, Huynen M, Jaeggi D, Schmidt S, Bork P, Snel B. (2003). STRING: a database of predicted functional associations between proteins. *Nucleic acids research, 31*, 258-261.

Vreven T et al. (2015). Updates to the Integrated Protein-Protein Interaction Benchmarks: Docking Benchmark Version 5 and Affinity Benchmark Version 2. *J Mol Biol, 427*, 3031-3041.

Waese J, Provart NJ. (2017). The Bio-Analytic Resource for Plant Biology. *Methods in molecular biology, 1533*, 119-148.

Wang L, Wang HF, Liu SR, Yan X, Song KJ. (2019). Predicting Protein-Protein Interactions from Matrix-Based Protein Sequence Using Convolution Neural Network and Feature-Selective Rotation Forest. *Sci Rep, 9*, 9848.

Warde-Farley D et al. (2010). The GeneMANIA prediction server: biological network integration for gene prioritization and predicting gene function. *Nucleic acids research, 38*, W214-220.

Wheeler DL et al. (2008). Database resources of the National Center for Biotechnology Information. *Nucleic acids research, 36*, D13-21.

Winter C, Henschel A, Kim WK, Schroeder M. (2006). SCOPPI: a structural classification of protein-protein interfaces. *Nucleic acids research, 34*, D310-314.

Wishart DS et al. (2018). DrugBank 5.0: a major update to the DrugBank database for 2018. *Nucleic acids research, 46*, D1074-D1082.

Wojcik J, Schachter V. (2001). Protein-protein interaction map inference using interacting domain profile pairs. *Bioinformatics, 17 Suppl 1*, S296-305.

Wong L, You Z-H, Li S, Huang Y-A, Liu G Detection of protein-protein interactions from amino acid sequences using a rotation forest model with a novel PR-LPQ descriptor. In: International Conference on Intelligent Computing, 2015. Springer, pp 713-720

Worth CL, Gong S, Blundell TL. (2009). Structural and functional constraints in the evolution of protein families. *Nature Reviews Molecular Cell Biology, 10*, 709-720.

Wu X et al. (2006). SPIDer: Saccharomyces protein-protein interaction database. *BMC bioinformatics, 7 Suppl 5*, S16.

Wuchty S, Barabási A-L, Ferdig MT. (2006). Stable evolutionary signal in a yeast protein interaction network. *BMC evolutionary biology, 6*, 8.

Xenarios I, Rice DW, Salwinski L, Baron MK, Marcotte EM, Eisenberg D. (2000). DIP: the database of interacting proteins. *Nucleic acids research, 28*, 289-291.

Xu Q, Canutescu AA, Wang G, Shapovalov M, Obradovic Z, Dunbrack Jr RL. (2008). Statistical analysis of interface similarity in crystals of homologous proteins. *Journal of molecular biology, 381*, 487-507.

Xue LC, Dobbs D, Bonvin AM, Honavar V. (2015). Computational prediction of protein interfaces: A review of data driven methods. *FEBS letters, 589*, 3516-3526.

Yachie N, Saito R, Sugiyama N, Tomita M, Ishihama Y. (2011). Integrative features of the yeast phosphoproteome and protein-protein interaction map. *PLoS computational biology, 7*, e1001064.

Yan Y, Zhang D, Zhou P, Li B, Huang S-Y. (2017). HDOCK: a web server for protein–protein and protein–DNA/RNA docking based on a hybrid strategy. *Nucleic acids research*, *45*, W365-W373.

Yanai I, Derti A, DeLisi C. (2001). Genes linked by fusion events are generally of the same functional category: a systematic analysis of 30 microbial genomes. *Proceedings of the National Academy of Sciences*, *98*, 7940-7945.

Yang L, Xia J-F, Gui J. (2010). Prediction of protein-protein interactions from protein sequence using local descriptors. *Protein and Peptide Letters*, *17*, 1085-1090.

Yang X, Yang S, Li Q, Wuchty S, Zhang Z. (2020). Prediction of human-virus protein-protein interactions through a sequence embedding-based machine learning method. *Computational and structural biotechnology journal*, *18*, 153-161.

Yook SH, Oltvai ZN, Barabasi AL. (2004). Functional and topological characterization of protein interaction networks. *Proteomics*, *4*, 928-942.

You Z-H, Lei Y-K, Zhu L, Xia J, Wang B Prediction of protein-protein interactions from amino acid sequences with ensemble extreme learning machines and principal component analysis. In: BMC bioinformatics, 2013. vol S8. Springer, p S10

Yu H, Paccanaro A, Trifonov V, Gerstein M. (2006). Predicting interactions in protein networks by completing defective cliques. *Bioinformatics*, *22*, 823-829.

Yu T, Liu Y, Zeng Y, Chen J, Yang G, Li Y. (2019). Triplet-Triplet Annihilation Upconversion for Photocatalytic Hydrogen Evolution. *Chemistry*, *25*, 16270-16276.

Yue J et al. (2017). PCPPI: a comprehensive database for the prediction of Penicillium-crop protein-protein interactions. *Database : the journal of biological databases and curation*, *2017*.

Zanzoni A, Montecchi-Palazzi L, Quondam M, Ausiello G, Helmer-Citterich M, Cesareni G. (2002). MINT: a Molecular INTeraction database. *FEBS letters*, *513*, 135-140.

Zhang QC, Petrey D, Garzón JI, Deng L, Honig B. (2012). PrePPI: a structure-informed database of protein–protein interactions. *Nucleic acids research*, *41*, D828-D833.

Zhang X, Jiao X, Song J, Chang S. (2016). Prediction of human protein–protein interaction by a domain-based approach. *Journal of Theoretical Biology*, *396*, 144-153.

Zhao S et al. (2010). Regulation of cellular metabolism by protein lysine acetylation. *Science*, *327*, 1000-1004.

Zhou H-X, Qin S. (2007). Interaction-site prediction for protein complexes: a critical assessment. *Bioinformatics*, *23*, 2203-2209.

Zhou YZ, Gao Y, Zheng YY (2011) Prediction of protein-protein interactions using local description of amino acid sequence. In: Advances in computer science and education applications. Springer, pp 254-262

Zhu D, Qin ZS. (2005). Structural comparison of metabolic networks in selected single cell organisms. *BMC bioinformatics*, *6*, 8.

Zhu G et al. (2016). PPIM: A Protein-Protein Interaction Database for Maize. *Plant physiology*, *170*, 618-626.

Zhu H, Domingues FS, Sommer I, Lengauer T. (2006). NOXclass: prediction of protein-protein interaction types. *BMC bioinformatics*, *7*, 27.

Zhu H, Sommer I, Lengauer T, Domingues FS. (2008). Alignment of non-covalent interactions at protein-protein interfaces. *PLoS One*, *3*, e1926.

Zhu X, Mitchell JC. (2011). KFC2: a knowledge-based hot spot prediction method based on interface solvation, atomic density, and plasticity features. *Proteins: Structure, Function, and Bioinformatics*, *79*, 2671-2683.

Zuiderweg ER. (2002). Mapping protein-protein interactions in solution by NMR spectroscopy. *Biochemistry*, *41*, 1-7.